Computational Methods for Option Pricing

siam FRONTIERS IN APPLIED MATHEMATICS

The SIAM series on Frontiers in Applied Mathematics publishes monographs dealing with creative work in a substantive field involving applied mathematics or scientific computation. All works focus on emerging or rapidly developing research areas that report on new techniques to solve mainstream problems in science or engineering.

The goal of the series is to promote, through short, inexpensive, expertly written monographs, cutting edge research poised to have a substantial impact on the solutions of problems that advance science and technology. The volumes encompass a broad spectrum of topics important to the applied mathematical areas of education, government, and industry.

BOOKS PUBLISHED IN FRONTIERS IN APPLIED MATHEMATICS

Achdou, Yves, and Pironneau, Olivier, *Computational Methods for Option Pricing*

Smith, Ralph C., *Smart Material Systems: Model Development*

Iannelli, M.; Martcheva, M.; and Milner, F. A., *Gender-Structured Population Modeling: Mathematical Methods, Numerics, and Simulations*

Pironneau, O. and Achdou, Y., *Computational Methods in Option Pricing*

Day, William H. E. and McMorris, F. R., *Axiomatic Consensus Theory in Group Choice and Biomathematics*

Banks, H. T. and Castillo-Chavez, Carlos, editors, *Bioterrorism: Mathematical Modeling Applications in Homeland Security*

Smith, Ralph C. and Demetriou, Michael, editors, *Research Directions in Distributed Parameter Systems*

Höllig, Klaus, *Finite Element Methods with B-Splines*

Stanley, Lisa G. and Stewart, Dawn L., *Design Sensitivity Analysis: Computational Issues of Sensitivity Equation Methods*

Vogel, Curtis R., *Computational Methods for Inverse Problems*

Lewis, F. L.; Campos, J.; and Selmic, R., *Neuro-Fuzzy Control of Industrial Systems with Actuator Nonlinearities*

Bao, Gang; Cowsar, Lawrence; and Masters, Wen, editors, *Mathematical Modeling in Optical Science*

Banks, H. T.; Buksas, M. W.; and Lin, T., *Electromagnetic Material Interrogation Using Conductive Interfaces and Acoustic Wavefronts*

Oostveen, Job, *Strongly Stabilizable Distributed Parameter Systems*

Griewank, Andreas, *Evaluating Derivatives: Principles and Techniques of Algorithmic Differentiation*

Kelley, C. T., *Iterative Methods for Optimization*

Greenbaum, Anne, *Iterative Methods for Solving Linear Systems*

Kelley, C. T., *Iterative Methods for Linear and Nonlinear Equations*

Bank, Randolph E., *PLTMG: A Software Package for Solving Elliptic Partial Differential Equations. Users' Guide 7.0*

Moré, Jorge J. and Wright, Stephen J., *Optimization Software Guide*

Rüde, Ulrich, *Mathematical and Computational Techniques for Multilevel Adaptive Methods*

Cook, L. Pamela, *Transonic Aerodynamics: Problems in Asymptotic Theory*

Banks, H. T., *Control and Estimation in Distributed Parameter Systems*

Van Loan, Charles, *Computational Frameworks for the Fast Fourier Transform*

Van Huffel, Sabine and Vandewalle, Joos, *The Total Least Squares Problem: Computational Aspects and Analysis*

Castillo, José E., *Mathematical Aspects of Numerical Grid Generation*

Bank, R. E., *PLTMG: A Software Package for Solving Elliptic Partial Differential Equations. Users' Guide 6.0*

McCormick, Stephen F., *Multilevel Adaptive Methods for Partial Differential Equations*

Grossman, Robert, *Symbolic Computation: Applications to Scientific Computing*

Coleman, Thomas F. and Van Loan, Charles, *Handbook for Matrix Computations*

McCormick, Stephen F., *Multigrid Methods*

Buckmaster, John D., *The Mathematics of Combustion*

Ewing, Richard E., *The Mathematics of Reservoir Simulation*

Computational Methods for Option Pricing

Yves Achdou
Université Denis Diderot
Paris, France

Olivier Pironneau
Université Pierre et Marie Curie
Institut Universitaire de France
Paris, France

siam
Society for Industrial and Applied Mathematics
Philadelphia

10 9 8 7 6 5 4 3 2 1

Library of Congress Cataloging-in-Publication Data

Achdou, Yves.
Computational methods for option pricing / Yves Achdou, Olivier Pironneau.
p. cm.--(Frontiers in applied mathematics)
Includes bibliographical references and index.
ISBN 0-89871-573-3 (pbk.)
1. Options (Finance)—Prices—Mathematical models. I. Pironneau, Olivier. II. Title. III. Series.

HG6024.A3A26 2005
322.64'53'01519--dc22

2005046506

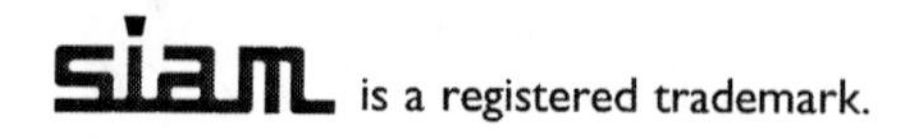

To Juliette and Raphaël

Firm theory, assured significance,
Appeared as frauds upon Time's credit bank
Or assets valueless in Truth's treasury.
From Sri Aurobindo's epic, Savitri

Contents

List of Algorithms

Preface

Mathematical finance is an old science but has become a major topic for numerical analysts since Merton [97], Black–Scholes [16] modeled financial derivatives. An excellent book for the mathematical foundation of option pricing is Lamberton and Lapeyre's [85]. Since the Black–Scholes model relies on stochastic differential equations, option pricing rapidly became an attractive topic for specialists in the theory of probability, and stochastic methods were developed first for practical applications, along with analytical closed formulas. But soon, with the rapidly growing complexity of the financial products, other numerical solutions became attractive. Applying the Monte-Carlo method to option pricing is very natural and not difficult, at least for European options, but speeding up the method by variance reduction may become tricky. Similarly, tree methods are very intuitive and fast but also rapidly become difficult as the complexity of the financial product grows.

Focusing on the Black–Scholes model, a partial differential equation is obtained by Itô's calculus. It can be approximated and integrated numerically by various methods, to which a very clear and concise introduction may be found in the book by Wilmott, Howison, and Dewynne [117]: the basic idea is to approximate the partial differential equation by a system of equations with a finite number of unknowns, which may be solved numerically to obtain a discrete solution. The discrete problems can be computationally intensive.

The aim of this book is neither to present financial models nor to discuss their validity; we must be very modest in this perspective, since our expertise is not here. This book is not a recipe book either, and although we have tried to be broad, many financial products such as bonds are not covered.

The purpose is rather to discuss some modern numerical techniques which we believe to be useful for simulations in finance. We are not going to dwell on Monte-Carlo and tree methods, because these have been studied very well elsewhere (see [60, 116]). We essentially focus on the finite difference (Chapter 3) and the finite element methods (Chapter 4) for the partial differential equation, trying to answer the following three questions:

- Are these methods reliable?
- How can their accuracy be controlled and improved?
- How can these methods be implemented efficiently?

Several applications to financial products with programs in C++ are proposed.

In this book, we stress the notions of error control and adaptivity: the aim is to control a posteriori the accuracy of the numerical method, and if the desired accuracy is not reached, to refine the discretization precisely where it is necessary, i.e., most often where the solution exhibits singularities. We believe that mesh adaption based on a posteriori estimates is an important and practical tool because it is the only existing way to certify that a numerical scheme will give the solution within a given error bound. It is therefore a road for software certification. Mesh adaption greatly speeds up computer programs because grid nodes are present only where they are needed; it is particularly important for American options, because the option price as a function of time and the spot price exhibits a singularity on a curve which is itself unknown. A posteriori error estimates are the subject of Chapter 5, and adaptive methods are also used for pricing American options in Chapter 6.

Controlling the accuracy of a numerical method requires a rather complete mathematical analysis of the underlying partial differential equation: this motivates partially the theoretical results contained in Chapters 2 and 6.

The Black–Scholes model is by many aspects too simple to fit the market prices, and several more elaborate models were proposed:

- *Assume that the volatility is itself a stochastic process* [70, 110, 51]. This is discussed in Chapters 2 and 4.
- *Generalize the Black–Scholes model by assuming that the spot price is a Lévy process* [31]. We discuss the pricing of Lévy driven options in Chapters 2 and 4.
- *Use local volatility*; i.e., assume that the volatility in the Black–Scholes model is a function of time and of the prices of the underlying assets.

In the latter case one has to *calibrate the volatility* from the market data, i.e., find a volatility function which permits one to recover the prices of the options available on the market. This means solving an inverse problem with a partial differential equation. Inverse problems are used in many fields of engineering: for instance, in the oil industry, people try to recover underground properties from seismic data. Such problems are usually ill-posed in the sense that a very small variation of the data may cause huge changes in the computed volatility. This unstable character of inverse problems is a well-known fact, and the best-known cure is *least square* with *Tychonoff regularization*: calling $(\bar{u}_i)_{i\in I}$ the prices of a family of options available on the market, σ the local volatility function, and $(u_i(\sigma))_{i\in I}$ the prices computed with the local volatility model, the least squares approach solves the minimization problem

$$\min_{\sigma\in\mathcal{K}} J(\sigma) := \sum_{i\in I} \omega_i |u_i(\sigma) - \bar{u}_i|^2,$$

where $\mathcal{K}$ is a closed subset of a suitable function space W. This problem is ill-posed; the Tychonoff regularization replaces J by

$$J(\sigma) + J_R(\sigma) \quad \text{with} \quad J_R(\sigma) = \|\sigma\|^2,$$

where $\|\cdot\|$ is a suitable norm on W. One has to choose properly the space W, the set $\mathcal{K}$, and the norm $\|\cdot\|$. For that, one has to study first the sensitivity of the options' prices with respect to the local volatility. Here also the careful analysis of the partial differential equations and inequalities done in Chapters 2 and 6 proves useful. We discuss volatility calibration with European options and American options. To our knowledge, calibration with American options is not discussed elsewhere. We also treat computational methods in order to evaluate the Greeks, i.e., the sensitivity of the options' prices with respect to various parameters: for that, we describe the method of automatic differentiation of computer codes. We think that this idea is also new in the field of computational finance.

The book is organized as follows: Chapter 1 contains an introduction to option pricing, and to the numerical methods not based on partial differential equations. Chapter 2 deals with the mathematical analysis of partial differential equations in finance: it is the cornerstone for the numerical analysis performed in the rest of the book. Chapter 3 contains the description and the numerical analysis of some finite difference methods. Chapter 4 is devoted to the finite element method: we insist on practical implementation rather than on numerical analysis. This chapter is a rather long one, since the finite element method is not too well known in finance. We apply the finite element method to many kinds of options, including Lévy driven options, basket options, and the case where the volatility is also a stochastic process. We also survey recent works where partial differential equation based methods have been applied successfully for pricing options on baskets with several assets (say ten or twenty). In Chapter 5, we discuss a posteriori error estimates and adaptive mesh refinement. American options are studied in the separate Chapter 6, which goes from mathematical and numerical analysis to computational algorithms. Chapter 7 deals with the use of automatic differentiation of computer programs for evaluating the Greeks. It is also an introduction to the calibration of volatility. In Chapter 8, we describe two ways of calibrating the volatility with European options: the first one is based on least squares and Dupire's equation, and the second one uses techniques of stochastic optimal control. Finally, Chapter 9 is devoted to calibration with American options.

This book is for anyone who wishes to become acquainted with the modern tools of numerical analysis for some computational problems arising in finance. Although some important aspects of finance modeling are reviewed, the main topic is numerical algorithms for fast and accurate computations of financial derivatives and for calibration of parameters.

While some parts of this book are written at a graduate level, this book aims also at being useful to Ph.D. students and professional scientists in the field of finance. In particular it contains rigorous results on the mathematical regularity of solutions, on modern algorithms with adaptive mesh refinement, and on the calibration with European and American options which do not seem to be available anywhere else. In particular, we believe that adaptive methods are not popular enough in the field of computational finance and that promoting them is useful.

Except in Chapter 1, the book deals mostly with partial differential equations. It goes from mathematical and numerical analysis to practical computational algorithms. It is application oriented and does not contain any theoretical development which is not used elsewhere for a practical purpose. The computer programs are given, partially in the book in order to point out the important ideas, fully on the web site www.ann.jussieu.fr/~achdou. They are written in C++ but do not require a thorough knowledge of the language. They run on most platforms.

Our interest in this topic originated from a numerical challenge posed to us by Nicolas Di Césaré and Jean-Claude Hontand at the bank Natexis-Banques Populaires. Later we offered two courses at the graduate level, one at University Pierre et Marie Curie (Paris 6) for students of numerical analysis wishing to know more about numerical mathematical finance, and the other at University Denis Diderot (Paris 7) for teaching numerical methods to students more familiar with finance and statistics. The book grew out of this experience.

Being computational applied mathematicians we are indebted to a number of colleagues for their counseling in mathematical finance: H. Berestycki, R. Cont, J. Di Fonseca, L. Élie, N. El Karoui, D. Gabay, M. Lezmi, C. Martini, G. Papanicolaou, C. Schwab, and A. Sulem. We thank also C. Bernardi for having revised part of the manuscript. The first author also acknowledges very fruitful cooperation with N. Tchou and B. Franchi. We have also borrowed some C++ classes from F. Hecht.

Chapter 1

Option Pricing

1.1 Orientation

After a brief introduction to options and option pricing, we briefly discuss two pricing methods which will not be used in the other chapters of the book: the Monte-Carlo simulations and the binomial tree methods. Since this book is devoted to deterministic techniques for partial differential equations in finance, we shall not dwell on these two methods; however, they are important alternatives used often in practice and so it is necessary to know them to appreciate the advantages (and limitations) of deterministic methods.

1.2 A Brief Introduction to Options

1.2.1 The European Vanilla Call Option

Quoting Cox and Rubinstein [36], *a European vanilla call option is a contract giving its owner the right to buy a fixed number of shares of a specific common stock at a fixed price at a certain date.* The act of making the transaction is referred to as *exercising the option*: note that the owner of the option may or may not exercise it.

- The specific stock is called the *underlying asset* or the *underlying security*. For simplicity, we will assume that the fixed number of shares is one. The price of the underlying asset will be referred to as the *spot price* and will be denoted by S or S_t.
- The fixed price is termed the *strike*, the *striking price*, or the *exercise price*. We will often use the notation K for the strike.
- The given date is termed the *maturity date*, the *maturity*, or the *date of expiration*. It will often be denoted by T.

The term *vanilla* is used to signify that this kind of option is the simplest one: there are more complicated contracts, some of which will be described later.

An option has a value. We wish to solve the following problem: is it possible to evaluate the market price C_t of the call option at time t, $0 \leq t \leq T$?

For that, we have to make assumptions on the market: we assume that the transactions have no cost and are instantaneous and that the market rules out *arbitrage*; it is impossible to make an instantaneous benefit without taking any risk.

Pricing the option at maturity is easy. If S_T is the spot price at maturity, then the owner of the call option will make a benefit of $(S_T - K)_+ = \max(S_T - K, 0)$ by exercising the option and immediately selling the asset. Thus, assuming that there is no arbitrage, the value of the call on the expiration date is

$$C(S_T, T) = (S_T - K)_+.$$

1.2.2 Option Pricing: The Black–Scholes Model

Definitions. Before describing the model, let us recall very briefly some notions of probability: Let Ω be a set, $\mathcal{A}$ a σ-algebra of subsets of Ω, and $\mathbb{P}$ a nonnegative measure on Ω such that $\mathbb{P}(\Omega) = 1$. The triple $(\Omega, \mathcal{A}, \mathbb{P})$ is called a *probability space*.

Recall that a *real-valued random variable* X on $(\Omega, \mathcal{A}, \mathbb{P})$ is an $\mathcal{A}$-measurable real-valued function on Ω; i.e., for each Borel subset B of $\mathbb{R}$, $X^{-1}(B) \in \mathcal{A}$. Also, a *real-valued stochastic process* $(X_t)_{t\geq 0}$ on $(\Omega, \mathcal{A}, \mathbb{P})$ assigns to each time t a random variable X_t on $(\Omega, \mathcal{A}, \mathbb{P})$. The process X_t is continuous if for $\mathbb{P}$-almost every $\omega \in \Omega$, the function $t \mapsto X_t(\omega)$ is continuous. More generally, it is possible to define similarly random variables and stochastic processes with values on separable complete metric spaces, for example, finite-dimensional vector spaces.

Recall that a *filtration* $F_t = (\mathcal{A}_t)_{t\geq 0}$ is an increasing family of σ-algebras $\mathcal{A}_t$; i.e., for $t > \tau$, we have $\mathcal{A}_\tau \subset \mathcal{A}_t \subset \mathcal{A}$. The σ-algebra $\mathcal{A}_t$ usually represents a certain past history available at time t.

A stochastic process $(X_t)_{t\geq 0}$ is F_t-adapted if for each $t \geq 0$, X_t is $\mathcal{A}_t$-measurable.

Following, e.g., [85], we will consider only filtrations F_t such that for all $t \geq 0$, $\mathcal{A}_t$ contains the set $\mathcal{N}$ of all the subsets $A \in \mathcal{A}$ with $\mathbb{P}(A) = 0$.

For a given stochastic process X_t on $(\Omega, A, \mathbb{P})$, it is possible to construct a filtration $F_t = (\mathcal{A}_t)_{t\geq 0}$ by taking $\mathcal{A}_t$ as the smallest σ-algebra such that X_τ for all $\tau \leq t$ and all the negligible subsets of A for $\mathbb{P}$ are $\mathcal{A}_t$-measurable (we say that $\mathcal{A}_t$ is the *σ-algebra generated by $(X_\tau)_{\tau\leq t}$ and by $\mathcal{N}$*). The obtained filtration is called the *natural filtration* of the process X_t.

For a filtration $F_t = (\mathcal{A}_t)_{t\geq 0}$, we call *stopping time* a random variable τ with value in $\mathbb{R}_+ \cup \{+\infty\}$ such that, for all t, the event $\{\tau \leq t\} \in \mathcal{A}_t$.

For a filtration $F_t(\mathcal{A}_t)_{t\geq 0}$, an F_t-adapted stochastic process $(M_t)_{t\geq 0}$ is called a *martingale* if

- $\mathbb{E}(|M_t|) < +\infty$,
- for all $\tau \leq t$, $\mathbb{E}(M_t | F_\tau) = M_\tau$.

Notions on Itô's stochastic integral and stochastic differential equations are necessary for the following. We refer the reader, for example, to [79, 85, 104, 48].

The Black–Scholes Model. The Black–Scholes model [16, 97] is a continuous-time model involving a risky asset (the underlying asset) whose price at time t is S_t and a risk-free asset whose price at time t is S_t^0: the evolution of S_t^0 is found by solving the ordinary differential equation

$$\frac{dS_t^0}{dt} = r(t)S_t^0,$$

where $r(t)$ is an instantaneous *interest rate*. Setting $S_0^0 = 1$, we find that

$$S_t^0 = e^{\int_0^t r(\tau)d\tau}, \quad \text{and} \quad S_t^0 = e^{rt} \quad \text{if } r \text{ is constant.}$$

For simplicity, we suppose here that r is constant.

The Black–Scholes model decomposes the return on the asset $\frac{dS_t}{S_t}$ as a sum of a deterministic term μdt (hence μ is an average rate of growth of the asset price), called the *drift*, and a random term which models the price variations in response to external effects. More precisely, the Black–Scholes model assumes that the price of the risky asset is a solution to the following stochastic differential equation:

$$dS_t = S_t(\mu dt + \sigma_t dB_t), \tag{1.1}$$

where B_t is a standard Brownian motion on a probability space $(\Omega, \mathcal{A}, \mathbb{P})$, i.e., a real-valued continuous stochastic process whose increments are independent and stationary, with $B_0 = 0$ $\mathbb{P}$-almost surely, $\mathbb{E}(B_t) = 0$, and $\mathbb{E}(B_t^2) = t$. It can be proved that, in the limit $\delta t \to 0$, the law of $B_{t+\delta t} - B_t$ is a Gaussian random variable with zero mean and variance δt. We call F_t the natural filtration associated to B_t. Here σ_t is a real process adapted to F_t. The number σ_t is called the *volatility*. It is assumed that $0 < \underline{\sigma} \le \sigma_t \le \bar{\sigma}$ for all $t \in [0, T]$.

The accurate mathematical meaning of (1.1) can be found in [79, 85, 104, 48].

For simplicity here, we assume that σ_t is constant: $\sigma_t = \sigma$, $0 \le t \le T$. The value of S_t can be deduced from that of B_t by

$$S_t = S_0 \exp\left(\left(\mu - \frac{\sigma^2}{2}\right)t + \sigma B_t\right).$$

Pricing the Option. The Black–Scholes model yields a formula for pricing the option at $t < T$. There are many ways to derive formula (1.4)–(1.5). We choose to follow the arguments of Harrison and Pliska [67] (see also Bensoussan [12]), presented in the book by Lamberton and Lapeyre [85]: Girsanov's theorem tells us that there exists a probability $\mathbb{P}^*$ equivalent to $\mathbb{P}$ such that the price S_t satisfies the stochastic differential equation

$$dS_t = S_t(rdt + \sigma dW_t), \tag{1.2}$$

where $W_t = B_t + \int_0^t \frac{\mu - r}{\sigma} ds$ is a standard Brownian motion under $\mathbb{P}^*$. In other words, under probability $\mathbb{P}^*$, the discounted price $\tilde{S}_t = S_t e^{-rt}$ satisfies $d\tilde{S}_t = \sigma \tilde{S}_t dW_t$, so it is a martingale; see [79, 85, 104, 48].

Then, it can be proved that it is possible to simulate the option by a self-financed portfolio containing H_t shares of the underlying asset and H_t^0 shares of the risk-free asset, i.e., to find a pair of adapted processes H_t^0 and H_t such that

$$\int_0^T |H_t^0| dt + \int_0^T H_t^2 dt < +\infty \quad \text{almost surely (a.s.),}$$

$$H_t^0 S_t^0 + H_t S_t = H_0^0 S_0^0 + H_0 S_0 + \int_0^t H_\tau^0 dS_\tau^0 + \int_0^t H_\tau dS_\tau \quad \text{a.s. } \forall t \in [0, T],$$

and

$$C_T = H_T^0 S_T^0 + H_T S_T.$$

Then since arbitrage is ruled out, the option's price must be given by

$$C_t = H_t^0 S_t^0 + H_t S_t. \tag{1.3}$$

It is also possible to see that for a self-financed portfolio, one has

$$\begin{aligned} H_t^0 + H_t \tilde{S}_t &= H_0^0 + H_0 \tilde{S}_0 + \int_0^t H_\tau d\tilde{S}_\tau \quad \text{a.s. } \forall t \in [0, T], \\ &= H_0^0 + H_0 \tilde{S}_0 + \int_0^t \sigma H_\tau \tilde{S}_\tau dW_\tau. \end{aligned}$$

This implies that $H_t^0 + H_t \tilde{S}_t$ is a square integrable martingale under $\mathbb{P}^*$, and from (1.3), we have

$$C_t = e^{rt}\mathbb{E}^*(C_T e^{-rT} | F_t) = \mathbb{E}^*(e^{-r(T-t)}(S_T - K)_+ | F_t). \tag{1.4}$$

The essential argument for proving that the option can be simulated by a self-financed portfolio is a representation theorem for Brownian martingales, which states that for every square integrable martingale M_t for the filtration F_t, there exists an adapted process H_t such that $dM_t = H_t dW_t$ and $\mathbb{E}^*(\int H_\tau^2 d\tau) < +\infty$.

All the arguments above can be generalized when $\sigma_t = \sigma(S_t, t)$ and when $r_t = r(t)$, with, for example, r and σ continuous functions such that $S \mapsto S\sigma(S, t)$ is a Lipschitz regular function of S with a Lipschitz constant independent of t, and σ is bounded from above and away from 0 uniformly in t, and the Black–Scholes formula is

$$C_t = \mathbb{E}^*(e^{-\int_t^T r(\tau)d\tau}(S_T - K)_+ | F_t). \tag{1.5}$$

When the volatility σ_t is a function of t and S_t, we will speak of *local volatility*. The Black–Scholes model with uniform coefficients is often too rough to recover the prices of the options on the market, so a current practice is to *calibrate the volatility surface* $(S, t) \mapsto \sigma(S, t)$ to fit the market prices.

Remark 1.1. *The argument used in the original paper by Black and Scholes* [16] *and presented in the book by Wilmott, Howison, and Dewynne* [117] *is slightly different. After postulating that $C_t = C(S_t, t)$, one constructs an instantaneously risk-free portfolio with one option and Δ_t shares of the underlying asset: the choice of Δ_t is called* hedging. *From Itô's formula, the hedging factor Δ_t is shown to be*

$$\Delta_t = -\frac{\partial C}{\partial S}(S_t, t).$$

Then one obtains a partial differential equation for C, with the infinitesimal generator of the Markov process S_t. Finally (1.5) *is obtained. This remark shows why precisely computing the derivatives of C is important.*

1.2.3 Other European Options

European Vanilla Put Options. A wide variety of contracts can be designed. Analogous to the vanilla call options are the vanilla put options: quoting [36], *a European vanilla put option is a contract giving its owner the right to sell a share of a specific common stock (whose price is S_t at time t) at a fixed price K at a certain date T*. It is clear that the value of the put option at maturity is

$$P_T = \max(K - S_T, 0) = (K - S_T)_+. \tag{1.6}$$

Using the Black–Scholes model for pricing the put option, we obtain

$$P_t = \mathbb{E}^*(e^{-\int_t^T r(\tau)d\tau}(K - S_T)_+|F_t). \tag{1.7}$$

The Put-Call Parity. Subtracting (1.7) from (1.5), we obtain that

$$C_t - P_t = \mathbb{E}^*(e^{-\int_t^T r(\tau)d\tau}(S_T - K)|F_t),$$

which yields

$$C_t - P_t = S_t - Ke^{-\int_t^T r(\tau)d\tau}, \tag{1.8}$$

since the discounted price $e^{-\int_0^t r(\tau)d\tau}S_t$ is a martingale under $\mathbb{P}^*$. The relation (1.8) is called the put-call parity.

Dividends. Pricing options on assets which yield dividends will be discussed in Chapter 2.

General European Options. Let $Q^\circ : \mathbb{R}_+ \to \mathbb{R}_+$ be a function bounded on the bounded subsets of $\mathbb{R}_+$ and such that $\frac{Q^\circ(S)}{S}$ is bounded as $S \to +\infty$. It is possible to design a contract that gives its owner the payoff $Q^\circ(S_T)$ at maturity T. The function Q° is called the *payoff function*. Following the arguments above, the value of this option at maturity is $Q^\circ(S_T)$, and at time t, $0 \le t < T$, the value of the option is

$$Q_t = \mathbb{E}^*(e^{-\int_t^T r(\tau)d\tau}Q^\circ(S_T)|F_t). \tag{1.9}$$

For example, the cash or nothing call option, with strike K, is the right to buy the asset at the price $S_T - E$ if $S_T > K$ at maturity T. The payoff function of this option is $Q^\circ(S) = E1_{S>K}$. Similarly, the cash or nothing put option is the right to sell the asset at the price $S_T + E$ if $S_T < K$ at maturity T. Its payoff function is $Q^\circ(S) = E1_{S<K}$.

Barrier Options. A barrier option with payoff Q° and maturity T is a contract which yields a payoff $Q^\circ(S_T)$ at maturity T, as long as the spot price S_t remains in the interval $(a(t), b(t))$ for all time $t \in [0, T]$. For simplicity, we assume that a and b do not depend on time. The option is extinguishable in the sense that its value vanishes as soon as S_t leaves the interval (a, b). With the Black–Scholes model, one obtains a formula for the option's price:

$$Q_t = \mathbb{E}^*(1_{\{\forall\tau\in[t,T], S_\tau\in(a,b)\}}e^{-\int_t^T r(\tau)d\tau}Q^\circ(S_T)|F_t), \tag{1.10}$$

where the expectation is computed under the risk neutral probability.

European Options on a Basket of Assets. Consider a basket containing I assets, whose prices S_{it}, $i = 1, \dots, I$, satisfy the system of stochastic differential equations

$$dS_{it} = S_{it}\left(\mu_i dt + \frac{\sigma_i}{\sqrt{1+\sum_{j\neq i}\rho_{i,j}^2}}\left(dB_{it} + \sum_{j\neq i}\rho_{i,j}dB_{jt}\right)\right), \qquad i = 1, \dots, I, \tag{1.11}$$

where $B_{1t}, \dots, B_{It}$ are I independent standard Brownian motions under probability $\mathbb{P}$, and where the correlation coefficients $\rho_{i,j}$ satisfy $\rho_{i,j} = \rho_{j,i}$, $1 \le i < j \le I$. The process $\frac{1}{\sqrt{1+\sum_{j\neq i}\rho_{i,j}^2}}(B_{it} + \sum_{j\neq i}\rho_{i,j}B_{jt})$ is a standard Brownian motion, so

$$S_{it} = S_{i0}\exp\left(\left(\mu_i - \frac{\sigma_i^2}{2}\right)t + \frac{\sigma_i}{\sqrt{1+\sum_{j\neq i}\rho_{i,j}^2}}\left(B_{it} + \sum_{j\neq i}\rho_{i,j}B_{jt}\right)\right).$$

For a function $Q^\circ : (\mathbb{R}_+)^I \to \mathbb{R}_+$, the European option on this basket of assets of maturity T and payoff Q° can be exercised at $t = T$ for a payoff of $Q^\circ(S_{1T}, \dots, S_{IT})$.

As for the options on a single asset, it is possible to find a risk neutral probability $\mathbb{P}^*$ under which the price of the option is

$$Q_t = \mathbb{E}^*(e^{-\int_t^T r(\tau)d\tau}Q^\circ(S_{1T}, \dots, S_{IT})|F_t).$$

1.3 Constant Coefficients. The Black–Scholes Formula

Calling $Q(S, t)$ the price of an option with maturity T and payoff function Q°, and assuming that r and σ are constant, the Black–Scholes formula is

$$Q(S, t) = e^{-r(T-t)}\mathbb{E}^*(Q^\circ(Se^{r(T-t)}e^{\sigma(W_T-W_t)-\frac{\sigma^2}{2}(T-t)})), \tag{1.12}$$

and since under P^*, $W_T - W_t$ is a centered Gaussian distribution with variance $T - t$,

$$Q(S, t) = \frac{1}{\sqrt{2\pi}}e^{-r(T-t)}\int_{\mathbb{R}} Q^\circ(Se^{(r-\frac{\sigma^2}{2})(T-t)+\sigma x\sqrt{T-t}})e^{-\frac{x^2}{2}}dx. \tag{1.13}$$

When the option is a vanilla European option (noting C the price of the call and P the price of the put), a more explicit formula can be deduced from (1.13). Take, for example, a call

$$\begin{aligned} C(S, t) &= \frac{1}{\sqrt{2\pi}}\int_{-d_2}^{+\infty}(Se^{-\frac{\sigma^2}{2}(T-t)+\sigma x\sqrt{T-t}} - Ke^{-r(T-t)})e^{-\frac{x^2}{2}}dx \\ &= \frac{1}{\sqrt{2\pi}}\int_{-\infty}^{d_2}(Se^{-\frac{\sigma^2}{2}(T-t)-\sigma x\sqrt{T-t}} - Ke^{-r(T-t)})e^{-\frac{x^2}{2}}dx, \end{aligned} \tag{1.14}$$

where

$$d_1 = \frac{\log(\frac{S}{K}) + (r + \frac{\sigma^2}{2})(T-t)}{\sigma\sqrt{T-t}} \quad \text{and} \quad d_2 = d_1 - \sigma\sqrt{T-t}. \tag{1.15}$$

Finally, introducing the upper tail of the Gaussian function

$$N(d) = \frac{1}{\sqrt{2\pi}} \int_{-\infty}^{d} e^{-\frac{x^2}{2}} dx, \tag{1.16}$$

and using the new variable $x + \sigma\sqrt{T-t}$ instead of x, we obtain the *Black–Scholes formula.*

Proposition 1.1. *When σ and r are constant, the price of the call is given by*

$$C(S,t) = SN(d_1) - Ke^{-r(T-t)}N(d_2), \tag{1.17}$$

and the price of the put is given by

$$P(S,t) = -SN(-d_1) + Ke^{-r(T-t)}N(-d_2), \tag{1.18}$$

where d_1 and d_2 are given by (1.15) *and N is given by* (1.16).

Remark 1.2. *If r is a function of time,* (1.15) *must be replaced by*

$$d_1 = \frac{\log(\frac{S}{K}) + \int_t^T r(\tau)d\tau + \frac{\sigma^2}{2}(T-t)}{\sigma\sqrt{T-t}} \quad \text{and} \quad d_2 = d_1 - \sigma\sqrt{T-t}. \tag{1.19}$$

Remark 1.3. *For vanilla barrier options with constant volatility, and when $b = \infty$ or $a = 0$ (see the paragraph on barrier options above), there are formulas similar to* (1.17) *and* (1.18)*; see, for example,* [117].

The Black–Scholes Formula Programmed with the GSL. There is a function in the GNU Scientific Library (GSL) [59] for computing $N(d)$: it is based on approximate formulas that can be found in Abramowitz and Stegun; see [1]. The GSL is a very nice public domain package written in C for scientific computing. It includes programs for special functions, linear algebra, approximate integration, fast Fourier transform, polynomials and interpolation, random variables, etc. Here, we shall make use of two of these.

ALGORITHM 1.1. GSL function calls.

```
int gsl_sf_erf_Q_e(double x, gsl_sf_result * result);
double gsl_sf_erf_Q(double x);
```

The program for computing a vanilla European call is as follows.

ALGORITHM 1.2. Black–Scholes formula using the GSL.

```
                                                  //    file exactBSgsl.cpp
#include <iostream>
#include<gsl/gsl_sf_erf.h>
using namespace std;

                                                      //   !  Call premium
double Call(double S, double K, double r, double Vol, double theta)
```

```
            //    !S: Stock price -- K: Strike -- r:  riskless interest rate
                              //    !Vol:  volatility -- theta:  time to maturity
{
 if(S>0)
   {
     double standard_deviation= Vol*sqrt(theta);
     double d1 = (log(S/K)+r*theta)/standard_deviation
             + 0.5*standard_deviation;
     double d2 = d1 -  standard_deviation;
     return S*gsl_sf_erf_Q(-d1) - K*exp(-r*theta)* gsl_sf_erf_Q(-d2);
   }
 else
   return 0;
}

int main(){ cout<<Call(111,100,0.1,0.1,1)<<endl;}
```

On Linux machines (and cygwin and MacOS X) this is compiled, linked, and executed by

```
g++ -c exactBSgsl.cpp
g++ exactBSgsl.o -lgsl
.../a.out
```

Remark 1.4. *The program below also computes the value of the Black–Scholes formula. Integrals are replaced by sums on quadrature points. The reader can check that this is much slower than with the GSL, which uses special functions.*

ALGORITHM 1.3. Black–Scholes formula by a naive method.

```
double BSformula(double t, double S, double sig, double r, double K)
{ double dS=0.01, sig2=2*sig*sig*t, r2=r*t-sig2/4, aux=0;
   for(double Sp=0.001; Sp<500; Sp+=dS)
     aux += ((K-Sp)>0?K-Sp:0)*exp(-pow(log(Sp/S)-r2,2)/sig2)*dS/Sp;
   return aux * exp(-r*t)/sqrt(8*atan(1)*t)/sig;
}
```

1.4 Monte-Carlo Methods

For a complete survey of Monte-Carlo methods in finance, a good reference is the book by Glasserman [60]. The book by Lyuu [91] also contains a lot of practical information.

1.4.1 Numerical Random Variables

In the C-library `stdlib.h`, there is a function `rand()` which returns an integer value (of type `long int`) uniformly distributed in `[0,RAND_MAX]`. To obtain a Gaussian random variable one does the following.

First make the change of scale

$$w \to \tilde{w} := \frac{w}{\text{RAND\_MAX}} \tag{1.20}$$

so that $\tilde{w} \in [0, 1]$.

Now let w_1, w_2 be two independent random variables uniformly distributed in $[0, 1]$; then

$$x = \sqrt{-2\log(w_1)}\cos(2\pi w_2) \tag{1.21}$$

is a Gaussian random variable $\mathcal{N}(0, 1)$ (with zero mean value, unit variance, and density $\frac{1}{\sqrt{2\pi}}e^{-\frac{x^2}{2}}$). Therefore $x\sqrt{\delta t}$ may be used to simulate $W_{t+\delta t} - W_t$.

The C program (with a touch of C++) which uses these two ideas for S_t and C_t is as follows.

ALGORITHM 1.4. Pricing by Monte-Carlo.

```
#include <iostream>
#include <math.h>
#include <stdlib.h>
#include <fstream.h>
using namespace std;

const int M=500;                        //  nb of time steps of size dt
const int N=50000;                      //  nb of stochastic realization
const int L=40;                         //  nb of sampling point for S
const double K = 100;                                    //  the strike
const double left=0, right=130;                          //  the barriers
const double  sigmap=0.2, r=0.1;                         //  vol., rate
const double pi2 =8*atan(1), dt=1./M, sdt =sqrt(dt), eps=1.e-50;
const double er=exp(-r);

double gauss();
double EDOstoch(const double x, int m);
double EDOstoch_barrier(const double x, int m,
const double Smin, const double Smax);
double payoff(double s);

double gauss()
{
  return sqrt(eps-2.*log(eps+rand()/(double)RAND_MAX))
    *cos(rand()*pi2/RAND_MAX);
}

double EDOstoch(const double x, int m)
{
  double  S= x;
  for(int i=0;i<m;i++)
    S += S*(sigmap*gauss()*sdt+r*dt);
  return S;                                  //  gives S(x, t=m*dt)
}

double EDOstoch_barrier(const double x, int m,
const double Smin, const double Smax)
{
```

```
  if ((x<=Smin)||(x>=Smax))
    return -1000;
  double  S= x;
  for(int i=0;i<m;i++)
    {
      if ((S<=Smin)||(S>=Smax))
return -1000;
      S += S*(sigmap*gauss()*sdt+r*dt);
    }
  return S;
}

double payoff(double s)
{   if(s>K) return s-K; else return 0;}

int main( void )
{
  ofstream ff("stoch.dat");
  for(double x=0.;x<2*K;x+=2*K/L)
    {                                              //    sampling values for x=S
      double value =0;
      double y,S ;
      for(int i=0;i<N;i++)
{
  S=EDOstoch(x,M);
         //   S=EDOstoch_barrier(x,M, left, right); //for barrier options
  double y=0;
  if (S>= 0)
    y = er*payoff(S);
  value += y;
}
      ff << x <<"\t" << value/N << endl;
    }
  return 0;
}
```

The program stores the result in a file called stoch.dat in a format that `gnuplot` can use for graphics (see www.gnuplot.org). For barrier options, one has just to replace the line `EDOstoch(x,M);` by `EDOstoch-barrier(x,M, left, right);`. In Figure 1.1, we have plotted the price of the vanilla call option with payoff $K = 100$ one year to maturity, as a function of $\frac{20S}{K}$. The prices have been computed using Algorithm 1.4. In Figure 1.2, we have plotted the price of the barrier vanilla call option one year to maturity with $a = 0$, $b = 130$ in Algorithm 1.4.

Project 1.1. *Adapt the Monte-Carlo program above to compute a basket option with two assets and run some tests. Implement one variance reduction method (see below) to speed up the program.*

One can estimate the accuracy of the Monte-Carlo method by using the law of large numbers.

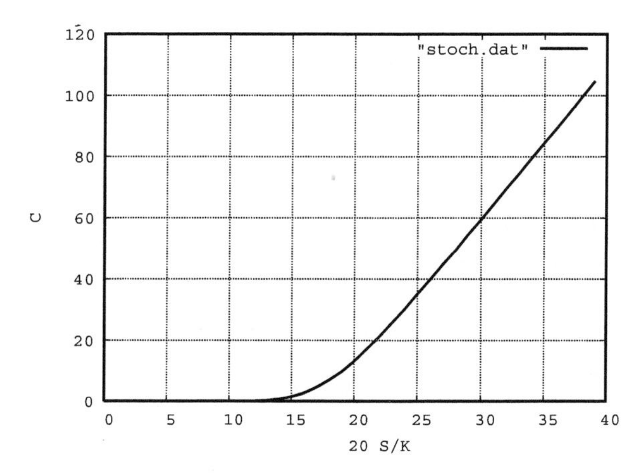

Figure 1.1. *Computation of the call price one year to maturity by using the Monte-Carlo algorithm above. The curve displays C versus S. It has been obtained from the file "stoch.dat," which contains the values shown in Table* 1.1, *by using the public domain program* `gnuplot` *(and the command* `plot "stoch.dat" w l`*).*

Table 1.1. *Content of the file* `stoch.dat` *generated by Algorithm 1.4. For clarity we have broken the single column into 5, so the numbers must be read left to right and down as usual.*

0	0	0	0	0
0	0	0	0	7.02358e-05
0.00657661	0.0295531	0.0975032	0.298798	0.748356
1.54013	2.77273	4.62055	7.06252	9.73593
13.1792	17.1289	21.1632	25.7196	30.1016
35.0933	39.847	44.8094	49.4797	54.7059
59.5088	64.5122	69.5124	74.3284	79.4387
84.3937	89.3514	94.3428	99.6568	104.677

Theorem 1.2 (central limit). *Let x be a random variable with probability density p, expectation $\mathbb{E}(x)$, and variance*

$$\operatorname{var}(x) = \mathbb{E}((x - \mathbb{E}(x))^2).$$

The approximation

$$\mathbb{E}(x) := \int_{-\infty}^{+\infty} p(x)x dx \approx \mathbb{E}_N(x) := \frac{1}{N}\sum_1^N x_i \tag{1.22}$$

satisfies, for all $c_1 < 0 < c_2$,

$$\lim_{N\to\infty} \mathcal{P}\left(\mathbb{E}(x) - \mathbb{E}_N(x) \in \left(c_1\sqrt{\frac{\operatorname{var}(x)}{N}}, c_2\sqrt{\frac{\operatorname{var}(x)}{N}}\right)\right) = \frac{1}{\sqrt{2\pi}}\int_{c_1}^{c_2} e^{-x^2/2}dx, \tag{1.23}$$

where $\mathcal{P}(y \in Y)$ stands for the probability that y belongs to Y.

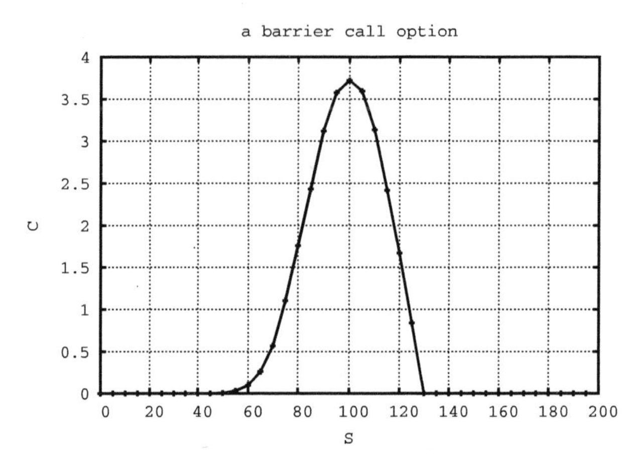

Figure 1.2. *The price of a barrier vanilla call option one year to maturity ($K = 100$, $a = 0$, $b = 130$, $\sigma = 0.2$, $r = 0.1$), computed by the Monte-Carlo method.*

1.4.2 Random Variables Using the GSL

Users should know that `rand()` is never perfect and that after many calls there is some periodicity in the answer. There are dedicated libraries which do a better and quicker job in the case of Gaussian variables. The GSL [59] implements a number of different functions to generate random variables. We give below a better implementation of the C-function `gauss()` by the GSL with a slight modification in that it returns a Gaussian variable of variance dt instead of variance 1.

ALGORITHM 1.5. Gaussian function using the GSL.

```
#include <gsl/gsl_rng.h>
#include <gsl/gsl_randist.h>
const gsl_rng_type *Tgsl=gsl_rng_default;
gsl_rng_env_setup();
gsl_rng *rgsl=gsl_rng_alloc(Tgsl);
.....
double gauss(double dt) { return gsl_ran_gaussian(rgsl, dt);}
.....
```

1.4.3 Variance Reduction

From the central limit theorem we see that the accuracy of the Monte-Carlo method for simulating the expectation of X is controlled by $\sqrt{\text{var}(X)/N}$. Thus the computing time for a given accuracy ϵ behaves like $\frac{\text{var}(X)}{\epsilon^2}$. Therefore, any change of variables which decreases σ will be valuable.

There are several such transformations but no really general technique.

Control Variates. To compute $\mathbb{E}(X)$ with a better precision one may instead compute $\mathbb{E}(X - X')$, where X' is a random variable with $\mathbb{E}(X')$ known and $\text{var}(X - X') < \text{var}(X)$. Indeed $\mathbb{E}(X) = \mathbb{E}(X - X') + \mathbb{E}(X')$ and $\mathbb{E}(X - X')$ will be known with better precision. To achieve $\text{var}(X - X') < \text{var}(X')$ one must choose X' "close" to X. The difficulty is to choose X'; we present below a method proposed in Glasserman [60].

Obviously, the problem of estimating the mean $\bar{X}$ of a random variable X with variance $\text{var}(X)$ can be shifted to that of estimating the mean $\bar{Z}$ of $Z = X - b(Y - \bar{Y})$, for a given constant b, and for any given random variable Y for which $\bar{Y}$ is known.

A quick calculation shows that

$$\text{var}(Z) = \mathbb{E}((Z - \bar{Z})^2) = \text{var}(X) + b^2\text{var}(Y) - 2b\mathbb{E}((X - \bar{X})(Y - \bar{Y}))$$

is minimized when

$$b = \frac{\mathbb{E}((X - \bar{X})(Y - \bar{Y}))}{\text{var}(Y)} \approx \frac{\sum_1^n (X_i - \bar{X})(Y_i - \bar{Y})}{\sum_1^n (Y_i - \bar{Y})^2},$$

where $\{X_i\}_1^n$ (resp., $\{Y_i\}_1^n$) is a set of n samples of X (resp., Y). Then

$$\frac{\text{var}(Z)}{\text{var}(X)} = 1 - \frac{\mathbb{E}\left((X - \bar{X})(Y - \bar{Y})^2\right)}{\mathbb{E}((X - \bar{X})^2)\mathbb{E}((Y - \bar{Y})^2)}.$$

To apply this idea to the pricing of a European vanilla call at time $t = 0$ by Black–Scholes formula (1.5), we take

$$X_T = e^{-rT}(S_T - K)_+, \quad Y_T = S_T \text{ and notice that } \mathbb{E}^*(Y_T) = e^{rT}S_0.$$

Let $\{S_i\}_1^n$ be n samples of S_T obtained by integrating (1.1); then $C_0 = \mathbb{E}^*(X_T) = \mathbb{E}^*(Z_T)$ with $Z_T = X_T - b(Y_T - \mathbb{E}^*(Y_T))$, which yields

$$C_0 = \mathbb{E}^*(Z_T) \approx \frac{1}{n}\sum_1^n (e^{-rT}(S_i - K)_+ - b(S_i - e^{rT}S_0)).$$

As seen in Figure 1.3, the convergence of the Monte-Carlo method is much improved with b chosen as above, when $K \ll S_0$, and not so much otherwise (see Glasserman [60]).

Antithetic Variates. Consider the stochastic ordinary differential equation with a change of sign in the term containing the Brownian motion and the process ζ_t given by

$$\begin{aligned} dS_t &= S_t(r dt + \sigma_t dW_t), \quad S_0 = x, \\ dS_t^- &= S_t^-(r dt - \sigma_t dW_t), \quad S_0^- = x, \\ \zeta_t &= \frac{e^{-rt}}{2}[(S_t - K)_+ + (S_t^- - K)_+]. \end{aligned} \tag{1.24}$$

Then

$$\mathbb{E}^*(\zeta_t) = \mathbb{E}^*(e^{-rt}(S_t - K)_+) \quad \text{and} \quad C_0 = \mathbb{E}^*(\zeta_T),$$

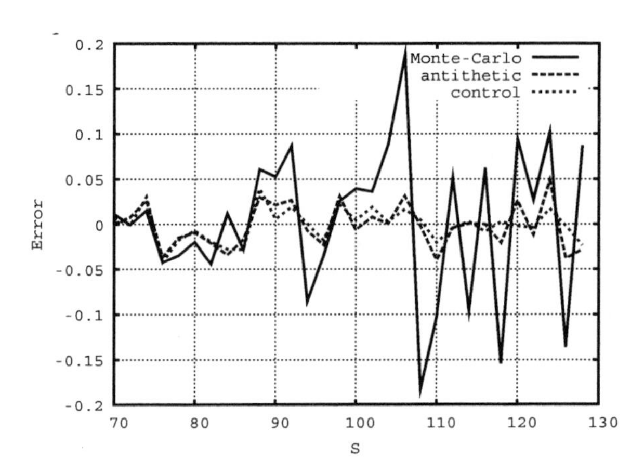

Figure 1.3. *Errors on the price of the vanilla call option one year to maturity ($K = 100, r = 0.1, \sigma = 0.2$) with the standard Monte-Carlo method and with variance reduction methods.*

because S_t^- and S_t have the same law, and

$$\begin{aligned}
\operatorname{var}(\zeta_t) &= \frac{e^{-2rt}}{4}\operatorname{var}((S_t - K)_+) + \frac{e^{-2rt}}{4}\operatorname{var}((S_t^- - K)_+) \\
&\quad + \frac{e^{-2rt}}{2}\mathbb{E}^*\left(\left((S_t - K)_+ - \mathbb{E}^*((S_t - K)_+)\right)\left((S_t^- - K)_+ - \mathbb{E}^*((S_t^- - K)_+)\right)\right) \\
&= \frac{e^{-2rt}}{2}\operatorname{var}((S_t - K)_+) \\
&\quad + \frac{e^{-2rt}}{2}\mathbb{E}^*\left(\left((S_t - K)_+ - \mathbb{E}^*((S_t - K)_+)\right)\left((S_t^- - K)_+ - \mathbb{E}^*((S_t^- - K)_+)\right)\right) \\
&\le \frac{1}{2}\operatorname{var}(e^{-rt}(S_t - K)_+),
\end{aligned}$$

since the function $x \mapsto (x - K)_+$ is monotone and the two variables S_t and S_t^- are negatively correlated. Therefore we have obtained a new process with the same expectation and with a smaller variance. To simulate S_t^- one reuses the same random variables used for S_t, so the additional operation count is small. Note that this technique needs a monotone payoff function.

A Program. The control variates (variable Y below) and the antithetic variates (variable Z below) are implemented and compared to the standard Monte-Carlo method.

ALGORITHM 1.6. Variance reduction by control variate.

```
int main( void )
{
 const int M=100;                               //    nb of time steps of size dt
 const int N=10000;                             //   nb of stochastic realization
```

```
const int L=100;                              //    nb of sampling point for S
const double K = 100, sigmap=0.2,  r=0.1;
const double dt=1./M, sdt =sqrt(dt), er = exp(r);
double X, Y, Z;
ofstream ff("comp.dat");
for(double x=70.;x<130;x+=2*K/L)
  {
    double meanY=x*er, meanX =0,meanZ=0, barY=0;
    double varX=0, varY=0, Exy=0;
    for(int i=0;i<N;i++)
      {
        double  S1=x,S2= x;
        for(int ii=0;ii<M;ii++)
          {
            double y = sigmap*gauss()*sdt;
            S1 += S1*(y+r*dt);
            S2 += S2*(-y+r*dt);
          }
        Y = S1; Z =(phi_T(S1)+phi_T(S2))/(2*er);
        X = phi_T(S1)/er;
        meanX += X; meanZ += Z; barY+=Y;
        Exy+=X*Y; varX+=X*X; varY+=Y*Y;
      }
    meanX /=N; meanZ /= N; barY/=N;
    varX = varX/N - meanX*meanX;
    varY = varY/N-2*meanY*barY + meanY*meanY;
    Exy = Exy/N - meanX*barY;
    double b = Exy/varY, C=meanX -b*(barY-meanY);
    double exact = Call(x, K, r, sigmap, 1.);
    ff << x <<"\t" << meanX-exact<< '\t'
        << meanZ-exact <<'\t'<<C-exact <<endl;
  }
 return 0;
}
```

1.5 Other Options

1.5.1 American Options

An American vanilla call (resp., put) option is a contract giving its owner the right to buy (resp., sell) a share of a specific common stock at a fixed price K before a certain date T. More generally, for a payoff function Q°, the American option with payoff Q° and maturity T can be exercised at any $t < T$, yielding the payoff $Q^\circ(S_t)$. In contrast to European options, American options can be exercised anytime before maturity. Since the American option gives its owner more rights than the corresponding European option, its price should be higher.

Consider, for example, an American vanilla put: if P_t were less than $K - S_t$, then one could buy a put and a share of the underlying asset and immediately exercise the option, making a risk-free immediate benefit of $K - S_t - P_t > 0$; this is ruled out by the no-arbitrage assumption, so we see that $P_t \geq K - S_t$. More generally, the value at time t of an American option with payoff Q° is always larger than $Q^\circ(S_t)$.

Using the notion of strategy with consumption, the Black–Scholes model leads to the following formula for pricing an American option with payoff Q°: under the risk neutral probability,

$$Q_t = \sup_{\tau \in \mathcal{T}_{t,T}} \mathbb{E}^*(e^{-\int_t^\tau r(u)du} Q^\circ(S_\tau)|F_t), \tag{1.25}$$

where $\mathcal{T}_{t,T}$ denotes the set of stopping times in $[t, T]$ (see [85] for the proof of this formula). For an American vanilla put, with σ and r constant, this gives

$$P_t = \sup_{\tau \in \mathcal{T}_{t,T}} \mathbb{E}^*(e^{-r(\tau-t)}(K - Se^{(r-\frac{\sigma^2}{2})(\tau-t)+\sigma(W_\tau - W_t)})_+). \tag{1.26}$$

It can be seen that for an American vanilla call, the formula (1.25) coincides with (1.5), so American and European vanilla calls have the same price. This means that an American vanilla call should not be exercised before maturity.

For a Monte-Carlo simulation of an American option, one performs the same type of Monte-Carlo simulation as for European options but then takes the sup for all times $\tau \in [t, T]$ so as to obtain a realization of P_t. Then by doing this N times and taking the average, one obtains P_t for one value of S. In practice this is quite expensive and one may prefer binomial trees (see below) or finite difference or finite element methods (see Chapter 6).

Project 1.2. *Program a Monte-Carlo method for American options. Study the influence of N: plot the results. Apply a method of variance reduction and program it (see, for example, Lapeyre and Sulem* [87]*).*

1.5.2 Asian Options

A typical example of an Asian option is a contract giving its owner the right to buy an asset for its average price over some prescribed period. Depending on how the average is computed, there can be many kinds of Asian options. For example, one can use arithmetic averaging

$$I_T = \int_0^T S_\tau d\tau, \quad A_T = \frac{1}{T} I_T,$$

or geometric averaging

$$A_T = \exp\left(\frac{1}{T}\int_0^T \log(S_\tau) d\tau\right).$$

A simple example of an Asian option is the call (resp., put) with floating strike which gives its holder the right to buy (resp., sell) the underlying asset at $A(T)$ at maturity T. At maturity, the price of the call (resp., put) option is $(S_T - A_T)_+$ (resp., $(S_T - A_T)_-$).

Another example is that of a call (resp., put) which gives its holder the right to buy (resp., sell) the underlying asset at $K - A_T + S_T$ at maturity T, for a fixed strike K. At maturity, the price of the call (resp., put) option is $(A_T - K)_+$ (resp., $(K - A_T)_-$). Such an option is termed *Asian option with fixed strike*.

More generally, for a function $Q^\circ\colon \mathbb{R}_+^2 \to \mathbb{R}_+$, it is possible to define the Asian option with payoff $Q^\circ(S_T, A_T)$.

The Black–Scholes model yields the formula

$$Q_t = \mathbb{E}^*(e^{-\int_t^T r(u)du} Q^\circ(S_T, A_T)|F_t)$$

for the price of the Asian option at time t, where the expectation is computed under the risk neutral probability. A Monte-Carlo simulation to price an Asian option with arithmetic averaging is as follows.

- S_0 given, set $A = 0$.
- For $(t = 0\,; t \le T;\ t = t + \delta t)$
  - call the random generator to simulate $W_{t+\delta t} - W_t$;
  - compute $S_{t+\delta t} = S_t(1 + r\delta t + W_{t+\delta t} - W_t)$;
  - do $A = A + \frac{\delta t}{2T}(S_t + S_{t+\delta t})$.
- Compute $e^{-rT} Q^\circ(S, A)$.
- Repeat the above to obtain M values and average them.

Another class of Asian options involves the extremal values of the asset price for $t \le T$: they are called lookback options. The floating strike lookback call has a payoff of $(S_T - \min_{0\le\tau\le T} S_\tau)_+$, whereas the lookback put has a payoff of $(\max_{0\le\tau\le T} S_\tau - S_T)_+$. Similarly, the fixed strike lookback call (resp., put) has a payoff of $(\max_{0\le\tau\le T} S_\tau - K)_+$ (resp., $(K - \min_{0\le\tau\le T} S_\tau)_+$). One can also define lookback options on averages. For a function $Q^\circ : \mathbb{R}_+^4 \to \mathbb{R}_+$, for $m_T = \min_{0\le\tau\le T} S_\tau$ and $M_T = \max_{0\le\tau\le T} S_\tau$, it is possible to define Asian options with payoff $Q^\circ(S_T, A_T, m_T, M_T)$, and the price of the option at t is

$$\mathbb{E}^*(e^{-\int_t^T r(u)du} Q^\circ(S_T, A_T, m_T, M_T)|F_t).$$

1.6 Complement: Binomial Trees

The second most popular numerical method for pricing options is akin to Bellman's dynamic programming and uses a tree of possible events, the so-called *binomial option pricing model*. It leads to quite a good numerical method which is also easy to understand.

1.6.1 The Tree for S

Consider the very simple situation where the underlying asset (i.e., $S_n = S_t,\ t = n\delta t$) can evolve in only two ways:

- either it goes up by a factor $u \ge 1$: $S_{n+1} = uS_n$ with probability p, or
- it goes down by a factor $d \le 1$: $S_{n+1} = dS_n$ with probability $1 - p$.

So if we denote by S_n^m one of the possible values of S at stage n, at the next stage we can have

$$S_n^m \begin{array}{l} \nearrow\ S_{n+1}^{m+1} = uS_n^m \\ \searrow\ S_{n+1}^m = dS_n^m \end{array}.$$

Note that at $n = 0$, S_0 is known; then at $n = 1$, $S_1 \in \{uS_0, dS_0\}$, at $n = 2$, $S_2 \in \{u^2S_0, udS_0, d^2S_0\}$ with probability $p^2, 2p(1-p), (1-p)^2$, and so forth.

At $n = 2$ the mean value of S_2 is

$$E(S_2) = S_0(u^2p^2 + 2ud(1-p)p + d^2(1-p)^2).$$

The factor 2 is because the middle state can be reached either by $S_0 \to uS_0 \to udS_0$ or by $S_0 \to dS_0 \to udS_0$ with the same probability $p(1-p)$.

Similarly, the variance of S_2 is

$$E(S_2^2 - E(S_2)^2) = S_0^2\left[u^4p^2 + 2u^2d^2(1-p)p + d^4(1-p)^2 - (up + d(1-p))^2\right].$$

After N steps, we have

$$\frac{S_N}{S_0} \in \left\{u^N, u^{N-1}d, u^{N-2}d^2, \dots, u^{N-j}d^j, \dots, d^N\right\},$$

the state $\frac{S_N}{S_0} = u^{N-j}d^j$ occurring with probability $\binom{j}{N}p^{N-j}(1-p)^j$, where $\binom{j}{N} = \frac{j!(N-j)!}{N!}$ are the binomial factors.

The problem now is to see what are the conditions necessary on d, u, p for $S_n \to S_t$ with $t = n\delta t$, $N = T/\delta t \to \infty$, and $\mathrm{d}S_t = S_t(r\mathrm{d}t + \sigma \mathrm{d}W_t)$.

Note first that the expectation of S_{n+1} knowing S_n is $(up + d(1-p))S_n$, and its variance is

$$E(S_{n+1}^2 - E(S_n)^2) = S_n^2(pu^2 + (1-p)d^2 - (up + d(1-p))^2) = p(1-p)(u-d)^2S_n^2.$$

When $\mathrm{d}S_t = S_t(r\mathrm{d}t + \sigma\mathrm{d}W_t)$, under the risk neutral probability $\mathbb{P}^*$, the mean and variance of S_t are $S_ne^{r\delta t}$ and $S_n^2e^{(2r+\sigma^2)\delta t}$, and therefore it is necessary that

$$up + d(1-p) = e^{r\delta t}$$

and

$$p(1-p)(u-d)^2 = e^{(2r+\sigma^2)\delta t}.$$

We have two equations for three variables, so there remains to add a third arbitrary equation: a popular choice is

$$p = \frac{1}{2},$$

which implies

$$u = e^{r\delta t}(1 + \sqrt{e^{\sigma^2\delta t} - 1}), \qquad d = e^{r\delta t}(1 - \sqrt{e^{\sigma^2\delta t} - 1}).$$

Another choice is

$$u = \frac{1}{d},$$

which yields

$$d = \alpha - \sqrt{\alpha^2 - 1}, \quad d = \alpha + \sqrt{\alpha^2 - 1}, \quad \text{with} \quad \alpha = \frac{1}{2}(e^{-rt} + e^{(r+\sigma^2)\delta t}),$$

and

$$p = \frac{e^{r\delta t} - d}{u - d}.$$

1.6.2 The Tree for C

By definition, the price of the European vanilla call option can be approximated by

$$C_N e^{rN\delta t} = \mathbb{E}^*((S_N - K)_+) = \sum_{j=0}^{N} \binom{j}{N} p^{N-j}(1-p)^j (u^{N-j} d^j S_0 - K)_+. \qquad (1.27)$$

Although (1.27) could be programmed directly, it is much faster to give a name to the intermediate states of S and C and use a backward induction relation.

Let $S_n^m = d^m u^{n-m} S_0$ be the mth possible value of S_n. Then

$$S_{n+1}^m = u S_n^m, \quad m = 0, \dots, n, \quad \text{and} \quad S_{n+1}^{n+1} = d S_n^n.$$

It is not difficult to see that $C_n = (S_n - K)_+$ at $t = n\delta t$ also has only two possible changes, a growth or a decrease, and that a two-stage change "up-down" is the same as a "down-up." Thus let C_n^m, $m = 0, \dots, n$, be the possible values of C_n. Because C_n^m increases to C_{n+1}^{m+1} with probability p or decreases to C_{n+1}^m with probability $1 - p$,

$$C_n^m \begin{array}{l} \nearrow C_{n+1}^{m+1} \\ \searrow C_{n+1}^{m} \end{array},$$

the expected value of C_{n+1} knowing $C_n = C_n^m$ is $pC_{n+1}^{m+1} + (1-p)C_{n+1}^m$, so the analogue to (1.5) is

$$e^{-r\delta t}(pC_{n+1}^{m+1} + (1-p)C_{n+1}^m) = C_n^m. \qquad (1.28)$$

Since the values C_N^m are known from the payoff function, this recurrence can be used with decreasing n.

Notice that it is not necessary to store all the intermediate values, and one can use the same memory for C_n^m and C_{n+1}^m.

ALGORITHM 1.7. European call by binomial tree.

```
double binomial(const double S0)
{
   double disc = exp(-r*dt);
   double u = (1+sqrt(exp(sigmap*sigmap*dt)-1))/disc;
   double d=(1-sqrt(exp(sigmap*sigmap*dt)-1))/disc, p=0.5;

   S[0] = S0;
   for(int m=1; m<M; m++)
   {
       for(int n=m; n>0; n--)
           S[n] = u*S[n-1];
       S[0] = d*S[0];
   }
   for(int n=0;n<M;n++)
       C[n] = S[n]>K?S[n]-K:0;
   for(int m=M-1;m>0;m--)
       for(int n=0; n<m;n++)
           C[n] = (p*C[n+1]+(1-p)*C[n])*disc;
   return C[0];
}
```

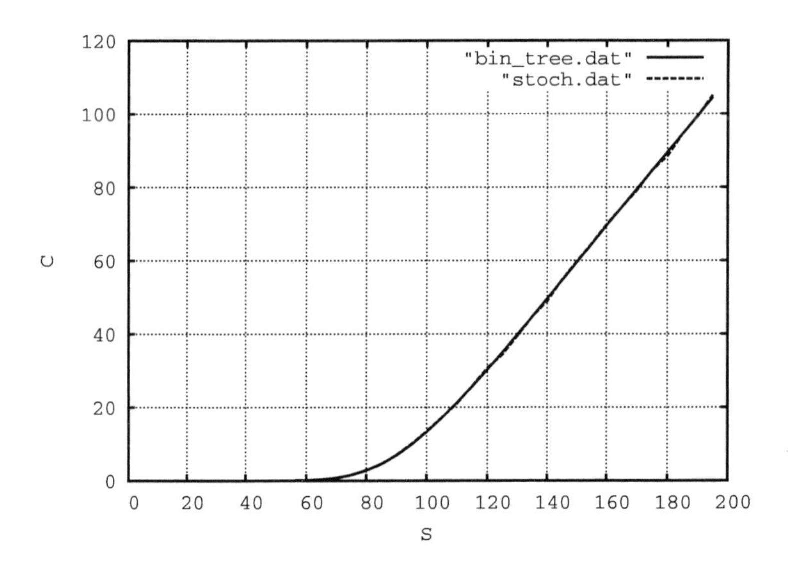

Figure 1.4. *Results for the call price one year to maturity by using the binomial tree method with $p = 0.5$ and 500 time steps. The curve displays C versus S. It has been obtained by Algorithm 1.7. For comparison we have also plotted the results with a stochastic method using only 500 samples (the dashed line).*

```
void main(){
 for(int i=0;i<nx;i++)
 {     double x=(2.*K*i)/nx;
       cout  << "C_0("<<x<<")  =  "<<binomial(x)<<endl;
 }
}
```

In Figure 1.4 we have plotted the price of a vanilla call option with strike $K = 100$ one year to maturity, as a function of the spot price. The prices have been computed by the binomial tree method of Algorithm 1.7.

Remark 1.5. *Note that the algorithm has $O(M^2)$ operations and can be slow for large M. There are implementations which are $O(M)$ (see, e.g.,* [91]*).*

Trees can be used for pricing American options; see [36, 117, 91]. The principle is the same but in the backward induction one must check that the expected profit is higher than the current one $S_0 u^{n-m} d^m$. An American put can be implemented as follows.

ALGORITHM 1.8. Binary tree for an American put.

```
double binomial(const double S0)
{
   double disc = exp(-r*dt);
   double u =  (1+sqrt(exp(sigmap*sigmap*dt)-1))/disc;
   double d=(1-sqrt(exp(sigmap*sigmap*dt)-1))/disc, p=0.5;
```

```
    S[0] = S0;um[0]=1; du[0]=1;
    for(int m=1; m<M; m++)
    {
        for(int n=m; n>0; n--)
            S[n] = u*S[n-1];
        S[0] = d*S[0];
        um[m]=u*um[m-1]; du[m]=d*du[m-1]/u;
    }
    for(int n=0;n<M;n++)
        P[n] = S[n]>K?0:K-S[n];
    for(int m=M-1;m>0;m--)
        for(int n=0; n<m;n++)
        {   P[n] = (p*P[n+1]+(1-p)*P[n])*disc;
            double gain=K-S0*um[m]*du[n];
                                                  //  pow(u,m-n)*pow(d,n);
            if(P[n]<aux) P[n]=gain;
        }
    return P[0];
}
```

Notice that to save computing time, u^m and d^m are precomputed and stored and $u^{m-n}d^n$ is written as $u^m(d/u)^n$. There are ways to reduce the computing time by eliminations of trivial branches in the algorithms, but the method remains inherently $O(M^2)$, i.e., slow for large values of M.

Project 1.3. *Study the influence of the choice of p, u, d on the results. Plot a convergence curve for the L2-norm of the error as a function of δt. Present the results of error estimations (search the literature,* [87] *in particular). Study the case of American options and adapt the programs to the case where σ depends on S. Adapt the programs to the case of a dividend paying stock (see §2.3).*

Chapter 2

The Black–Scholes Equation: Mathematical Analysis

2.1 Orientation

The Black–Scholes model introduced in Chapter 1 gives the option prices as expectations of random variables computed from the solution to a stochastic differential equation.

Itô's calculus allows the derivation of a partial differential equation of parabolic type for such mean quantities. We shall first recall how the partial differential equation associated to the Black–Scholes model is derived; then we shall study it in detail by variational methods and solve the following problems:

- What are the conditions for the parabolic problem to be well-posed?
- What are the qualitative properties of the solutions to the parabolic boundary value problems? In particular, is there a singularity at the origin $S = 0$? What is the regularity of the solutions? Is there a comparison principle between solutions to the partial differential equation?

Next, we consider the price of a vanilla European option as a function of the strike K and of maturity T, for a fixed spot price: it turns out this function is a solution to a partial differential equation in the variables K and T, known as Dupire's equation [41]. We will use this equation later for calibration.

The chapter ends with similar considerations for other options:

- barrier options,
- basket options,
- options with stochastic volatilities,
- options built on Lévy processes.

The Black–Scholes model involves a risk-free asset whose price S_t° satisfies the differential equation $dS_t^\circ = r(t)S_t^\circ$, and a risky asset under a probability $\mathbb{P}$ and a filtration F_t, whose price S_t satisfies the stochastic differential equation $dS_t = S_t(\mu dt + \sigma dB_t)$, where B_t is a standard Brownian motion. Here the volatility σ may depend on time and on the price of the underlying asset, i.e., $\sigma = \sigma(S_t, t)$. We assume that the functions $\sigma(S, t)$ and $r(t)$ are continuous nonnegative and bounded, respectively, on $\mathbb{R}_+ \times \mathbb{R}_+$ and $\mathbb{R}_+$, and that $S \mapsto S\sigma(S, t)$ is Lipschitz continuous with a constant independent of t, which ensure the existence and uniqueness of a square integrable S_t. As seen in Chapter 1, under the Black–Scholes assumption, there exists a probability $\mathbb{P}^*$ equivalent to $\mathbb{P}$ (the risk neutral probability) such that $W_t = B_t + \int_0^t \frac{\mu - r(s)}{\sigma_s} ds$ is a standard Brownian motion, and the price of a European option with payoff P_0 and maturity T is given by

$$\phi(S_t, t) = \mathbb{E}^* \left(e^{-\int_t^T r(s)ds} P_\circ(S_T) | F_t \right) \tag{2.1}$$

(F_t is the natural filtration of W_t). In the case of vanilla calls and puts with constant interest rate r, we have

$$\begin{aligned} C_t &= e^{-r(T-t)} \mathbb{E}^*((S_T - K)_+ | F_t), \\ P_t &= e^{-r(T-t)} \mathbb{E}^*((K - S_T)_+ | F_t). \end{aligned} \tag{2.2}$$

2.2 The Partial Differential Equation

2.2.1 Infinitesimal Generator of the Diffusion Process

It is possible to relate the function ϕ in (2.1) to the solution of a parabolic partial differential equation. The operator corresponding to this partial differential equation appears in a natural way in the following result.

Proposition 2.1. *Assume that the functions $\sigma(S, t)$ and $r(t)$ are continuous nonnegative and bounded, respectively, on $\mathbb{R}_+ \times [0, T]$ and $[0, T]$, and that $S \mapsto S\sigma(S, t)$ is Lipschitz continuous with a constant independent of t. Then, for any function $u : (S, t) \mapsto u(S, t)$ continuous in $\mathbb{R}_+ \times [0, T]$, C^1-regular with respect to t and C^2-regular with respect to S in $\mathbb{R}_+ \times [0, T)$, and such that $|S \frac{\partial u}{\partial S}| \le C(1 + S)$ with C independent of t, the process*

$$\begin{aligned} M_t &= e^{-\int_0^t r(\tau)d\tau} u(S_t, t) \\ &\quad - \int_0^t e^{-\int_0^\tau r(v)dv} \left(\frac{\partial u}{\partial t}(S_\tau, \tau) + L_\tau u(S_\tau, \tau) - r(\tau) u(S_\tau, \tau) \right) d\tau \end{aligned}$$

is a martingale under F_t, where L_t is the differential operator

$$L_t f(S) = \frac{\sigma^2(S, t) S^2}{2} \frac{d^2}{dS^2} f(S) + r(t) S \frac{d}{dS} f(S).$$

The differential operator L_t is called the infinitesimal generator of the Markov process S_t.

Proof. From the assumptions on r and σ, we know that S_t is a square integrable process, i.e., $\mathbb{E}^*(S_t^2) < +\infty$. Thanks to Itô's formula,

$$\begin{aligned} & d(e^{-\int_0^t r(\tau)d\tau} u(S_t, t)) \\ &= \left(-r(t) u(S_t, t) + \frac{\partial}{\partial t} u(S_t, t) + r S_t \frac{\partial}{\partial S} u(S_t, t) + \frac{\sigma^2(S_t, t) S_t^2}{2} \frac{\partial^2}{\partial S^2} u(S_t, t) \right) e^{-\int_0^t r(\tau)d\tau} dt \\ &\quad + \sigma(S_t, t) S_t \frac{\partial}{\partial S} u(S_t, t) e^{-\int_0^t r(s)ds} dW_t, \end{aligned}$$

which yields, from the definition of M_t,

$$dM_t = \sigma(S_t, t)S_t \frac{\partial}{\partial S} u(S_t, t)e^{-\int_0^t r(s)ds} dW_t.$$

Therefore, M_t is a martingale, because from the assumptions on σ and u,

$$\mathbb{E}^* \left(\int_0^t \left| \sigma(S_\tau, \tau)S_\tau \frac{\partial}{\partial S} u(S_\tau, \tau) \right|^2 d\tau \right) < +\infty$$

and $\mathbb{E}^*(M_T|F_t) = M_t$. □

Theorem 2.2. *Assume that σ and r are continuous nonnegative and bounded, and that $S \mapsto S\sigma(S, t)$ is Lipschitz continuous with a constant independent of t. Consider a function $P : \mathbb{R}_+ \times [0, T] \to \mathbb{R}$, continuous in $\mathbb{R}_+ \times [0, T]$ and C^1-regular with respect to t and C^2-regular with respect to S in $\mathbb{R}_+ \times [0, T)$, such that $|S\frac{\partial P}{\partial S}| \le C(1+S)$ with C independent of t. Assume that P satisfies*

$$\frac{\partial}{\partial t} P(S, t) + \frac{\sigma^2(S, t)S^2}{2} \frac{\partial^2}{\partial S^2} P(S, t) + r(t)S \frac{\partial}{\partial S} P(S, t) - r(t)P(S, t) = 0 \tag{2.3}$$

and

$$P(S, T) = P_0(S), \quad S \in \mathbb{R}_+; \tag{2.4}$$

then, with ϕ given by (2.1), *we have $\phi = P$.*

Proof. Applying Proposition 2.1 to $u = P$, the solution to (2.3), (2.4), we get

$$\mathbb{E}^*(e^{-\int_0^T r(s)ds} P_\circ(S_T)|F_t) = e^{-\int_0^t r(s)ds} P(S_t, t),$$

which is exactly the desired result. □

The problem (2.3), (2.4) is a backward-in-time parabolic boundary value problem, with a terminal Cauchy condition. Henceforth, we shall refer to (2.3) as the Black–Scholes equation. We are going to study in detail a weak formulation of the boundary value problem. Before that, we give some considerations on the asymptotics of the solutions to (2.3), (2.4) for large values of S, and on the Black–Scholes equation written in the variables $(\log(S), t)$.

2.2.2 Vanilla Options: Behavior for Extremal Values of S

We give here heuristic considerations which will be fully justified in §2.3. For simplicity, we assume that the interest rate r is constant. By construction we have that

$$C(S, T) = (S - K)_+ \quad \text{and} \quad P(S, T) = (K - S)_+ \quad \forall S > 0. \tag{2.5}$$

Furthermore if the financial model is reasonable, the call option should always be less than the underlying asset: $C_t < S_t$. Applied at $S = 0$ this gives $C(0, t) = 0$ for all t. Then the put-call parity implies $P(0, t) = Ke^{-r(T-t)}$.

On the other hand, when S is very large, the put option becomes useless, so we expect $P(S, t)$ to vanish as $S \to \infty$, and by the put-call parity, $C \approx S - Ke^{-r(T-t)}$.

From a mathematical point of view, it is important to understand that the behavior of P or C for small and large values of S need not be imposed in order to have a well-posed problem. We shall see later that along with the Black–Scholes partial differential equation, the terminal condition at T and a very weak growth condition for large values of S (namely, $C(S,t)$ is negligible compared to $e^{\eta \log^2(S)}$ for any $\eta > 0$) suffice to determine completely the price of the option. Similarly, in the numerical simulations, it will not be necessary to impose explicitly the value of the option at $S = 0$.

2.2.3 The Black–Scholes Equation in Logarithmic Prices

We consider the vanilla European call. It is convenient to set

$$\theta = T - t, \qquad x = \log S, \qquad \text{and} \quad \varphi(x,\theta) = C(e^x, T-\theta). \tag{2.6}$$

Notice that

$$S\frac{\partial C}{\partial S} = \frac{\partial \varphi}{\partial x} \quad \text{and} \quad S^2\frac{\partial^2 C}{\partial S^2} = S\frac{\partial}{\partial S}\left(S\frac{\partial C}{\partial S}\right) - S\frac{\partial C}{\partial S} = \frac{\partial^2 \varphi}{\partial x^2} - \frac{\partial \varphi}{\partial x}.$$

So, for a call, (2.3) becomes

$$\begin{aligned}
&\frac{\partial \varphi}{\partial \theta} - \frac{1}{2}\sigma^2\frac{\partial^2 \varphi}{\partial x^2} - \left(r - \frac{\sigma^2}{2}\right)\frac{\partial \varphi}{\partial x} + r\varphi = 0 \quad \text{in} \quad \mathbb{R}\times(0,T],\\
&\varphi(x,0) = (e^x - K)_+,\\
&\varphi(x,\theta) \sim 0 \quad \text{as} \quad x \to -\infty,\\
&\varphi(+\infty,\theta) = e^x - Ke^{-r\theta} \quad \text{as} \quad x \to +\infty.
\end{aligned} \tag{2.7}$$

The advantage of (2.7) is that it has constant coefficients, so we shall be able to recast it into the heat equation, after suitable changes of variables.

We set $\varphi(x,\theta) = \psi(x,\theta)e^{a\theta + bx}$, with $b = \frac{1}{2} - \frac{r}{\sigma^2}$ and $a = -r - \frac{\sigma^2}{2}b^2$. We obtain that ψ satisfies

$$\begin{aligned}
&\frac{\partial \psi}{\partial \theta} - \frac{\sigma^2}{2}\frac{\partial^2 \psi}{\partial x^2} = 0 \quad \text{in} \quad \mathbb{R}\times(0,T],\\
&\psi(x,0) = (e^{(\frac{1}{2}+\frac{r}{\sigma^2})x} - Ke^{(-\frac{1}{2}+\frac{r}{\sigma^2})x})_+,
\end{aligned} \tag{2.8}$$

and the growth of ψ at infinity is known thanks to (2.7). Therefore,

$$\psi(x,\theta) = \frac{1}{\sqrt{2\pi\sigma^2 t}}\int_{y\in\mathbb{R}} e^{-\frac{y^2}{2\sigma^2 t}}\psi(x-y,0)dy,$$

because

$$\frac{e^{-\frac{x^2}{2\sigma^2 t}}}{\sqrt{2\pi\sigma^2 t}}$$

is the fundamental solution to (2.7). This representation formula is another way of obtaining (1.17).

2.3 Mathematical Analysis of the Black–Scholes Equation with Local Volatility

We consider the Black–Scholes equation for a European option with a local volatility, i.e., σ is a function of S and t, and with a variable interest rate. In this section, it will be convenient to replace the time variable t by the time to maturity $T - t$; doing so, we get a forward parabolic equation: for $S > 0$ and $t \in (0, T]$,

$$\frac{\partial}{\partial t}P(S,t) - \frac{\sigma^2(S,t)S^2}{2}\frac{\partial^2}{\partial S^2}P(S,t) - r(t)S\frac{\partial}{\partial S}P(S,t) + r(t)P(S,t) = 0, \tag{2.9}$$

with the Cauchy data

$$P(S,0) = P_0(S), \quad S \in \mathbb{R}_+, \tag{2.10}$$

where P_0 is the payoff function.

Remark 2.1. *It is easy to see that for all $a \in \mathbb{R}$, $b \in \mathbb{R}$, the function $P(S,t) = aS + be^{-\int_0^t r(\tau)d\tau}$ is a solution to* (2.9)*, and it is also clear that such a function cannot satisfy the Cauchy condition* (2.10) *when P_0 is not an affine function.*

Dividends. Discretely paid dividends cause jumps on the price of the underlying asset: if a dividend D is paid at time to maturity t, then the process S_t satisfies

$$S_{(T-t)_+} = S_{(T-t)_-} - D,$$

because arbitrage is ruled out. On the other hand, the option price must not jump at t, because the option's owner does not get any benefit from the dividend, and because the dividend and date are known in advance. Therefore, the pricing function $P(S,t)$ (here t is the time to maturity) should satisfy

$$P(S,t_+) = P(S+D,t_-). \tag{2.11}$$

This means that the pricing function jumps at t: one has to integrate (2.9) in the time interval $(0,t)$, implement the jump condition (2.11), and integrate again (2.9) till the next dividend. This means that, when using a discrete method to compute P with a subdivision $(t_i)_{i\in\{0,\dots,M\}}$ of the time interval $[0,T]$, then the date of a discretely paid dividend should coincide with some t_i, so that (2.11) above can be implemented.

Note that if, for each time t, the asset pays out a dividend $q(t)S_t dt$ in dt, then the equation becomes

$$\frac{\partial}{\partial t}P(S,t) - \frac{\sigma^2(S,t)S^2}{2}\frac{\partial^2}{\partial S^2}P(S,t) - (r(t)-q(t))S\frac{\partial}{\partial S}P(S,t) + r(t)P(S,t) = 0. \tag{2.12}$$

This equation is of the same nature as (2.9), and if q is sufficiently well behaved, then (2.12) does not imply any additional mathematical difficulties.

Thus, in most cases, we will assume that $q = 0$, and that there are no discretely paid dividends.

Strong Solutions to (2.9), (2.10). The Cauchy problem (2.9), (2.10) has been very much studied: it is proved, for example, in [55] that if

- the function $(S, t) \mapsto S\sigma(S, t)$ is Lipschitz continuous on $\mathbb{R}_+ \times [0, T]$ (this condition can be weakened by considering only Hölder regularity),
- the function $\sigma(S, t)$ is bounded on $\mathbb{R}_+ \times [0, T]$ and bounded from below by a positive constant,
- the function $t \mapsto r(t)$ is bounded and Lipschitz continuous (this assumption can be relaxed),
- the Cauchy data P_0 satisfies $0 \le P_0(S) \le C(1 + S)$ for a given constant C,

then there exists a unique function $P \in \mathcal{C}^0(\mathbb{R}_+ \times [0, T])$, $\mathcal{C}^1$-regular with respect to t in $\mathbb{R}_+ \times (0, T]$ and $\mathcal{C}^2$-regular with respect to S in $\mathbb{R}_+ \times (0, T]$, which is solution to (2.9), (2.10) and which satisfies $0 \le P(S, t) \le C'(1 + S)$ for a given constant C'.

The function P is called a *strong solution* to the Cauchy problem. In what follows, we present the concept of *weak solutions* to (2.9), (2.10), which is fundamental for the development of many numerical methods presented in this book.

General Orientation. In what follows, we are going to present mathematical results about the Cauchy problem (2.9), (2.10): the concept of weak solutions to parabolic equations, as in [90, 21], will be central. The idea behind it is that there is a natural energy associated to (2.9). This leads us to introduce a suitable Sobolev space (the space of functions for which the above-mentioned energy is defined), and to define the concept of weak solutions to (2.9), (2.10). Then the abstract theory of [90] will yield the existence and uniqueness of such a weak solution, yet without giving much information on its regularity. The next step will consist in obtaining regularity results depending on the regularity of the payoff function. We will also present the weak maximum principle, which will permit us to compare various solutions to (2.9), and to obtain bounds and qualitative results. For instance, the put-call parity will be proved as a consequence of the maximum principle. Finally, we shall treat the case of an option on a basket of assets, showing that the theory extends very naturally to this case.

Note also that in addition to being a very general mathematical tool, the weak or variational formulation to (2.9), (2.10) will be the ground for the finite element method for the numerical approximation of (2.9), (2.10).

2.3.1 Some Function Spaces

We denote by $L^2(\mathbb{R}_+)$ the Hilbert space of square integrable functions on $\mathbb{R}_+$, endowed with the norm $\|v\|_{L^2(\mathbb{R}_+)} = (\int_{\mathbb{R}_+} v(x)^2 dx)^{\frac{1}{2}}$ and the inner product $(v, w)_{L^2(\mathbb{R}_+)} = \int_{\mathbb{R}_+} v(x)w(x)dx$. Calling $\mathcal{D}(\mathbb{R}_+)$ the space of the smooth functions with compact support in $\mathbb{R}_+$, we know that $\mathcal{D}(\mathbb{R}_+)$ is dense in $L^2(\mathbb{R}_+)$.

Let us introduce the space

$$W = \left\{ w \text{ continuous on } [0, +\infty) : \ w(x) = \int_0^x \phi(s)ds, \ \text{with } \phi \in L^2(\mathbb{R}_+) \right\}. \quad (2.13)$$

It is clear that the functions of W vanish at $x = 0$. The space W endowed with the norm $\|w\|_W = \|\frac{dw}{dx}\|_{L^2(\mathbb{R}_+)}$ is a Hilbert space, which is topologically isomorphic to $L^2(\mathbb{R}_+)$. Thus W is separable.

Lemma 2.3. *The space $\mathcal{D}(\mathbb{R}_+)$ is dense in W.*

Proof. Consider a function $w \in W$: $w(x) = \int_0^x \phi(s)ds$, with $\phi \in L^2(\mathbb{R}_+)$. We know that $\mathcal{D}(\mathbb{R}_+)$ is dense in $L^2(\mathbb{R}_+)$, so we can find a sequence of functions $(\phi_m)_{m\in\mathbb{N}}$, $\phi_m \in \mathcal{D}(\mathbb{R}_+)$, converging to ϕ in $L^2(\mathbb{R}_+)$. Therefore, the sequence $(w_m)_{m\in\mathbb{N}} : w_m(x) = \int_0^x \phi_m(s)ds$ converges to w in W. The function w_m is smooth but its support may not be compact in $\mathbb{R}_+$. Let us modify it slightly by introducing a smooth nonnegative function ψ on $\mathbb{R}_+$, with total mass 1, and supported in the interval $(1, 2)$, and by setting, for a small positive parameter ϵ, $\phi_{m,\epsilon}(x) = \phi_m(x) - \epsilon\psi(\epsilon x)\int_{\mathbb{R}_+}\phi_m(s)ds$: this function is contained in $\mathcal{D}(\mathbb{R}_+)$ and its total mass is 0. Now, if $w_{m,\epsilon}(x) = \int_0^x \phi_{m,\epsilon}(s)ds$,

$$\|w - w_{m,\epsilon}\|_W^2 \le 2\|w - w_m\|_W^2 + 2\epsilon^2\left|\int_{\mathbb{R}_+}\phi_m(s)ds\right|^2 \int_{\mathbb{R}_+}\psi^2(\epsilon s)ds,$$

and $\int_{\mathbb{R}_+}\psi^2(\epsilon s)ds \le \frac{C}{\epsilon}$, so choosing m large enough, then ϵ small enough, $\|w - w_{m,\epsilon}\|_W$ can be made as small as desired. The result is proved. □

Lemma 2.4 (Hardy's inequality). *If $w \in W$, then $\frac{w}{x} \in L^2(\mathbb{R}_+)$, and*

$$\left\|\frac{w}{x}\right\|_{L^2(\mathbb{R}_+)} \le 2\|w\|_W. \tag{2.14}$$

Proof. From Lemma 2.3, it is enough to prove (2.14) for $w \in \mathcal{D}(\mathbb{R}_+)$. Clearly, $\|\frac{w}{x}\|^2_{L^2(\mathbb{R}_+)} = \int_{\mathbb{R}_+}\frac{1}{x^2}(\int_0^x w'(s)ds)^2dx$, and integrating by parts, we obtain that

$$\begin{aligned}\left\|\frac{w}{x}\right\|^2_{L^2(\mathbb{R}_+)} &= \int_{\mathbb{R}_+}\frac{2}{x}w'(x)\left(\int_0^x w'(s)ds\right)dx \\ &= \int_{\mathbb{R}_+}\frac{2}{x}w'(x)w(x)dx \le 2\left\|\frac{w}{x}\right\|_{L^2(\mathbb{R}_+)}\|w\|_W,\end{aligned}$$

where the last estimate comes from the Cauchy–Schwarz inequality. □

From Lemma 2.4, we can define the space V:

$$V = \left\{\frac{w}{x} : w \in W\right\} = \left\{v \in L^2(\mathbb{R}_+) : v = \frac{w}{x}, \text{ with } w \in W\right\}. \tag{2.15}$$

It is clear that a function $v \in V$ is continuous on $\mathbb{R}_+$, for xv is continuous on $[0, +\infty)$.

Lemma 2.5. *We have the identity*

$$V = \left\{v \in L^2(\mathbb{R}_+) : x\frac{dv}{dx} \in L^2(\mathbb{R}_+)\right\}, \tag{2.16}$$

where the derivative must be understood in the sense of the distributions on $\mathbb{R}_+$.

Proof. A function $v \in L^2(\mathbb{R}_+)$ satisfies $xv \in W$ if and only if $\frac{d}{dx}(xv) \in L^2(\mathbb{R}_+)$, by the definition of W. This is equivalent to $x\frac{dv}{dx} \in L^2(\mathbb{R}_+)$ because $\frac{d}{dx}(xv) = v + x\frac{dv}{dx}$. □

From Lemma 2.5, we can endow V with the inner product $(v, w)_V = (v, w) + (x\frac{dv}{dx}, x\frac{dw}{dx})$, and with the Euclidean norm $\|v\|_V = \sqrt{(v, v)_V}$, and one can check easily that V is a Hilbert space.

Lemma 2.6. *The space $\mathcal{D}(\mathbb{R}_+)$ is dense in V.*

Proof. A function $v \in V$ if and only if $xv \in W$. Using Lemma 2.3, let $(w_m)_{m\in\mathbb{N}}$ be a sequence of functions in $\mathcal{D}(\mathbb{R}_+)$ converging to xv in W. The functions $v_m = \frac{w_m}{x}$ belong to $\mathcal{D}(\mathbb{R}_+)$, and converge to v in $L^2(\mathbb{R}_+)$ by Lemma 2.4. Furthermore, $x\frac{dv_m}{dx} = \frac{dw_m}{dx} - \frac{w_m}{x}$. The first term of this sum converges to $\frac{d}{dx}(xv)$ in $L^2(\mathbb{R}_+)$, whereas the second one converges to v in $L^2(\mathbb{R}_+)$. Therefore, $x\frac{dv_m}{dx}$ converges to $x\frac{dv}{dx}$ in $L^2(\mathbb{R}_+)$. □

Lemma 2.7 (Poincaré's inequality). *If $v \in V$, then*

$$\|v\|_{L^2(\mathbb{R}_+)} \le 2\left\|x\frac{dv}{dx}\right\|_{L^2(\mathbb{R}_+)}. \tag{2.17}$$

Proof. From Lemma 2.6, it is enough to prove (2.17) for $v \in \mathcal{D}(\mathbb{R}_+)$: we have

$$2\int_{\mathbb{R}_+} xv(x)\frac{dv}{dx}(x)dx = -\int_{\mathbb{R}_+} v^2(x)dx.$$

Using the Cauchy–Schwarz inequality on the left-hand side of this identity, we deduce that

$$\|v\|^2_{L^2(\mathbb{R}_+)} \le 2\|v\|_{L^2(\mathbb{R}_+)}\left\|x\frac{dv}{dx}\right\|_{L^2(\mathbb{R}_+)},$$

which yields the desired estimate. □

From Lemma 2.7, we see that the seminorm $|v|_V = \|x\frac{dv}{dx}\|_{L^2(\mathbb{R}_+)}$ is in fact a norm on V, which is equivalent to $\|\cdot\|_V$. In the same manner, using the density of $\mathcal{D}(\mathbb{R}_+)$ in $L^2(\mathbb{R}_+)$, one can prove the following result.

Lemma 2.8. *If $w \in L^2(\mathbb{R}_+)$, then the function v, $v(x) = \frac{1}{x}\int_0^x w(s)ds$ belongs to V, and there exists a positive constant C independent of w such that $\|v\|_V \le \|w\|_{L^2(\mathbb{R}^+)}$.*

We denote by V' the topological dual space of V. For simplicity, we also denote by $(\cdot,\cdot)$ the duality pairing between V' and V, and we define $\|\cdot\|_{V'}$ by

$$\|w\|_{V'} = \sup_{v\in V\setminus\{0\}} \frac{(w, v)}{|v|_V}. \tag{2.18}$$

2.3.2 The Weak Formulation of the Black–Scholes Equation

Let us multiply (2.9) by a smooth real-valued function ϕ on $\mathbb{R}_+$ and integrate in S on $\mathbb{R}_+$. Assuming that integrations by parts are permitted, we obtain

$$\begin{aligned}
0 = &\frac{d}{dt}\int_{\mathbb{R}_+} P(S,t)\phi(S)dS \\
&+\int_{\mathbb{R}_+} \frac{S^2\sigma^2(S,t)}{2}\frac{\partial P}{\partial S}(S,t)\frac{\partial \phi}{\partial S}(S)\,dS \\
&+\int_{\mathbb{R}_+} \left(-r(t)+\sigma^2(S,t)+S\sigma(S,t)\frac{\partial \sigma}{\partial S}(S,t)\right) S\frac{\partial P}{\partial S}(S,t)\phi(S)\,dS \\
&+r(t)\int_{\mathbb{R}_+} P(S,t)\phi(S)\,dS.
\end{aligned}$$

This leads us to introduce the bilinear form a_t,

$$\begin{aligned}
a_t(v,w) = &\int_{\mathbb{R}_+} \frac{S^2\sigma^2(S,t)}{2}\frac{\partial v}{\partial S}\frac{\partial w}{\partial S}\,dS \\
&+\int_{\mathbb{R}_+} \left(-r(t)+\sigma^2(S,t)+S\sigma(S,t)\frac{\partial \sigma}{\partial S}(S,t)\right) S\frac{\partial v}{\partial S}w\,dS \\
&+r(t)\int_{\mathbb{R}_+} vw\,dS.
\end{aligned} \tag{2.19}$$

We make some assumptions on σ and r: we assume that the coefficients σ and $r \geq 0$ are continuous (only for simplicity), and that σ is sufficiently regular so that the following conditions make sense.

Assumption 2.1.

1. *There exist two positive constants, $\underline{\sigma}$ and $\overline{\sigma}$, such that for all $t \in [0,T]$ and all $S \in \mathbb{R}_+$,*

$$0 < \underline{\sigma} \leq \sigma(S,t) \leq \overline{\sigma}. \tag{2.20}$$

2. *There exists a positive constant C_σ such that for all $t \in [0,T]$ and all $S \in \mathbb{R}_+$,*

$$\left| S\frac{\partial \sigma}{\partial S}(S,t)\right| \leq C_\sigma. \tag{2.21}$$

Lemma 2.9. *Under Assumption 2.1, the bilinear form a_t is continuous on V; i.e., there exists a positive constant μ such that for all $v, w \in V$,*

$$|a_t(v,w)| \leq \mu |v|_V |w|_V. \tag{2.22}$$

Proof. If $v, w \in V$, then using (2.17), (2.20), (2.21) and calling $R = \max_{t\in[0,T]} r(t)$,

$$\left|\int_{\mathbb{R}_+} \frac{S^2\sigma^2(S,t)}{2}\frac{\partial v}{\partial S}\frac{\partial w}{\partial S}\,dS\right| \le \frac{\overline{\sigma}^2}{2}|v|_V|w|_V,$$

$$\left|\int_{\mathbb{R}_+}\left(-r(t)+\sigma^2(S,t)+S\sigma(S,t)\frac{\partial\sigma}{\partial S}(S,t)\right)S\frac{\partial v}{\partial S}w\,dS\right|$$

$$\le (R+\overline{\sigma}^2+C_\sigma\overline{\sigma})|v|_V\|w\|_{L^2(\mathbb{R}_+)} \le 2(R+\overline{\sigma}^2+C_\sigma\overline{\sigma})|v|_V|w|_V,$$

$$r(t)\left|\int_{\mathbb{R}_+} vw\,dS\right| \le 4R|v|_V|w|_V,$$

which yields (2.22) with $\mu = \frac{5}{2}\overline{\sigma}^2 + 2C_\sigma\overline{\sigma} + 6R$. □

It is possible to associate with the bilinear form a_t the linear operator A_t: $V \to V'$; for all $v, w \in V$, $(A_t v, w) = a_t(v, w)$. The operator A_t is bounded from V to V'.

We define $C^0([0,T]; L^2(\mathbb{R}_+))$ as the space of continuous functions on $[0,T]$ with values in $L^2(\mathbb{R}_+)$, and $L^2(0,T;V)$ as the space of square integrable functions on $(0,T)$ with values in V. Assuming that $P_0 \in L^2(\mathbb{R}_+)$, and following [90], it is easy to write a weak formulation for (2.9), (2.10).

Weak Formulation of (2.9), (2.10). Find $P \in C^0([0,T]; L^2(\mathbb{R}_+)) \cap L^2(0,T;V)$, such that $\frac{\partial P}{\partial t} \in L^2(0,T;V')$, satisfying

$$P_{|t=0} = P_0 \quad \text{in } \mathbb{R}_+ \text{ and for a.e. } t \in (0,T), \tag{2.23}$$

$$\forall v \in V, \quad \left(\frac{\partial P}{\partial t}(t), v\right) + a_t(P(t), v) = 0. \tag{2.24}$$

In order to apply the abstract theory of Lions and Magenes [90], we need the following estimate.

Lemma 2.10 (Gårding's inequality). *Under Assumption* 2.1, *there exists a nonnegative constant* λ *such that for all* $v \in V$,

$$a_t(v,v) \ge \frac{\underline{\sigma}^2}{4}|v|_V^2 - \lambda\|v\|^2_{L^2(\mathbb{R}_+)}. \tag{2.25}$$

Proof. If $v \in V$, then using (2.17), (2.20), (2.21) and calling $R = \max_{t\in[0,T]} r(t)$,

$$\left|\int_{\mathbb{R}_+} \frac{S^2\sigma^2(S,t)}{2}\left(\frac{\partial v}{\partial S}\right)^2 dS\right| \ge \frac{\underline{\sigma}^2}{2}|v|_V^2,$$

$$\left|\int_{\mathbb{R}_+}\left(-r(t)+\sigma^2(S,t)+S\sigma(S,t)\frac{\partial\sigma}{\partial S}(S,t)\right)S\frac{\partial v}{\partial S}v\,dS\right|$$

$$\le (R+\overline{\sigma}^2+C_\sigma\overline{\sigma})|v|_V\|v\|_{L^2(\mathbb{R}_+)} \le \frac{\underline{\sigma}^2}{4}|v|_V^2 + \lambda\|v\|^2_{L^2(\mathbb{R}_+)},$$

where $\lambda = (R+\overline{\sigma}^2+C_\sigma\overline{\sigma})^2/(\underline{\sigma}^2)$. This achieves the proof. □

Theorem 2.11. *If $P_0 \in L^2(\mathbb{R}_+)$, and under Assumption* 2.1, *the weak formulation* (2.23), (2.24) *has a unique solution, and we have the estimate, for all t, $0 < t < T$,*

$$e^{-2\lambda t}\|P(t)\|^2_{L^2(\mathbb{R}_+)} + \frac{1}{2}\underline{\sigma}^2 \int_0^t e^{-2\lambda\tau}|P(\tau)|^2_V d\tau \leq \|P_0\|^2_{L^2(\mathbb{R}_+)}. \tag{2.26}$$

Proof. The proof is given in [90]. The estimate (2.26) is obtained by taking $v = P(t)e^{-2\lambda t}$ in (2.24), using (2.25) and integrating in time between 0 and t. □

Note that Theorem 2.11 applies for any European option with a payoff function in $L^2(\mathbb{R}_+)$, in particular to vanilla puts. It does not apply to vanilla calls, and we will come back to this later.

2.3.3 Regularity of the Weak Solutions

If the interest rate, the volatility, and the payoff are smooth enough, then it is possible to prove additional regularity for the solution to (2.23), (2.24).

Calling A_t the unbounded operator in $L^2(\mathbb{R}_+)$,

$$A_t v = -\frac{\sigma^2(S,t)S^2}{2}\frac{\partial^2 v}{\partial S^2} - r(t)S\frac{\partial v}{\partial S} + r(t)v,$$

it can be checked that for all $t \in [0, T]$ and for λ given in Lemma 2.10, the domain of $A_t + \lambda$ is

$$D = \left\{ v \in V;\, S^2\frac{\partial^2 v}{\partial S^2} \in L^2(\mathbb{R}_+) \right\}. \tag{2.27}$$

Assumption 2.2. *There exist a positive constant C and $0 \leq \alpha \leq 1$ such that for all $t_1, t_2 \in [0, T]$ and $S \in \mathbb{R}_+$,*

$$|r(t_1) - r(t_2)| + |\sigma(S,t_1) - \sigma(S,t_2)| + S\left|\frac{\partial \sigma}{\partial S}(S,t_1) - \frac{\partial \sigma}{\partial S}(S,t_2)\right| \leq C|t_1 - t_2|^\alpha. \tag{2.28}$$

With Assumptions 2.1 and 2.2, it is possible to prove what is called a smoothing effect: the solution to (2.23), (2.24) belongs to D at any time $t > 0$, for any Cauchy data $P_0 \in L^2(\mathbb{R}_+)$. More precisely, Assumptions 2.1 and 2.2 ensure that

- the domain of A_t is D, which is dense in $L^2(\mathbb{R}_+)$ and independent of t;
- we have the Gårding's inequality (2.25);
- if $\tilde{A}_t = A_t + \lambda I$, there exists a constant L such that

$$\|(\tilde{A}_t - \tilde{A}_s)\tilde{A}_\tau^{-1}\|_{\mathcal{L}(L^2(\mathbb{R}_+))} \leq L|t-s|^\alpha. \tag{2.29}$$

With these three facts, we can apply a general result of Kato on parabolic evolution equations (see [98, Theorem 5.6.8] and [25]) and obtain the following result.

Theorem 2.12. *Under Assumptions* 2.1 *and* 2.2, *for all s*, $0 < t \le T$, *the solution* P *of* (2.23), (2.24) *satisfies* $P \in \mathcal{C}^0([t,T];D)$ *and* $\frac{\partial P}{\partial t} \in \mathcal{C}^0([t,T];L^2(\mathbb{R}_+))$, *and there exists a constant* C *such that for all* t, $0 < t \le T$,

$$\|A_t P(t)\|_{L^2(\mathbb{R}_+)} \le \frac{C}{t}.$$

If $P_0 \in D$, *then the solution* P *of* (2.23), (2.24) *belongs to* $\mathcal{C}^0([0,T];D)$ *and* $\frac{\partial P}{\partial t} \in \mathcal{C}^0([0,T];L^2(\mathbb{R}_+))$.

Remark 2.2. *Note that for the second part of Theorem* 2.12, *it is possible to relax Assumption 2.2.*

Let us give a mild regularity result when $P_0 \in V$.

Proposition 2.13. *If Assumption* 2.1 *is satisfied and if* $P_0 \in V$, *then the solution to* (2.23), (2.24) *belongs to* $\mathcal{C}^0([0,T];V) \cap L^2(0,T;D)$, $\frac{\partial P}{\partial t} \in L^2(0,T;L^2(\mathbb{R}_+))$, *and there exists a nonnegative constant* $\tilde{\lambda}$ *such that the estimate*

$$e^{-2\tilde{\lambda}t}\left\|S\frac{\partial P}{\partial S}(t)\right\|^2_{L^2(\mathbb{R}_+)} + \frac{\underline{\sigma}^2}{2}\int_0^t e^{-2\tilde{\lambda}\tau}\left|S\frac{\partial P}{\partial S}(\tau)\right|^2_V d\tau \le \left\|S\frac{\partial P_0}{\partial S}\right\|^2_{L^2(\mathbb{R}_+)} \tag{2.30}$$

holds.

Proof. Consider first the case when the coefficients of (2.9) satisfy Assumptions 2.1 and 2.2: calling $Q = \frac{\partial P}{\partial S}$ and taking the derivative of (2.9) with respect to S, one obtains in the sense of distributions

$$\frac{\partial}{\partial t}Q(S,t) - \frac{\partial}{\partial S}\left(\frac{\sigma^2(S,t)S^2}{2}\frac{\partial}{\partial S}Q(S,t)\right) - r(t)S\frac{\partial}{\partial S}Q(S,t) = 0, \tag{2.31}$$

and by multiplying by S, we obtain with $W = SQ$

$$\frac{\partial}{\partial t}W(S,t) - \frac{\partial}{\partial S}\left(\frac{\sigma^2(S,t)S^2}{2}\frac{\partial}{\partial S}W(S,t)\right) + \left(\frac{\sigma^2(S,t)}{2} - r(t)\right)S\frac{\partial}{\partial S}W(S,t) + \left(r(t) - \frac{\sigma^2(S,t)}{2}\right)W(S,t) = 0, \tag{2.32}$$

which can be written in the shorter form

$$\frac{d}{dt}W + B_t W = 0, \tag{2.33}$$

where B_t is given by

$$B_t v = -\frac{\partial}{\partial S}\left(\frac{\sigma^2(S,t)S^2}{2}\frac{\partial v}{\partial S}\right) + \left(\frac{\sigma^2(S,t)}{2} - r(t)\right)S\frac{\partial v}{\partial S} + \left(r(t) - \frac{\sigma^2(S,t)}{2}\right)v.$$

Under Assumption 2.1, B_t is a linear operator from V to V', bounded uniformly with respect to t, and it is possible to prove Gårding's inequality: there exists a nonnegative constant $\tilde{\lambda}$ such that, for all $v \in V$,

$$(B_t v, v) \ge \frac{\underline{\sigma}^2}{4}|v|^2_V - \tilde{\lambda}\|v\|^2_{L^2(\mathbb{R}_+)}.$$

Under the assumptions of Theorem 2.12, the terms in (2.33) belong to $L^2(0,T;V')$, so it is possible to take the duality product with $W(S,t)e^{-2\tilde{\lambda}t}$ and to integrate in time. One obtains

$$\frac{1}{2}(e^{-2\tilde{\lambda}t}\|W(t)\|^2_{L^2(\mathbb{R}_+)} - \|W(0)\|^2_{L^2(\mathbb{R}_+)}) + \int_0^t e^{-2\tilde{\lambda}\tau}((\tilde{\lambda} + B_\tau)W(\tau), W(\tau))d\tau = 0,$$

and using Gårding's inequality, and the fact that $W = S\frac{\partial P}{\partial S}$, we find (2.30), i.e.,

$$e^{-2\tilde{\lambda}t}\left\|S\frac{\partial P}{\partial S}(t)\right\|^2_{L^2(\mathbb{R}_+)} + \frac{\underline{\sigma}^2}{2}\int_0^t e^{-2\tilde{\lambda}\tau}\left|S\frac{\partial P}{\partial S}(\tau)\right|^2_V d\tau \le \left\|S\frac{\partial P_0}{\partial S}\right\|^2_{L^2(\mathbb{R}_+)}.$$

With some technical arguments that can be skipped, it is possible to prove that the estimate (2.30) holds if $P_0 \in V$ and if the volatility satisfies only Assumption 2.1. Indeed,

- D is a dense subspace of V so it is possible to approximate P_0 in V by a sequence of functions $P_{0,\epsilon} \in D$;
- for $p > 1$ large enough, it is possible to approximate the interest rate r by nonnegative smooth functions r_ϵ, uniformly bounded and such that $r_\epsilon \to r$ in $L^p((0,T))$;
- it is possible to approximate the volatility σ by nonnegative smooth functions σ_ϵ, obeying Assumption 2.1 uniformly and Assumption 2.2 with a constant C_ϵ, and such that $\sigma_\epsilon \to \sigma$ and $S\frac{\partial \sigma_\epsilon}{\partial S} \to S\frac{\partial \sigma}{\partial S}$ in $L^p(\omega)$, for all compact ω of $\mathbb{R}_+ \times [0,T]$ and for all $p < +\infty$.

We call P_ϵ the solution to the Black–Scholes equation with payoff $P_{0,\epsilon}$, interest rate r_ϵ, and volatility σ_ϵ. Thanks to (2.30), it is possible to extract a subsequence still called P_ϵ that converges weakly * in $L^\infty(0,T;V)$ and weakly in $L^2(0,T;D)$ and such that $\frac{\partial P_\epsilon}{\partial t}$ converges weakly in $L^2(\mathbb{R}_+ \times (0,T))$. The limit satisfies (2.30) a.e. in t. On the other hand, by passing to the limit as $\epsilon \to 0$, one sees that the limit of P_ϵ must satisfy (2.9) in the sense of distributions and (2.10). Therefore, the limit is the unique solution P to (2.23), (2.24). ☐

Remark 2.3. *As a consequence of Proposition* 2.13, *the solution to* (2.23), (2.24) *is continuous if Assumption* 2.1 *is satisfied and* $P_0 \in V$.

Note that Proposition 2.13 and Remark 2.3 apply to a European vanilla put: indeed, $P_0(S) = (K - S)_+ \in V$.

2.3.4 The Maximum Principle

We are going to give a maximum principle for weak solutions of (2.9). The solutions of (2.9) may not vanish for $S \to +\infty$. Therefore, we are going to state the maximum principle for a class of functions much larger than V; see [102] for a reference book on the maximum principle. We define

$$\mathcal{V} = \{v : \forall\epsilon > 0,\ v(S)e^{-\epsilon\log^2(S+2)} \in V\}. \tag{2.34}$$

Note that the polynomial functions belong to $\mathcal{V}$.

We are going to use the truncation method of Stampacchia. With this aim, we define, for $v \in \mathcal{V}$, $v_+(S) = v(S)1_{v(S)>0}$.

Lemma 2.14. *If $v \in \mathcal{V}$, then $v_+ \in \mathcal{V}$ and $\frac{d(v_+)}{dS}(S) = \frac{dv}{dS}(S)1_{v(S)>0}$.*

In the same manner $v_-(S) = -v(S)1_{v(S)<0}$ and $\frac{d(v_-)}{dS}(S) = -\frac{dv}{dS}(S)1_{v(S)<0}$.

Theorem 2.15 (weak maximum principle). *Let $u(S,t)$ be such that for all positive numbers ϵ,*

- $ue^{-\epsilon\log^2(S+2)} \in \mathcal{C}^0([0,T]; L^2(\mathbb{R}_+)) \cap L^2(0,T;V)$,
- $\frac{\partial u}{\partial t}e^{-\epsilon\log^2(S+2)} \in L^2(\mathbb{R}_+ \times (0,T))$,
- $u|_{t=0} \geq 0$ *a.e.*,
- $\frac{\partial u}{\partial t} + A_t u \geq 0$ *(in the sense of distributions);*

then $u \geq 0$ a.e.

Proof. Consider the function $H(s) = \frac{1}{2}s_-^2$ (its derivative is $H'(s) = -s_-$), and for two positive parameters z and ζ, the function

$$\psi(t) = \int_{\mathbb{R}_+} H(u(S,t))e^{-\phi(S,t)}dS,$$

where

$$\phi(S,t) = \frac{\zeta}{2T-t}\log^2(S) + zt.$$

It is easy to see that $\psi \in \mathcal{C}^0([0,T];\mathbb{R})$, and for all t, $\psi(t) \geq 0$. We have also $\psi(0) = 0$.

Assume that $\frac{\partial u}{\partial t}e^{-\epsilon\log^2(S+2)} \in L^2(\mathbb{R}_+ \times (0,T))$: in this case, $u_-e^{-\epsilon\log^2(S+2)}$ belongs to $L^2(0,T;V)$ and is such that $\frac{\partial u_-}{\partial t}e^{-\epsilon\log^2(S+2)} \in L^2(\mathbb{R}_+ \times (0,T))$. Therefore, $\psi' \in L^1(0,T;\mathbb{R})$ and a.e.,

$$\begin{aligned}\psi'(t) &= -\int_{\mathbb{R}_+} u_-(S,t)\frac{\partial u}{\partial t}(S,t)e^{-\phi(S,t)}dS - \frac{1}{2}\int_{\mathbb{R}_+} u_-^2(S,t)\frac{\partial\phi}{\partial t}e^{-\phi(S,t)}dS \\ &\leq -\left((A_t u(t), u_-(t)e^{-\phi(t)}) + \frac{1}{2}\int_{\mathbb{R}_+} u_-^2(S,t)\frac{\partial\phi}{\partial t}(S,t)e^{-\phi(S,t)}dS\right).\end{aligned}$$

Calling $\chi = e^{-\frac{\phi}{2}}u_-$, we have

$$\psi'(t) \leq -(A_t\chi(t),\chi(t)) - \int_{\mathbb{R}_+}\kappa(S,t)\chi^2(S,t)\,dS, \tag{2.35}$$

with

$$\kappa(S,t) = \frac{1}{2}z + \frac{\zeta^2}{2(2T-t)^2}\log^2(S)\left(\frac{1}{\zeta} - \sigma^2\right) + \left(\sigma^2 + S\sigma\frac{\partial\sigma}{\partial S} - r(t)\right)\frac{\zeta}{2T-t}\log(S).$$

We choose two values $S_1 < 1$ and $S_2 > 1$ such that, for $S \in (0, S_1) \cup (S_2, +\infty)$, $4T \max_{\mathbb{R}_+ \times [0,T]} |\sigma^2 + S\sigma \frac{\partial \sigma}{\partial S} - r| < \frac{1}{2} |\log(S)|$. We define $\mathcal{S}_1 = (0, S_1) \cup (S_2, +\infty)$ and $\mathcal{S}_2 = [S_1, S_2]$:

$$\kappa(S,t) \geq \frac{\zeta}{2(2T-t)^2} \log^2(S) \left(1 - \zeta\sigma^2 - 4T \sup_{[0,T]\times\mathcal{S}_1} \frac{|\sigma^2 + S\sigma\frac{\partial\sigma}{\partial S} - r|}{|\log(S)|}\right)$$

$$+ \frac{1}{2} z - \zeta \max_{\mathcal{S}_2 \times [0,T]} \frac{|\sigma^2 + S\sigma\frac{\partial\sigma}{\partial S} - r|}{2T - t} |\log(S)| .$$

But $4T \max_{\mathcal{S}_1 \times [0,T]} \frac{|\sigma^2 + S\sigma\frac{\partial\sigma}{\partial S} - r|}{|\log(S)|} \leq \frac{1}{2}$, so, for ζ small enough,

$$1 - \zeta\sigma^2 - 4T \max_{\mathcal{S}_1 \times [0,T]} \frac{|\sigma^2 + S\sigma\frac{\partial\sigma}{\partial S} - r|}{|\log(S)|} \geq 0.$$

Then, for z large enough, we have for all $\chi \in V$,

$$\langle a_t\chi, \chi\rangle + \left(\frac{1}{2} z - \zeta \max_{\mathcal{S}_2\times[0,T]} \frac{|\sigma^2 + S\sigma\frac{\partial\sigma}{\partial S} - r|}{2T - t} |\log(S)|\right) \|\chi\|^2_{L^2(\mathbb{R}_+)} \geq \alpha\|\chi\|^2_V$$

with $\alpha > 0$. With these choices of ζ and z, (2.35) implies

$$\psi'(t) \leq 0 \quad \text{a.e.},$$

and for all $t \in [0, T]$, $\psi(t) = 0$, i.e., $u \geq 0$. $\square$

Remark 2.4. *It is possible to generalize Theorem* 2.15*: for example, with Assumption* 2.1*, consider the weak solution to* (2.9), (2.10) *given by Theorem* 2.11*, with* $P_0 \in L^2(\mathbb{R}_+)$ *and* $P_0 \geq 0$. *We have* $P \geq 0$ *a.e. and for that we do not need any further assumptions on* $\frac{\partial P}{\partial t}$*, because we can approximate* P_0 *by a smooth function* $P_{0,\epsilon} \in V$*, use Proposition* 2.13 *then Theorem* 2.15 *for the solution to* (2.9) *with Cauchy data* $P_{0,\epsilon}$*, and finally pass to the limit as* $\epsilon \to 0$.

2.3.5 Consequences of the Maximum Principle

Various Bounds. The maximum principle is an extremely powerful tool for proving estimates on the solutions of elliptic and parabolic partial differential equations.

Proposition 2.16. *Under Assumption* 2.1*, let* P *be the weak solution to* (2.9), (2.10)*, with* $P_0 \in L^2(\mathbb{R}_+)$ *a bounded function, i.e.,* $0 \leq \underline{P_0} \leq P_0(S) \leq \overline{P_0}$*. Then, a.e.*

$$\underline{P_0} e^{-\int_0^t r(\tau)d\tau} \leq P(S,t) \leq \overline{P_0} e^{-\int_0^t r(\tau)d\tau}. \tag{2.36}$$

Proof. We know that $\underline{P_0} e^{-\int_0^t r(\tau)d\tau}$ and $\overline{P_0} e^{-\int_0^t r(\tau)d\tau}$ are two solutions of (2.9). Therefore, we can apply the maximum principle (see Remark 2.4) to both $P - \underline{P_0} e^{-\int_0^t r(\tau)d\tau}$ and $\overline{P_0} e^{-\int_0^t r(\tau)d\tau} - P$. $\square$

Remark 2.5. *In the case of a vanilla put option:* $P_0(S) = (K - S)_+$, *and Proposition* 2.16 *just says that* $0 \le P(S,t) \le Ke^{-\int_0^t r(\tau)d\tau}$, *which is certainly not a surprise for people with a background in finance.*

For the vanilla put option as in Remark 2.5, we have indeed more information.

Proposition 2.17. *Under Assumption* 2.1, *let* P *be the weak solution to* (2.9), (2.10), *with* $P_0(S) = (K - S)_+$*:*

$$(Ke^{-\int_0^t r(\tau)d\tau} - S)_+ \le P(S,t) \le Ke^{-\int_0^t r(\tau)d\tau} \tag{2.37}$$

and

$$P(0,t) = Ke^{-\int_0^t r(\tau)d\tau}. \tag{2.38}$$

Proof. Observe that the function $Ke^{-\int_0^t r(\tau)d\tau} - S$ is a solution to (2.9) and apply the maximum principle to $P(S,t) - (Ke^{-\int_0^t r(\tau)d\tau} - S)$. We have $Ke^{-\int_0^t r(\tau)d\tau} - S \le P(S,t)$. Then (2.37) is obtained by combining this estimate with the one given in Remark 2.5. □

Remark 2.6. *Note that we have not imposed* (2.38) *a priori. Similarly, when we look for discrete solutions either by finite differences or by finite elements, it will not be necessary to impose* (2.38) *as a boundary condition.*

The Super-Replication Principle. Take two European put options with the same maturity and two different payoff functions P_0 and Q_0. Call $P(S,t)$ and $Q(S,t)$ their respective prices, which both satisfy (2.9). One easy consequence of the maximum principle is that, if for all S, $P_0(S) \le Q_0(S)$, then for all t and S, $P(S,t) \le Q(S,t)$. This is called the super-replication principle in the finance community.

The Put-Call Parity. Take again a vanilla put option as in Remark 2.5, and consider the function $C(S,t)$ given by

$$C(S,t) = S - Ke^{-\int_0^t r(\tau)d\tau} + P(S,t). \tag{2.39}$$

From the fact that P and $S - Ke^{-\int_0^t r(\tau)d\tau}$ satisfy (2.9), it is clear that C is a solution to (2.9) with the Cauchy condition $C(0,S) = (S-K)_+$. This is precisely the boundary value problem for the European vanilla call option. On the other hand, from the maximum principle, we know that a well-behaved solution (in the sense of Theorem 2.15) to this boundary value problem is unique. Therefore, we can deduce the price of the call option from that of the put option: we recover the put-call parity seen in Chapter 1 and also well known to people trained in finance.

Convexity of P in the Variable S

Assumption 2.3. *There exists a positive constant* C *such that*

$$\left| S^2 \frac{\partial^2 \sigma}{\partial S^2}(S,t) \right| \le C \quad a.e. \tag{2.40}$$

Proposition 2.18. *Under Assumptions* 2.1 *and* 2.3, *let P be the weak solution to* (2.9), (2.10), *where $P_0 \in V$ is a convex function such that $\frac{\partial^2 P_0}{\partial S^2}$ has a compact support. Then, for all $t > 0$, $P(S,t)$ is a convex function of S.*

Proof. Assume first that the coefficients of (2.9) also satisfy Assumption 2.2, and that $P_0 \in \mathcal{D}(\mathbb{R}_+)$, so by Theorem 2.12, $Q = S^2 \frac{\partial^2 P}{\partial S^2} \in C^0([0,T]; L^2(\mathbb{R}_+))$. Deriving twice (2.9) with respect to S, and multiplying by S^2,

$$\frac{\partial}{\partial t} Q - \frac{\partial}{\partial S}\left(\frac{\sigma^2 S^2}{2}\frac{\partial Q}{\partial S}\right) - S(r-\sigma^2)\frac{\partial Q}{\partial S} + \left(r - S\frac{\partial \sigma^2}{\partial S} + \frac{1}{2}\frac{\partial^2}{\partial S^2}\left(S^2\sigma^2\right)\right) Q = 0. \quad (2.41)$$

Thanks to Assumptions 2.1 and 2.3, a maximum principle analogue to Theorem 2.15 applies, because Q satisfies (2.41) and $Q|_{t=0} \geq 0$. Therefore $Q \geq 0$, which says that $\frac{\partial^2 P}{\partial S^2} \geq 0$ for all $t \in [0,T]$ and a.e. in S.

To prove the result for nonsmooth coefficients and Cauchy data, we approximate P_0 by a smooth convex function with compact support, we approach the coefficients by ones satisfying Assumption 2.2, and we pass to the limit as in the proof of Proposition 2.13. □

As a consequence, we see that under Assumptions 2.1 and 2.3, the price of a vanilla European put option is convex with respect to S, and thanks to the call-put parity, this is also true for the vanilla European call.

Remark 2.7. *The assumption of the compact support of $\frac{\partial^2 P_0}{\partial S^2}$ in Proposition* 2.18 *can of course be relaxed.*

More Bounds. We focus on a vanilla put, i.e., the solution to (2.9), (2.10) with $P_0(S) = (K-S)_+$. It is possible to compare P with prices of vanilla puts with constant volatilities.

Proposition 2.19. *Under Assumption* 2.1, *we have, for all $t \in [0,T]$ and for all $x > 0$,*

$$\underline{P}(S,t) \leq P(S,t) \leq \overline{P}(S,t), \quad (2.42)$$

where $\underline{P}$ (resp., $\overline{P}$) is the solution to (2.9), (2.10) *with $\sigma = \underline{\sigma}$, (resp., $\overline{\sigma}$).*

Proof. Consider the difference $E = P - \overline{P}$. It satisfies $E|_{t=0} = 0$ and

$$\frac{\partial E}{\partial t} - \frac{\sigma^2 S^2}{2}\frac{\partial^2 E}{\partial S^2} - rS\frac{\partial E}{\partial S} + rE = \frac{S^2}{2}(\overline{\sigma}^2 - \sigma^2)\frac{\partial^2 \overline{P}}{\partial S^2} \geq 0,$$

because $\overline{P}$ is convex. Then, the maximum principle leads to the upper bound in (2.42) and the lower bound is obtained in the same manner. □

Localization. Again, we focus on a vanilla put. For computing numerically an approximation to P, one has to limit the domain in the variable S, i.e., consider only $S \in (0, \bar{S})$ for $\bar{S}$ large enough, and impose some artificial boundary condition at $S = \bar{S}$. Imposing that the new function vanishes on the artificial boundary, we obtain the new boundary value problem

$$\begin{aligned} \frac{\partial \tilde{P}}{\partial t} - \frac{\sigma^2 S^2}{2} \frac{\partial^2 \tilde{P}}{\partial S^2} - rS\frac{\partial \tilde{P}}{\partial S} + r\tilde{P} = 0, \quad t \in (0, T], \ S \in (0, \bar{S}), \\ \tilde{P}(\bar{S}, t) = 0, \quad t \in (0, T], \end{aligned} \tag{2.43}$$

with the Cauchy data $\tilde{P}(S, 0) = (K - S)_+$ in $(0, \bar{S})$. The theory of weak solutions applies to this new boundary value problem, but one has to work in the new Sobolev space

$$\tilde{V} = \left\{ v, S\frac{\partial v}{\partial S} \in L^2((0, \bar{S})), v(\bar{S}) = 0 \right\}.$$

The results in §2.3.1, §2.3.2, and §2.3.3 can be transposed to problem (2.43). The question is to estimate the error between P and $\tilde{P}$. For that, we use a version of the maximum principle adapted to (2.43), the proof of which is omitted for brevity.

Proposition 2.20. *Let $u(S, t)$ be a function such that*

- $u \in C^0([0, T]; L^2((0, \bar{S})), S\frac{\partial u}{\partial S} \in C^0((0, \bar{S}) \times (0, T))$, *and* $\frac{\partial u}{\partial t} \in L^2((0, \bar{S}) \times (0, T))$,
- $u(S, 0) \geq 0$ *in* $[0, \bar{S}]$,
- $\frac{\partial u}{\partial t} + A_t u \geq 0$ *a.e. in* $(0, \bar{S}) \times (0, T)$,
- $u(\bar{S}, t) \geq 0$ *in* $[0, T]$;

then $u \geq 0$ *in* $[0, \bar{S}] \times [0, T]$.

Proposition 2.21. *Under Assumption* 2.1, *the error* $\max_{t\in[0,T], S\in[0,\bar{S}]} |P(S, t) - \tilde{P}(S, t)|$ *decays faster than any exponential* $\exp(-\eta\bar{S})$ *($\eta > 0$) as* $\bar{S} \to \infty$.

Proof. From Proposition 2.20, we immediately see that $P \geq \tilde{P}$ in $(0, \bar{S}) \times (0, T)$, because $P(\bar{S}, t) \geq \tilde{P}(\bar{S}, t) = 0$. On the other hand, from Proposition 2.19, $P \leq \bar{P}$, which implies that $P(\bar{S}, t) \leq \bar{P}(\bar{S}, t)$. Call $\pi(\bar{S}) = \max_{t\in[0,T]} \bar{P}(\bar{S}, t)$, and consider the function $E(S, t) = \pi - P(S, t) + \tilde{P}(S, t)$. It is easy to check that the function E satisfies the assumptions of Proposition 2.20, so $\pi(\bar{S}) \geq P - \tilde{P}$ in $[0, \bar{S}] \times [0, T]$. At this point, we have proved that

$$0 \leq P - \tilde{P} \leq \pi(\bar{S}) \quad \text{in } [0, \bar{S}] \times [0, T].$$

But $\pi(\bar{S})$ can be computed semiexplicitly by the Black–Scholes formula (1.18), and it is easy to see that for all $\eta > 0$, $\lim_{\bar{S}\to\infty} \pi(\bar{S})e^{\eta\bar{S}} = 0$. Therefore, $\max_{t\in[0,T], S\in[0,\bar{S}]} |P(S, t) - \tilde{P}(S, t)|$ decays faster than any exponential $\exp(-\eta\bar{S})$ as $\bar{S} \to \infty$. □

2.3.6 Dupire's Equation

We reproduce the arguments in [6].

We consider a vanilla European call, subject to the Black–Scholes equation with local volatility,

$$
\begin{aligned}
\partial_t C + \frac{\sigma^2(S,t)S^2}{2}\frac{\partial^2 C}{\partial S^2} + (r-q)S\frac{\partial C}{\partial S} - rC = 0 \qquad & \text{in } [0,T)\times\mathbb{R}_+, \\
C(S,T) = (K-S)_+ \qquad & \text{in } \mathbb{R}_+,
\end{aligned}
\tag{2.44}
$$

where we have supposed that the underlying asset yields a distributed dividend.

Let us call $G(S,t,\xi,T)$ the Green's function, i.e., the solution to

$$
\begin{aligned}
\partial_t G + \frac{\sigma^2(S,t)S^2}{2}\frac{\partial^2 G}{\partial S^2} + (r-q)S\frac{\partial G}{\partial S} - rG = -(\delta_\xi\otimes\delta_T)(S,t) \quad & \text{in } \mathbb{R}_+\times\mathbb{R}, \\
G(S,t,\xi,T) = 0 \quad & \text{in } \mathbb{R}_+\times(t,+\infty).
\end{aligned}
\tag{2.45}
$$

As a function of T and ξ, G satisfies the adjoint equation (see [55])

$$
\begin{aligned}
\partial_T G - \frac{1}{2}\frac{\partial^2(\sigma^2(\xi,T)\xi^2 G)}{\partial\xi^2} + \frac{\partial((r-q)\xi G)}{\partial\xi} + rG = (\delta_S\otimes\delta_t)(\xi,T) \quad & \text{in } \mathbb{R}_+\times\mathbb{R}, \\
G(S,t,\xi,T) = 0 \quad & \text{in } \mathbb{R}_+\times(t,+\infty).
\end{aligned}
\tag{2.46}
$$

To obtain (2.46), it suffices to rewrite (2.45) in the abstract form

$$
L(S,t)G(S,t,\xi,T) = -(\delta_\xi\otimes\delta_T)(S,t),
$$

where $L(S,t)$ stands for the Black–Scholes operator. This implies that for any smooth function $\phi(S,t)$, $\psi(\xi,T)$ compactly supported in $\mathbb{R}_+\times\mathbb{R}$, and calling $\zeta(S,t)$ a solution to $L(S,t)^T\zeta(S,t) = \phi(S,t)$, we have

$$
\begin{aligned}
&\int_{\xi,T} L(\xi,T)\psi(\xi,T)\langle L(S,t)G(\cdot,\cdot,\xi,T),\zeta\rangle_{S,t}\,d\xi dT \\
&= \int_{S,t} L^T(S,t)\zeta(S,t)\langle L^T(\cdot,\cdot)G(S,t,\cdot,\cdot),\psi\rangle_{\xi,T}\,dSdt \\
&= \int_{S,t} \phi(S,t)\langle L^T(\cdot,\cdot)G(S,t,\cdot,\cdot),\psi\rangle_{\xi,T}\,dSdt.
\end{aligned}
$$

On the other hand,

$$
\begin{aligned}
&\int_{\xi,T} L(\xi,T)\psi(\xi,T)\langle L(S,t)G(\cdot,\cdot,\xi,T),\zeta\rangle_{S,t}\,d\xi dT \\
&= -\int_{\xi,T} L(\xi,T)\psi(\xi,T)\zeta(\xi,T)\,d\xi dT \\
&= -\int_{\xi,T} \psi(\xi,T)L^T(\xi,T)\zeta(\xi,T)\,d\xi dT = \int_{\xi,T}\psi(\xi,T)\phi(\xi,T)\,d\xi dT.
\end{aligned}
$$

From the two previous identities, we get that $-L^T(\xi,T)G(S,t,\xi,T) = (\delta_S\otimes\delta_t)(\xi,T)$, which is exactly (2.46).

The price of the call (solution to (2.44)) is given by the representation formula

$$C(S,t,K,T) = \int_0^{+\infty} G(S,t,\xi,T)(\xi - K)_+ \, d\xi = \int_K^{+\infty} G(S,t,\xi,T)(\xi - K)\, d\xi, \tag{2.47}$$

so it is possible to compute the derivatives of C with respect to K:

$$\frac{\partial C}{\partial K} = -\int_K^{+\infty} G(S,t,\xi,T)\, d\xi, \qquad \frac{\partial^2 C}{\partial K^2} = G(S,t,K,T)\, d\xi. \tag{2.48}$$

From the adjoint equation (2.46), we deduce

$$\partial_T \frac{\partial^2 C}{\partial K^2} - \frac{1}{2}\frac{\partial^2(\sigma^2(K,T)K^2 \frac{\partial^2 C}{\partial K^2})}{\partial K^2} + \frac{\partial((r-q)K\frac{\partial^2 C}{\partial K^2})}{\partial K} + r\frac{\partial^2 C}{\partial K^2} = 0 \quad \text{in } \mathbb{R}_+ \times (t,+\infty) \tag{2.49}$$

and integrating twice with respect to K, we obtain

$$\partial_T C - \frac{1}{2}\sigma^2(K,T)K^2\frac{\partial^2 C}{\partial K^2} + (r-q)K\frac{\partial C}{\partial K} + qC = A(T)K + B(T). \tag{2.50}$$

Following Dupire [41], we assume that all the terms on the left-hand side of (2.50) decay when K tends to $+\infty$, so $A(T) = B(T) = 0$. We have obtained that fixing the date t and the spot price S, the price of the European vanilla call satisfies the partial differential equation

$$\partial_T C - \frac{1}{2}\sigma^2(K,T)K^2\frac{\partial^2 C}{\partial K^2} + (r-q)K\frac{\partial C}{\partial K} + qC = 0 \tag{2.51}$$

with respect to the strike K and the maturity T. This equation is a forward parabolic equation resembling very much (2.44). It is known as Dupire's equation in the finance literature. We shall see later that (2.50) is very useful when trying to calibrate the local volatility from the prices of options on the market. Indeed, if the options for all strikes and maturities were on the market, then $C(K,T)$ would be known for all K and T, and the local volatility in (2.44) would be given by (2.51), and, at least formally,

$$\sigma^2(K,T) = 2\frac{\partial_T C(K,T) + (r-q)K\frac{\partial C}{\partial K}(K,T) + qC(K,T)}{K^2\frac{\partial^2 C}{\partial K^2}(K,T)}. \tag{2.52}$$

Dupire versus Black–Scholes. It is interesting to compare $P(S,0,K,T)$ as a function of K, obtained either by solving numerically Dupire's equation or the Black–Scholes equations (in the variables S and t) for different strikes K, for a given volatility function. To do so, we compute several vanilla European puts by solving the Black–Scholes equations (in the variables S and t) for different values of K, and we plot $K \mapsto P(S,0,K,T)$ and compare to the plot of the $K \mapsto v(S,0,K,T)$ solution to Dupire's equation (in the variables K and T); the results are shown in Figure 2.1. We see that there is a noticeable difference for large values of K, but this is due to the localization in Dupire's equation at $\bar{K} = 2S$ ($\bar{K} = 10S$, for instance, would have given a better result). In Chapter 4, we propose a better boundary condition for localizing, which applies when the local volatility is constant for large values of S.

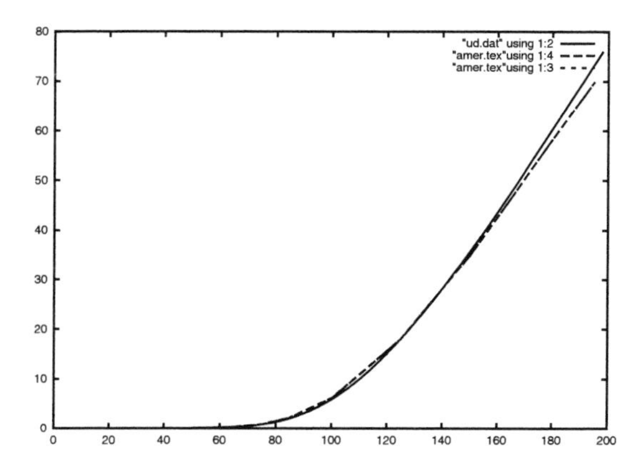

Figure 2.1. *$P(S, 0, K, T)$ versus K (S is fixed) for a family of European vanilla puts with a constant volatility computed by Dupire's equation, the Black–Scholes formula, and finite differences for the Black–Scholes partial differential equation.*

2.4 Barrier Options

As seen in Chapter 1, barrier options are options for which the right to exercise disappears out of a subregion of $\mathbb{R}_+ \times (0, T)$ (the option becomes worthless out of this region).

The boundary of this region is called a knockout boundary. The knockout boundary may or may not depend on time. Consider the case when the boundary does not depend on time: for example, a double knockout option is an option (a put or a call) which becomes worthless if the price of the underlying asset either rises to an upper barrier S_2 or falls below a lower value $S_1 \geq 0$.

In $(S_1, S_2) \times (0, T)$, the price of the option satisfies the Black–Scholes equation (possibly with local volatility), and we have the boundary conditions

$$P(S_1, t) = P(S_2, t) = 0.$$

If the volatility is constant, then it is possible to compute the price P semiexplicitly by the method of the images (see Wilmott, Howison, and Dewynne [117], for instance). In other cases, numerical computation is needed, and one can rely on a weak formulation: calling $\Omega = (S_1, S_2)$, and defining

$$\tilde{V} = \left\{ v \in L^2(\Omega) \; : \; S\frac{\partial v}{\partial S} \in L^2(\Omega) \right\},$$

which is a Hilbert space with the norm

$$\|v\|_{\tilde{V}} = \left(\|v\|^2_{L^2(\Omega)} + \left\| S\frac{\partial v}{\partial S} \right\|^2_{L^2(\Omega)} \right)^{\frac{1}{2}},$$

we can reproduce the arguments of §2.3.2, except that the space V is now defined as the closure of $\mathcal{D}(\Omega)$ into $\tilde{V}$ (where $\mathcal{D}(\Omega)$ is the space of smooth functions with compact support in Ω).

Note that if $S_1 \neq 0$ (i.e., no lower constraint), then

$$V = \left\{ v \in L^2(\Omega) \ : \ \frac{\partial v}{\partial S} \in L^2(\Omega);\ v(S_1) = v(S_2) = 0 \right\},$$

called $H_0^1(\Omega)$ in the mathematical literature.

On the contrary, if $S_1 = 0$, then

$$V = \left\{ v \in L^2(\Omega);\ S\frac{\partial v}{\partial S} \in L^2(\Omega);\ v(S_2) = 0 \right\}.$$

With these modifications, all the arguments in § 2.3.2 can be carried over to the present case.

If the barrier depends on time, then the boundary value problem can be seen as a parabolic problem in a domain which is not a cylinder in $\mathbb{R}_+ \times (0, T)$. In the simple cases, the mathematical analysis of such problems can be done by finding a change of variables which maps the noncylindrical domain to a cylindrical one, but this is beyond the scope of the present book.

2.5 Lévy Driven Assets

To improve the modeling one may consider assets described by exponentials of Lévy processes, i.e., processes that have stationary independent increments; see, for example, [26, 43, 42, 44, 92], the book by Cont and Tankov [30], and the references therein.

For a Lévy process X_t on a filtered probability space with probability $\mathbb{P}^*$, the Lévy–Khintchine formula says that there exists a function $\psi : \mathbb{R} \to \mathbb{C}$ such that

$$\mathbb{E}^*(e^{iuX_t}) = e^{-t\psi(u)} \tag{2.53}$$

and

$$\psi(u) = \frac{\sigma^2}{2} + i\alpha u + \int_{|y|<1} (1 - e^{-iuy} - iuy)\nu(dy) + \int_{|y|>1} (1 - e^{-iuy})\nu(dy) \tag{2.54}$$

for $\sigma, \alpha \in \mathbb{R}$ and a measure ν on $\mathbb{R}*$ such that $\int_{\mathbb{R}} \min(1, y^2)\nu(dy) < +\infty$. The measure ν is called the Lévy measure of X.

We consider the price of a financial asset S_t modeled as a stochastic process on a filtered probability space with probability $\mathbb{P}$. There exists an equivalent probability $\mathbb{P}^*$ under which the discounted price is a martingale. If S_t is modeled as an exponential of a Levy process, we have, under $\mathbb{P}^*$,

$$S_t = S_0 e^{(r+c-\frac{\sigma^2}{2})t + X_t},$$

and the correction parameter c is chosen such that the mean rate of return on the asset is risk neutrally r, i.e., $e^{(c+\frac{\sigma^2}{2})t} = \mathbb{E}^*(e^{X_t})$. The fact that the discounted price is a martingale is equivalent to

$$\int_{|y|>1} e^y \nu(dy) < \infty$$

and to

$$c = \int_{\mathbb{R}} (1 - e^y + y 1_{|y| \le 1}) \nu(dy).$$

We will assume also that $\int_{|y|>1} e^{2y} \nu(dy) < \infty$, so the discounted price is a square integrable martingale.

Consider an option with payoff $P_\circ$ and maturity T: its price at time t is

$$P(S_t, t) = \mathbb{E}^*(e^{-\int_t^T r(s)ds} P_\circ(S_T)|F_t), \tag{2.55}$$

and it can be proved, if the payoff function P_0 is in the domain of the infinitesimal generator of the semigroup (see [13]) when $\sigma \ge \underline{\sigma} > 0$, and [99, 33] without this assumption, that P is the solution to the partial integrodifferential equation

$$\begin{aligned} &\frac{\partial}{\partial t} P(S, t) + \frac{\sigma^2(S, t) S^2}{2} \frac{\partial^2}{\partial S^2} P(S, t) + r(t) S \frac{\partial}{\partial S} P(S, t) - r(t) P(S, t) \\ &+ \int_{\mathbb{R}} \left(P(Se^y, t) - P(S, t) - S(e^y - 1) \frac{\partial}{\partial S} P(S, t) \right) \nu(dy) = 0. \end{aligned} \tag{2.56}$$

We shall assume that the Lévy measure has a density $k(y)dy$, so (2.56) becomes, replacing t by the time to maturity,

$$\begin{aligned} &\frac{\partial}{\partial t} P(S, t) - \frac{\sigma^2(S, t) S^2}{2} \frac{\partial^2}{\partial S^2} P(S, t) - r(t) S \frac{\partial}{\partial S} P(S, t) + r(t) P(S, t) \\ &- \int_{\mathbb{R}} \left(P(Se^y, t) - P(S, t) - S(e^y - 1) \frac{\partial}{\partial S} P(S, t) \right) k(y) dy = 0. \end{aligned} \tag{2.57}$$

Many choices are possible for $k(y)$:

- The variance Gamma processes

$$k(y) = \begin{cases} \dfrac{1}{\nu|y|} e^{-\frac{|y|}{\eta_n}} & \text{if } y < 0, \\ \dfrac{1}{\nu|y|} e^{-\frac{|y|}{\eta_p}} & \text{if } y > 0, \end{cases} \tag{2.58}$$

  where η_n and η_p are positive and η_p is smaller than $1/2$.

- The CGMY processes are a generalization of the variance Gamma processes

$$k(y) = \begin{cases} \dfrac{C}{|y|^{1+Y}} e^{-G|y|} & \text{if } y < 0, \\ \dfrac{C}{|y|^{1+Y}} e^{-M|y|} & \text{if } y > 0, \end{cases} \tag{2.59}$$

  where $0 < Y < 2$ and $0 < G, 2 < M$.

- A simpler choice is Merton's model with Gaussian jump in log price with Lévy density (see [32])

$$k(y) = Ce^{-\frac{(y-\delta)^2}{2\gamma^2}}.$$

If the volatility is positive, it is possible to develop a theory on weak solutions to (2.57); see [95]. We will not cover this in the present book. When the volatility is 0, it is also possible to use semigroups (see [95]), but depending on k the semigroup may not be analytic. In this case, the notion of viscosity solutions (see [37]) is useful (see [32, 33, 99]), and it is possible to prove that even with nonsmooth payoffs, the function given by (2.55) is a viscosity solution to (2.57).

Remark 2.8. *Of course, it is possible to consider options on Lévy driven assets with barriers; see* [33].

2.6 Options on a Basket of Assets

We focus on a basket (see §1.2.3) containing two assets, whose prices are S_1 and S_2, but all that follows can be generalized. We assume that the prices of the underlying assets obey a system of stochastic differential equations:

$$\begin{aligned} dS_{1t} &= S_{1t}\left(\mu_1 dt + \frac{\sigma_1}{\sqrt{1+\rho^2}}(dW_{1t} + \rho dW_{2t})\right), \\ dS_{2t} &= S_{2t}\left(\mu_2 dt + \frac{\sigma_2}{\sqrt{1+\rho^2}}(\rho dW_{1t} + dW_{2t})\right), \end{aligned} \tag{2.60}$$

where W_{1t} and W_{2t} are two independent standard Brownian motions. For simplicity, we assume that σ_1 and σ_2 are positive constants, but generalization to functions $\sigma_1(S_1, S_2, t)$ and $\sigma_2(S_1, S_2, t)$ can be considered. The parameter ρ is the correlation factor: $-1 < \rho < 1$. Also for simplicity, we assume that the interest rate r of the risk-free asset is constant.

Consider a European option on this two-asset basket, whose payoff function is $P_0(S_1, S_2)$. As for the options on a single asset, it is possible to find a probability $\mathbb{P}^*$ under which the price of the option is

$$P(S_{1t}, S_{2t}, t) = e^{-r(T-t)}\mathbb{E}^*(P_0(S_{1T}, S_{2T})|F_t).$$

One can apply the two-dimensional Itô's formula, and find the partial differential equation for the price of the option $P(S_1, S_2, t)$: replacing the time with the time to maturity,

$$\begin{aligned} &\frac{\partial P}{\partial t} - \frac{1}{2}\sum_{k,l=1}^{2} \Xi_{k,l} S_k S_l \frac{\partial^2 P}{\partial S_k \partial S_l} - \sum_{k=1}^{2} r S_k \frac{\partial P}{\partial S_k} + rP = 0, \qquad t \in (0, T],\ S_1, S_2 > 0, \\ &P(S_1, S_2, 0) = P_0(S_1, S_2), \quad S_1, S_2 > 0. \end{aligned} \tag{2.61}$$

The tensor Ξ,

$$\Xi = \begin{pmatrix} \sigma_1^2 & \frac{2\rho}{1+\rho^2}\sigma_1\sigma_2 \\ \frac{2\rho}{1+\rho^2}\sigma_1\sigma_2 & \sigma_2^2 \end{pmatrix},$$

is clearly positive definite. Exactly as for the one-asset case, it is possible to study the weak solutions of (2.61). Noting $Q = \mathbb{R}_+^2$ and introducing the Hilbert space

$$V = \left\{v :\ v, S_1\frac{\partial v}{\partial S_1}, S_2\frac{\partial v}{\partial S_2} \in L^2(Q)\right\}, \tag{2.62}$$

with the norm

$$\|v\|_V = \left(\|v\|^2_{L^2(Q)} + \left\| S_1 \frac{\partial v}{\partial S_1} \right\|^2_{L^2(Q)} + \left\| S_2 \frac{\partial v}{\partial S_2} \right\|^2_{L^2(Q)} \right)^{\frac{1}{2}}, \tag{2.63}$$

one can check the following properties:

- The space $\mathcal{D}(Q)$ of smooth and compactly supported functions in Q is dense in V.
- V is separable.
- The seminorm $|\cdot|_V$ defined by $|v|^2_V = \|S_1 \frac{\partial v}{\partial S_1}\|^2_{L^2(Q)} + \|S_2 \frac{\partial v}{\partial S_2}\|^2_{L^2(Q)}$ is in fact a norm on V equivalent to $\|\cdot\|_V$ because $\|v\|_{L^2(Q)} \le 2|v|_V$.

Among the usual payoff functions, we can cite, for a put,

$$P_0(S_1, S_2) = (K - (S_1 + S_2))_+, \tag{2.64}$$

$$P_0(S_1, S_2) = (K - \max(S_1, S_2))_+, \tag{2.65}$$

$$P_0(S_1, S_2) = (K - \min(S_1, S_2))_+. \tag{2.66}$$

Note that the payoff functions given by (2.64) and (2.65) belong to $L^2(Q)$, which is not true if P_0 is given by (2.66). In what follows, we are going to outline the theory of the weak formulation to (2.61), which applies for $P_0 \in L^2(Q)$.

We introduce the bilinear form

$$a(v, w) = \int_Q \frac{1}{2} \sum_{k,l=1}^{2} \Xi_{k,l} S_k S_l \frac{\partial v}{\partial S_k} \frac{\partial w}{\partial S_l} + \int_Q \sum_{k=1}^{2} \left(-r + \sum_{l=1}^{2} \frac{1}{2} \Xi_{k,l} \right) S_k \frac{\partial v}{\partial S_k} w + r \int_Q vw. \tag{2.67}$$

One can prove that there exist two positive constants $\underline{\sigma} < \overline{\sigma}$ and a nonnegative constant λ such that for all $v, w \in V$,

$$a(v, w) \le \overline{\sigma}^2 |v|_V |w|_V \tag{2.68}$$

and

$$a(v, v) \ge \underline{\sigma}^2 |v|^2_V - \lambda \|v\|^2_{L^2(Q)}. \tag{2.69}$$

Assuming that $P_0 \in L^2(Q)$, the weak formulation of (2.61) consists in finding

$P \in C^0([0, T]; L^2(Q)) \cap L^2(0, T; V)$ such that $\frac{\partial P}{\partial t} \in L^2(0, T; V')$, satisfying

$$P_{|t=0} = P_0 \quad \text{in } Q, \tag{2.70}$$

and for a.a. $t \in (0, T)$,

$$\forall v \in V, \quad \left(\frac{\partial P}{\partial t}(t), v \right) + a(P(t), v) = 0. \tag{2.71}$$

The machinery of §2.3.2 applies and it is possible to prove that if $P_0 \in L^2(Q)$, the problem (2.70), (2.71) has a unique solution P and we have the estimate, for all t, $0 < t < T$,

$$e^{-2\lambda t} \|P(t)\|^2_{L^2(Q)} + 2\underline{\sigma}^2 \int_0^t e^{-2\lambda\tau} |P(\tau)|^2_V d\tau \le \|P_0\|^2_{L^2(Q)}. \tag{2.72}$$

Furthermore, analyzing the domain of the operator in (2.61), one can prove that

- for all $t > 0$, $P \in \mathcal{C}^0([t,T]; D) \cap \mathcal{C}^1([t,T]; L^2(Q))$, where
$$D = \left\{ v \in V,\ S_k S_l \frac{\partial^2 P}{\partial S_l \partial S_k} \in L^2(Q) \right\};$$
- if $P_0 \in D$, then $P \in \mathcal{C}^0([0,T]; D) \cap \mathcal{C}^1([0,T]; L^2(Q))$;
- if $P_0 \in V$, then $P \in \mathcal{C}^0([0,T]; V) \cap L^2(0,T; D)$.

Also, it is possible to prove a maximum principle analogue to Theorem 2.15. This is the main tool for establishing that if P_0 is given by (2.64) or by (2.65), then $P(S_1, 0, t) = (Ke^{-rt} - S_1)_+$ and $P(0, S_2, t) = (Ke^{-rt} - S_2)_+$.

Naturally, there may be barrier options on baskets of several assets. For two assets and a barrier independent of time, pricing the option then amounts to solving the boundary value problem

$$\begin{aligned}
&\frac{\partial P}{\partial t} - \frac{1}{2}\sum_{k,l=1}^{2} \Xi_{k,l} S_k S_l \frac{\partial^2 P}{\partial S_k \partial S_l} - \sum_{k=1}^{2} r S_k \frac{\partial P}{\partial S_k} + rP = 0, \qquad t \in (0,T],\ (S_1,S_2) \in \Omega,\\
&P(S_1,S_2,0) = P_0(S_1,S_2), \quad (S_1,S_2) \in \Omega, \qquad P(S_1,S_2,t) = 0 \quad \text{on } \partial\Omega
\end{aligned} \tag{2.73}$$

for a domain Ω of $\mathbb{R}_+^2$. We restrict ourselves to domains whose boundaries are locally the graph of Lipschitz continuous functions. Then, the Sobolev space to work with is the closure of $\mathcal{D}(\Omega)$ in the space $\{v \in L^2(\Omega);\ S_1 \frac{\partial v}{\partial S_1} \in L^2(\Omega);\ S_2 \frac{\partial v}{\partial S_2} \in L^2(\Omega)\}$ equipped with the norm

$$\left(\|v\|^2_{L^2(\Omega)} + \|S_1 \frac{\partial v}{\partial S_1}\|^2_{L^2(\Omega)} + \|S_2 \frac{\partial v}{\partial S_2}\|^2_{L^2(\Omega)} \right)^{\frac{1}{2}}.$$

If the domain is complex, simulations need some efforts, and computing techniques enabling one to describe the domain, like the finite element method, must be used.

2.7 Stochastic Volatility

We consider a financial asset whose price is given by the stochastic differential equation

$$dS_t = \mu S_t dt + \sigma_t S_t dW_t, \tag{2.74}$$

where $\mu S_t dt$ is a drift term, (W_t) is a Brownian motion, and (σ_t) is the volatility. The simplest models take a constant volatility, but these models are generally too coarse to match real market prices. A more realistic model consists in assuming that (σ_t) is a function of a mean reverting Orstein–Uhlenbeck process:

$$\begin{aligned}
\sigma_t &= f(Y_t),\\
dY_t &= \alpha(m - Y_t)dt + \beta d\hat{Z}_t,
\end{aligned} \tag{2.75}$$

where α, m, and β are positive constants, and where $(\hat{Z}_t)$ is a Brownian motion. As explained in [51], the law of Y_t knowing Y_0 is $\mathcal{N}(m + (Y_0 - m)e^{-\alpha t}, \frac{\beta^2}{\alpha}(1 - e^{-2\alpha t}))$. Therefore, m is the limit of the mean value of Y_t as $t \to +\infty$, and $\frac{1}{\alpha}$ is the characteristic time of mean reversion. The parameter α is called the *rate of mean reversion*. The ratio $\frac{\beta^2}{\alpha}$ is the limit of the variance of Y_t as $t \to +\infty$. For convenience, we introduce the parameter ν

$$\nu^2 = \frac{\beta^2}{2\alpha}. \tag{2.76}$$

The Brownian motion $\hat{Z}_t$ may be correlated with W_t: it can be written as a linear combination of (W_t) and an independent Brownian motion (Z_t),

$$\hat{Z}_t = \rho W_t + \sqrt{1 - \rho^2} Z_t, \tag{2.77}$$

where the *correlation factor* ρ lies in $[-1, 1]$.

Consider a European derivative on this asset, with expiration date T and payoff function $h(S_T)$. Its price at the time t will depend on t, on the price of the underlying asset S_t, and on Y_t. We denote by $P(S_t, Y_t, t)$ the price of the derivative, and by $\tilde{r}(t)$ the interest rate. By using the no-arbitrage principle and the two-dimensional Itô's formula, it is possible to prove that there exists a function $\tilde{\gamma}$ such that the pricing function P satisfies the partial differential equation

$$\begin{aligned} &\frac{\partial P}{\partial t} + \frac{1}{2} f(y)^2 S^2 \frac{\partial^2 P}{\partial S^2} + \rho \beta S f(y) \frac{\partial^2 P}{\partial S \partial y} + \frac{1}{2} \beta^2 \frac{\partial^2 P}{\partial y^2} \\ &+ \tilde{r}(t) \left(S \frac{\partial P}{\partial S} - P \right) + \left(\alpha(m - y) - \beta \tilde{\Lambda}(S, y, t) \right) \frac{\partial P}{\partial y} = 0, \quad \left\{ \begin{array}{l} 0 \le t < T, \\ S > 0,\ y \in \mathbb{R}, \end{array} \right. \end{aligned} \tag{2.78}$$

where

$$\tilde{\Lambda}(S, y, t) = \rho \frac{\mu - \tilde{r}(t)}{f(y)} + \sqrt{1 - \rho^2} \tilde{\gamma}(S, y, t), \tag{2.79}$$

with the terminal condition $P(S, y, T) = h(S)$. The function $\tilde{\gamma}(S, y, t)$ can be chosen arbitrarily.

The no-arbitrage argument can be summarized as follows: We look for the pricing function P by trying to construct a hedged portfolio of assets. It is not sufficient to hedge only with the underlying asset because there are two independent sources of randomness: dW_t and dZ_t. So the idea is to take a self-financing hedged portfolio containing a_t shares of the underlying asset, one option with expiration date T_1 whose price is

$$P_t^{(1)} = P^{(1)}(S_t, Y_t, t),$$

and b_t options with a larger expiration date $T_2 > T_1$ whose price is

$$P_t^{(2)} = P^{(2)}(S_t, Y_t, t).$$

The value of the portfolio is c_t. The no-arbitrage principle yields that for $t < T_1$,

$$dc_t = a_t dS_t + dP_t^{(1)} + b_t dP_t^{(2)} = \tilde{r}_t c_t dt = \tilde{r}_t (a_t S_t + P_t^{(1)} + b_t P_t^{(2)}) dt. \tag{2.80}$$

The two-dimensional Itô formula permits us to write $dP_t^{(1)}$ and $dP_t^{(2)}$ as combinations of dt, dW_t, and dZ_t. The left-hand side of (2.80) does not contain dZ_t so

$$b_t = -\frac{\frac{\partial P^{(2)}}{\partial y}}{\frac{\partial P^{(1)}}{\partial y}}.$$

From the last equation and since the left-hand side of (2.80) does not contain dW_t, we have also

$$a_t + \frac{\partial P^{(1)}}{\partial S} + b_t \frac{\partial P^{(2)}}{\partial S} = 0.$$

Comparing the dt terms in (2.80) and substituting the values of a_t and b_t, we obtain that

$$\frac{1}{\frac{\partial P^{(1)}}{\partial y}}\left(\frac{\partial P^{(1)}}{\partial t} + \frac{1}{2} f(y)^2 S^2 \frac{\partial^2 P^{(1)}}{\partial S^2} + \rho\beta S f(y) \frac{\partial^2 P^{(1)}}{\partial S \partial y} + \frac{1}{2}\beta^2 \frac{\partial^2 P^{(1)}}{\partial y^2} + \tilde{r}(t)\left(S\frac{\partial P^{(1)}}{\partial S} - P^{(1)}\right)\right)$$
$$= \frac{1}{\frac{\partial P^{(2)}}{\partial y}}\left(\frac{\partial P^{(2)}}{\partial t} + \frac{1}{2} f(y)^2 S^2 \frac{\partial^2 P^{(2)}}{\partial S^2} + \rho\beta S f(y) \frac{\partial^2 P^{(2)}}{\partial S \partial y} + \frac{1}{2}\beta^2 \frac{\partial^2 P^{(2)}}{\partial y^2} + \tilde{r}(t)\left(S\frac{\partial P^{(2)}}{\partial S} - P^{(2)}\right)\right).$$

In the equation above, the left-hand side does not depend on T_2 and the right-hand side does not depend on T_1, so there exists a function of $g(S, y, t)$ such that

$$\frac{1}{\frac{\partial P}{\partial y}}\left(\frac{\partial P}{\partial t} + \frac{1}{2} f(y)^2 S^2 \frac{\partial^2 P}{\partial S^2} + \rho\beta S f(y) \frac{\partial^2 P}{\partial S \partial y} + \frac{1}{2}\beta^2 \frac{\partial^2 P}{\partial y^2} + \tilde{r}(t)\left(S\frac{\partial P}{\partial S} - P\right)\right)$$
$$= g(S, y, t).$$

Choosing to write $g(S, y, t) = \alpha(y - m) + \beta\tilde{\lambda}(S, y, t)$ permits us to make the infinitesimal generator of the Orstein–Uhlenbeck process appear explicitly in the last equation and to obtain (2.78), so that as explained in [51] , we can group the differential operator in (2.78) as follows:

$$\underbrace{\frac{\partial P}{\partial t} + \frac{1}{2} f(y)^2 S^2 \frac{\partial^2 P}{\partial S^2} + \tilde{r}(t)\left(S\frac{\partial P}{\partial S} - P\right)}_{BS_{f(y)}} + \underbrace{\rho\beta S f(y) \frac{\partial^2 P}{\partial S \partial y}}_{\text{correlation}}$$
$$+ \underbrace{\frac{1}{2}\beta^2 \frac{\partial^2 P}{\partial y^2} + \alpha(m - y)\frac{\partial P}{\partial y}}_{\text{Orstein–Uhlenbeck}} \underbrace{- \beta\tilde{\Lambda}(S, y, t)\frac{\partial P}{\partial y}}_{\text{premium}}. \quad (2.81)$$

The premium term is the market price of the volatility risk: the reason to decompose $\tilde{\Lambda}$ as in (2.79) is that, in the perfectly correlated case ($|\rho| = 1$), it is possible to find the equation satisfied by P by a simpler no-arbitrage argument with a hedged portfolio containing only the option and shares of the underlying assets. In this case, the equation found for P is

$$\frac{\partial P}{\partial t} + \frac{1}{2} f(y)^2 S^2 \frac{\partial^2 P}{\partial S^2} + \rho\beta S f(y) \frac{\partial^2 P}{\partial S \partial y} + \frac{1}{2}\beta^2 \frac{\partial^2 P}{\partial y^2}$$
$$+ \tilde{r}(t)\left(S\frac{\partial P}{\partial S} - P\right) + \left(\alpha(m - y) - \beta\rho\frac{\mu - \tilde{r}(t)}{f(y)}\right)\frac{\partial P}{\partial y} = 0, \quad \begin{cases} 0 \le t < T, \\ S > 0, \ y \in \mathbb{R}. \end{cases} \quad (2.82)$$

The term $\frac{\mu - \tilde{r}(t)}{f(y)}$ is called the *excess return-to-risk ratio*.

Finally, with (2.78), the Itô formula, and (2.79)

$$dP(S_t, Y_t, t) = \left(Sf(Y_t)\frac{\partial P}{\partial S} + \beta\rho\frac{\partial P}{\partial y}\right)\left(\frac{\mu - \tilde{r}}{f(Y_t)}dt + dW_t\right) + \beta\sqrt{1-\rho^2}\frac{\partial P}{\partial y}(\tilde{\gamma}dt + dZ_t),$$

so we see that the function $\tilde{\gamma}$ is the contribution of the second source of randomness dZ_t to the risk premium.

There remains to choose the function f. In [110], E. Stein and J. Stein have considered the case when

$$f(y) = |y|, \tag{2.83}$$

but it is also possible to make other choices; see [51, 74]. A closed form for the price of a vanilla call has been given in [70] for $f(y) = |y|$ and $\tilde{\Lambda} = \lambda y^2$.

The partial differential equation is studied in [5, 3]. Here, we focus on the case (2.83), and for simplicity we consider only $\rho = 0$. Note that only in the other case, we may have to impose in addition the condition

$$P(S, 0, t) = g(S, t), \quad 0 \le t < T, \ S > 0, \tag{2.84}$$

where

$$\begin{aligned} &\frac{\partial g}{\partial t} + \tilde{r}(t)\left(S\frac{\partial g}{\partial S} - g\right) = 0, \quad 0 \le t < T, \ S > 0, \\ &g(S, t = T) = h(S, 0). \end{aligned} \tag{2.85}$$

To obtain a forward parabolic equation, we work with the time to maturity, i.e., $T - t \to t$. Also, in order to use a variational method, we make the change of unknown

$$u(S, y, t) = P(S, y, T - t)e^{-(1-\eta)\frac{(y-m)^2}{2\upsilon^2}}, \tag{2.86}$$

where η is a parameter such that $0 < \eta < 1$, because it can be seen very easily that if $\tilde{\Lambda} = 0$, then the function $e^{\frac{(y-m)^2}{2\upsilon^2}}$ satisfies (2.78), and we want to avoid such a behavior for large values of y. The parameter η will not be important for practical computations, because in any case, we have to truncate the domain and suppress large values of y.

With the notation $r(t) = \tilde{r}(T - t)$, $\gamma(t) = \tilde{\gamma}(T - t)$, and $\Lambda(t) = \tilde{\Lambda}(T - t)$ the new unknown u satisfies the degenerate parabolic partial differential equation

$$\begin{aligned} &\frac{\partial u}{\partial t} - \frac{1}{2}y^2 S^2 \frac{\partial^2 u}{\partial S^2} - r(t)\left(S\frac{\partial u}{\partial S} - u\right) - \frac{1}{2}\beta^2\frac{\partial^2 u}{\partial y^2} \\ &+ (-\alpha(y-m) + \beta\Lambda(S, y, t))\frac{\partial u}{\partial y} + \left(2\frac{\alpha}{\beta}\Lambda(S, y, t)(y-m) - \alpha\right)u \\ &+ \eta\left(2\alpha(y-m)\frac{\partial u}{\partial y} + 2\frac{\alpha^2}{\beta^2}(1-\eta)(y-m)^2 u - 2\frac{\alpha}{\beta}\Lambda(y-m)u + \alpha u\right) = 0, \end{aligned} \tag{2.87}$$

or, by expanding Λ and by denoting by $\mathcal{L}_t$ the linear partial differential operator

$$\begin{aligned}
&\mathcal{L}_t(v) \\
&= -\frac{1}{2}y^2S^2\frac{\partial^2 v}{\partial S^2} - \frac{1}{2}\beta^2\frac{\partial^2 v}{\partial y^2} - r(t)S\frac{\partial v}{\partial S} + (-(1-2\eta)\alpha(y-m) + \beta\gamma(S,y,t))\frac{\partial v}{\partial y} \\
&+ \left(r(t) + 2\frac{\alpha^2}{\beta^2}\eta(1-\eta)(y-m)^2 + 2(1-\eta)\frac{\alpha}{\beta}(y-m)(\gamma(S,y,t)) - \alpha(1-\eta)\right)v,
\end{aligned} \tag{2.88}$$

we obtain

$$\frac{\partial u}{\partial t} + \mathcal{L}_t u = 0. \tag{2.89}$$

We denote by Q the open half plane $Q = \mathbb{R}_+ \times \mathbb{R}$. Let us consider the weighted Sobolev space V:

$$V = \left\{ v : \left(\sqrt{1+y^2}v, \frac{\partial v}{\partial y}, S|y|\frac{\partial v}{\partial S} \right) \in (L^2(Q))^3 \right\}. \tag{2.90}$$

This space with the norm

$$|||v|||_V = \left(\int_Q (1+y^2)v^2 + \left(\frac{\partial v}{\partial y}\right)^2 + S^2y^2\left(\frac{\partial v}{\partial S}\right)^2 \right)^{\frac{1}{2}} \tag{2.91}$$

is a Hilbert space, and it has the following properties:

1. V is separable.
2. Calling $\mathcal{D}(Q)$ the space of smooth functions with compact support in Q, $\mathcal{D}(Q) \subset V$ and $\mathcal{D}(Q)$ is dense in V.
3. V is dense in $L^2(Q)$.

The crucial point is point 2, which can be proved by an argument due to Friedrichs (Theorem 4.2 in [58]). We also have the following lemma.

Lemma 2.22. *Let v be a function in V. Then*

$$\int_Q y^2v^2 \leq 4\int_Q y^2S^2\left(\frac{\partial v}{\partial S}\right)^2, \tag{2.92}$$

so the seminorm

$$\|v\|_V = \left(\int_Q v^2 + \left(\frac{\partial v}{\partial y}\right)^2 + S^2y^2\left(\frac{\partial v}{\partial S}\right)^2 \right)^{\frac{1}{2}} \tag{2.93}$$

is in fact a norm in V, equivalent to $|||\cdot|||$.

We call V' the dual of V. For using the general theory, we need to prove first the following lemma.

Lemma 2.23. *The operator $v \to \beta S \frac{\partial v}{\partial S}$ is continuous from V into V'.*

Proof. Call X and Y the differential operators

$$X(v) = Sy\frac{\partial v}{\partial S} + \beta\frac{\partial v}{\partial y}, \quad Y(v) = Sy\frac{\partial v}{\partial S} - \beta\frac{\partial v}{\partial y}. \tag{2.94}$$

The operators X and Y are continuous operators from V into $L^2(Q)$ and their adjoints are

$$\begin{aligned} X^T(v) &= -Sy\frac{\partial v}{\partial S} - \beta\frac{\partial v}{\partial y} - yv = -X(v) - yv, \\ Y^T v &= -Sy\frac{\partial v}{\partial S} + \beta\frac{\partial v}{\partial y} - yv = -Y(v) - yv. \end{aligned} \tag{2.95}$$

Consider the commutator $[X, Y] = XY - YX$: it can be checked that

$$[X, Y](v) = 2\beta S\frac{\partial v}{\partial S}. \tag{2.96}$$

Therefore, for $v \in V$ and $w \in \mathcal{D}(Q)$,

$$\left(2\beta S\frac{\partial v}{\partial S}, w\right) = -\int_Q Y(v)(X(w) + yw) + \int_Q X(v)(Y(w) + yw), \tag{2.97}$$

and from (2.92), there exists a constant C such that

$$\left|\left(2\beta S\frac{\partial v}{\partial S}, w\right)\right| \leq C\|v\|_V\|w\|_V.$$

To conclude, we use the density of $\mathcal{D}(Q)$ into V. $\square$

Lemma 2.23 implies the following proposition.

Proposition 2.24. *Assume that r is a bounded function of time and that γ is bounded by a constant. The operator $\mathcal{L}_t$ is a bounded linear operator from V into V' with a constant independent of t.*

We need also a Gårding inequality.

Proposition 2.25. *Assume that r is a bounded function of time and that γ is bounded by a constant Γ. Assume that $\alpha > \beta$; then there exist two positive constants C and c independent of t and two constants $0 < \eta_1 < \eta_2 < 1$ such that, for $\eta_1 < \eta < \eta_2$ and for any $v \in V$,*

$$(\mathcal{L}_t v, v) \geq C\|v\|_V^2 - c\|v\|_{L^2(Q)}^2. \tag{2.98}$$

From Propositions 2.24 and 2.25, we can prove the existence and uniqueness of weak solutions to the Cauchy problem with (2.89).

Theorem 2.26. *Assume that $\alpha > \beta$ and that η has been chosen as in Proposition 2.25. Then, for any $u_0 \in L^2(Q)$, there exists a unique u in $L^2(0,T;V) \cap C^0([0,T];L^2(Q))$, with $\frac{\partial u}{\partial t} \in L^2(0,T;V')$ such that, for a.e. $t \in (0,T)$,*

$$\left(\frac{\partial u}{\partial t}, v\right) + (\mathcal{L}_t u, v) = 0 \quad \forall v \in V \tag{2.99}$$

and

$$u(t=0) = u_0. \tag{2.100}$$

The mapping $u_0 \mapsto u$ is continuous from $L^2(Q)$ to $L^2(0,T;V) \cap C^0([0,T];L^2(Q))$.

Remark 2.9. *The ratio $\frac{\alpha^2}{\beta^2}$ is exactly the ratio between the rate of mean reversion and the asymptotic variance of the volatility. The assumption in Theorem 2.26 says that the rate of mean reversion should not be too small compared with the asymptotic variance of the volatility. This condition is usually satisfied in practice, since α is often much larger than the asymptotic variance $\frac{\beta^2}{\alpha}$.*

It is possible to prove a maximum principle similar to Lemma 2.14: as a consequence, in the case of a vanilla put, we see that the weak solution to Theorem 2.26 yields a solution to (2.82) with a financially correct behavior.

Proposition 2.27. *Assume that the coefficients are smooth and bounded, and that $\alpha > \beta$. If $P_0(S,y) = (K-S)_+$, then the function*

$$P = u(T-t,x,y)e^{(1-\eta)\frac{(y-m)^2}{2\nu^2}},$$

where u is the solution to (2.99), (2.100) with $u_0 = e^{(1-\eta)\frac{(y-m)^2}{2\nu^2}} P_0$, satisfies

$$(x - Ke^{-r(T-t)})_- \le P(t,x,y) \le Ke^{-r(T-t)}, \tag{2.101}$$

and we have a put-call parity.

In [3], it is shown that the domain of $\mathcal{L}_t$ does not depend on t; more precisely, the following result is proved.

Theorem 2.28. *If for all t, $r(t) > 0$, the domain D_t of $\mathcal{L}_t$ does not depend on t: $D_t = D$.*

Moreover, if there exists a constant $r_0 > 0$ such that $r(t) > r_0$ a.e., and if $\frac{\alpha^2}{\beta^2} > 2$, then for well-chosen values of η (in particular such that $2\frac{\alpha^2}{\beta^2}\eta(1-\eta) > 1$),

$$D = \left\{ v \in V;\, y^2x^2\frac{\partial^2 v}{\partial x^2}, \frac{\partial^2 v}{\partial y^2}, yx\frac{\partial^2 v}{\partial x \partial y}, x\frac{\partial v}{\partial x}, y\frac{\partial v}{\partial y}, y^2 v \in L^2(Q) \right\}. \tag{2.102}$$

Then we can prove stronger regularity results on the solution to (2.99), (2.100).

Theorem 2.29. *Assume that there exists* $\zeta, 0 < \zeta \leq 1$, *such that* $\gamma \in C^{\zeta}([0,T], L^{\infty}(Q))$ *and* r *is a Hölder function of time with exponent* ζ. *Assume also that* $r(t) > r_0$ *for a positive constant* r_0 *and that* $\frac{\alpha^2}{\beta^2} > 2$. *Then for* η *chosen as in Proposition 2.25 and Theorem 2.28, if* u_0 *belongs to* D *defined by (2.102), then the solution of (2.99), (2.100) given by Theorem 2.25 belongs also to* $\mathcal{C}^1((0,T); L^2(Q)) \cap \mathcal{C}^0([0,T]; D)$ *and satisfies the equation*

$$\frac{d}{dt}u - \mathcal{L}_t u = 0 \tag{2.103}$$

for each $t \in [0,T]$.

Furthermore, for $u_0 \in L^2(Q)$, *the weak solution of (2.99), (2.100) given by Theorem 2.26 belongs also to* $\mathcal{C}^1((\tau,T); L^2(Q)) \cap \mathcal{C}^0([\tau,T]; D)$ *for all* $\tau > 0$ *and we have that* $\|\frac{du}{dt}(t)\|_{L^2(Q)} + \|\mathcal{L}_t u(t)\|_{L^2(Q)} \leq \frac{C}{t}$ *for* $t > 0$.

Project 2.1. *Adapt the Monte-Carlo program in §1.4 to compute*

1. *a European put with a stochastic volatility model;*

2. *a basket put option with two assets.*

Run some tests with realistic data in both cases. Implement a variance reduction method to speed up the program.

Chapter 3

Finite Differences

Historically (see Lax and Richtmyer [88], Richtmyer and Morton [105], and Courant, Friedrichs, and Lewy [35]), the finite difference method is the first family of local methods for discretizing partial differential equations. Arguably, it can be attributed to Richardson in the beginning of the twentieth century. We are going to present several finite difference schemes for solving (2.9). There are many choices:

1. Should it be done in the actual price variable S or should we use the logarithmic price?
2. Should we use a method explicit in time?
3. Should the first order terms be upwinded?

To answer these we must first make sure that the question is understood. So we introduce first the simplest explicit scheme, then discuss two implicit schemes. As we shall see, a uniform mesh in logarithmic price is not the most accurate and efficient choice, so we will introduce a third scheme in the primitive variable with different proofs.

3.1 Finite Differences in Logarithmic Prices

3.1.1 Basic Estimates from Taylor Expansions

Lemma 3.1. *Let u be a C^4 function defined on a closed bounded interval I of $\mathbb{R}$. Then, for z and h such that $z, z+h, z-h \in I$,*

$$\left| \frac{u(z+h) - 2u(z) + u(z-h)}{h^2} - u''(z) \right| \leq \frac{h^2}{12} \|u^{(4)}\|_{L^\infty(I)}. \tag{3.1}$$

Proof. A Taylor expansion gives

$$u(z+h) = u(z) + u'(z)h + \frac{h^2}{2}u''(z) + \frac{h^3}{6}u^{(3)}(z) + \frac{h^4}{24}u^{(4)}(z+\theta h),$$

$$u(z-h) = u(z) - u'(z)h + \frac{h^2}{2}u''(z) - \frac{h^3}{6}u^{(3)}(z) + \frac{h^4}{24}u^{(4)}(z-\theta' h),$$

with $\theta, \theta' \in (0,1)$. Adding the two identities above,

$$u(z+h)+u(z-h)=2u(z)+h^2u''(z)+\frac{h^4}{24}(u^{(4)}(z+\theta h)+u^{(4)}(z-\theta' h)).$$

The conclusion is straightforward. □

Remark 3.1. *It can be proved in the same manner that if u is only C^3-regular, then for z and h such that $z, z+h, z-h \in I$,*

$$\left|\frac{u(z+h)-2u(z)+u(z-h)}{h^2}-u''(z)\right| \le \frac{h}{3}\|u^{(3)}\|_{L^\infty(I)}. \tag{3.2}$$

If u is only C^2-regular, then for all z in the interior of I,

$$\lim_{h\to 0}\left|\frac{u(z+h)-2u(z)+u(z-h)}{h^2}-u''(z)\right| = 0. \tag{3.3}$$

It is equally easy to establish the following result.

Lemma 3.2. *Let u be a C^2 function defined on a closed bounded interval I of $\mathbb{R}$. Then, for z and h such that $z, z+h \in I$,*

$$\begin{aligned}\left|\frac{u(z+h)-u(z)}{h}-u'(z)\right| &\le \frac{h}{2}\|u''\|_{L^\infty(I)},\\ \left|\frac{u(z+h)-u(z)}{h}-u'(z+h)\right| &\le \frac{h}{2}\|u''\|_{L^\infty(I)},\end{aligned} \tag{3.4}$$

and if u is C^3-regular,

$$\left|\frac{u(z+h)-u(z)}{h}-u'\left(z+\frac{h}{2}\right)\right| \le \frac{h^2}{24}\|u^{(3)}\|_{L^\infty(I)}. \tag{3.5}$$

3.1.2 Euler Explicit Scheme

Performing the change of variables $x = \log S$, $t = T - t$, leads to an equation of the form

$$\frac{\partial u}{\partial t}-\frac{1}{2}\sigma^2(x,t)\frac{\partial^2 u}{\partial x^2}-\beta(x,t)\frac{\partial u}{\partial x}+r(t)u=0.$$

We have seen in the previous chapter that this partial differential equation has a semi-analytical solution (2.7) when the volatility and the interest rate are constant.

Localization. For simplicity, we assume that r and σ are smooth functions, asymptotically constant when x is large. Consider, for example, a put of strike K. Let x_0 be a real number such that $-x_0 \ll \log K \ll x_0$ and let $\psi(x,t)$ be a function such that

1. $\psi(x,t) = Ke^{-\int_0^t r(\tau)d\tau} - e^x$ in $[-\infty, -x_0] \times [0,T]$,
2. $\psi(x,t) = 0$ in $[x_0, \infty] \times [0,T]$,
3. ψ is smooth.

Setting $\varphi = u - \psi$, one obtains

$$\frac{\partial \varphi}{\partial t} - \frac{1}{2}\sigma^2(x,t)\frac{\partial^2 \varphi}{\partial x^2} - \beta(x,t)\frac{\partial \varphi}{\partial x} + r(t)\varphi = f(x,t)$$

and $\lim_{|x|\to\infty} \varphi(x,t) = 0$, with a fast decay (see Proposition 2.21 for a partial justification). This permits us to truncate the domain in the variable x.

Choose $\bar{x} > x_0$, and consider the boundary value problem

$$\begin{aligned} \frac{\partial \varphi}{\partial t} - \frac{1}{2}\sigma^2(x,t)\frac{\partial^2 \varphi}{\partial x^2} - \beta(x,t)\frac{\partial \varphi}{\partial x} + r(t)\varphi = f(x,t), \quad & x \in (-\bar{x}, \bar{x}),\ t \in (0,T], \\ \varphi(-\bar{x},t) = \varphi(\bar{x},t) = 0, \quad & t \in (0,T], \\ \varphi(x,0) = \varphi_0(x), \quad & x \in (-\bar{x},\bar{x}). \end{aligned} \tag{3.6}$$

Discretization with the Euler Explicit Scheme. Given two positive integers N and M, we set $h = \frac{2\bar{x}}{N+1}$ and $\Delta t = \frac{T}{M}$, and we consider the real numbers

$$x_j = -\bar{x} + jh, \quad 0 \le j \le N+1, \qquad \text{and} \qquad t_m = m\Delta t, \quad 0 \le m \le M.$$

The points (x_j, t_m) are the nodes of a uniform grid in the rectangle $[-\bar{x}, \bar{x}] \times [0,T]$.

Using Lemmas 3.1 and 3.2, we may consider the *explicit scheme*:

Find φ_j^m, $m \in \{1,\dots,M\}$, $j \in \{1,\dots,N\}$, such that

$$\frac{1}{\Delta t}(\varphi_j^{m+1} - \varphi_j^m) - \frac{1}{2h^2}(\sigma_j^m)^2(\varphi_{j+1}^m - 2\varphi_j^m + \varphi_{j-1}^m) - \frac{\beta_j^m}{2h}(\varphi_{j+1}^m - \varphi_{j-1}^m) + r^m\varphi_j^m = f_j^m \tag{3.7}$$

for $0 \le m < M,\ 1 \le j \le N$, and

$$\begin{aligned} \varphi_0^m = \varphi_{N+1}^m = 0, \quad & m \in \{1,\dots,M\}, \\ \varphi_j^0 = \varphi_0(x_j), \quad & j \in \{0,\dots,N+1\}, \end{aligned} \tag{3.8}$$

where $\sigma_j^m = \sigma(x_j, t_m)$, $\beta_j^m = \beta(x_j, t_m)$, $r^m = r(t_m)$, and f_j^m is either $f(x_j, t_m)$ or an approximation of it.

It is crucial to understand that the value φ_j^m is not $\varphi(x_j, t_m)$; it is an approximation of $\varphi(x_j, t_m)$ for well-chosen values of parameters h and Δt.

For $0 \le m < M$, the values $\{\varphi_j^{m+1}\}_0^{N+1}$ can be computed from (3.7) in an explicit manner from $\{\varphi_j^m\}_0^{N+1}$. This is why this scheme is called an explicit scheme.

Abstract Results on Finite Difference Schemes. Let us reformulate what we have done so far in a general abstract setting with the following notation: let us call

$$I_{h,\Delta t} : C^0([-\bar{x}, \bar{x}] \times [0,T]) \to \mathbb{R}^{(N+2)\times(M+1)}$$

the operator which maps a function ϕ to its values at the nodes of the grid,

$$\mathcal{L}(\varphi) = \mathcal{G}$$

the boundary value problem (3.6), and

$$V_{h,\Delta t} = \mathbb{R}^{(N+2)\times(M+1)}$$

the space of unknowns after discretization.

We write

$$\mathcal{L}_{h,\Delta t}(\varphi_{h,\Delta t}) = \mathcal{G}_{h,\Delta t},$$

the discrete system (3.7). So $\mathcal{G}_{h,\Delta t}$ is obtained by taking the values of $\mathcal{G}$ at $(N+2)\times(M+1)$ points related to the nodes of the grid: $\mathcal{G}_{h,\Delta t} = \tilde{I}_{h,\Delta t}(\mathcal{G})$.

Remark 3.2. *In the case of the boundary value problem* (3.6),

$$\mathcal{L}(\varphi) = \left(\frac{\partial \varphi}{\partial t} - \frac{1}{2}\sigma^2 \frac{\partial^2 \varphi}{\partial x^2} - \beta \frac{\partial \varphi}{\partial x} + r\varphi, \varphi_{|x=-\bar{x}}, \varphi_{|x=\bar{x}}, \varphi_{|t=0}\right)$$

and

$$\mathcal{G} = (f, 0, 0, \varphi_0) \in \mathcal{C}^0([-\bar{x}, \bar{x}] \times (0, T]) \times (\mathcal{C}^0([0, T]))^2 \times \mathcal{C}^0([-\bar{x}, \bar{x}]).$$

With the Euler explicit scheme (3.7), (3.8), $\mathcal{G}_{h,\Delta t} = \tilde{I}_{h,\Delta t}(\mathcal{G}) = (g_j^m)_{0\le j\le N+1, 0\le m\le M}$, *where*

$$\begin{aligned} g_j^m &= f(x_j, t_{m-1}) \quad \text{for } 1 \le m \le M,\ 1 \le j \le N, \\ g_0^m &= g_{N+1}^m = 0 \quad \text{for } 1 \le m \le M, \\ g_j^0 &= \varphi_0(x_j) \quad \text{for } 0 \le j \le N+1, \end{aligned}$$

and

$$\begin{aligned} [\mathcal{L}_{h,\Delta t}(\phi)]_j^m = {} & \frac{1}{\Delta t}(\phi_j^m - \phi_j^{m-1}) - \frac{1}{2h^2}(\sigma_j^{m-1})^2(\phi_{j+1}^{m-1} - 2\phi_j^{m-1} + \phi_{j-1}^{m-1}) \\ & - \frac{\beta_j^{m-1}}{2h}(\phi_{j+1}^{m-1} - \phi_{j-1}^{m-1}) + r^{m-1}\phi_j^{m-1} \quad \text{for } m = 1, \dots, M,\ 1 \le j \le N, \\ [\mathcal{L}_{h,\Delta t}(\phi)]_j^m = {} & \phi_j^m \quad \text{if } j = 0,\ j = N+1, \text{ or } m = 0. \end{aligned}$$

Definition 3.3. *The scheme* $\mathcal{L}_{h,\Delta t}(\varphi_{h,\Delta t}) = \tilde{I}_{h,\Delta t}(\mathcal{G})$ *is said to be* consistent *for approximating* $\mathcal{L}(\varphi) = \mathcal{G}$ *if there exists a functional space W (containing smooth functions of t and x), such that* $I_{h,\Delta t}$ *and* $\tilde{I}_{h,\Delta t} \circ \mathcal{L}$ *are well defined on W and, for all* $\phi \in W$,

$$\lim_{h\to 0, \Delta t \to 0} \|\mathcal{L}_{h,\Delta t} \circ I_{h,\Delta t}(\phi) - \tilde{I}_{h,\Delta t} \circ \mathcal{L}(\phi)\|_\infty = 0.$$

The error $\mathcal{L}_{h,\Delta t} \circ I_{h,\Delta t}(\phi) - \tilde{I}_{h,\Delta t} \circ \mathcal{L}(\phi)$ *is called the consistency error of the scheme. If the scheme is consistent, let* k_t *and* k_x *be the largest nonnegative numbers such that, for any smooth function* ϕ, *there exists a positive constant* $C(\phi)$ *with*

$$\|\mathcal{L}_{h,\Delta t} \circ I_{h,\Delta t}(\phi) - \tilde{I}_{h,\Delta t} \circ \mathcal{L}(\phi)\|_\infty \le C(\phi)(h^{k_x} + \Delta t^{k_t});$$

then the scheme is said to be of order k_x *with respect to the variable x, and* k_t *with respect to the variable t.*

Definition 3.4. *We denote by* $\|\cdot\|$ *a family (depending on M and N) of norms on* $\mathbb{R}^{(N+2)\times(M+1)}$. *The scheme* $\mathcal{L}_{h,\Delta t}(\varphi_{h,\Delta t}) = \tilde{I}_{h,\Delta t}(\mathcal{G})$ *is said to be* stable *with respect to the norms* $\|\cdot\|$ *if there exists a constant C independent of h and* Δt *such that* $\|\mathcal{L}_{h,\Delta t}^{-1}\| \le C$.

Definition 3.5. *The scheme* $\mathcal{L}_{h,\Delta t}(\varphi_{h,\Delta t}) = \tilde{I}_{h,\Delta t}(\mathcal{G})$ *is said to be* convergent *with respect to the norms* $\|\cdot\|$ *if*

$$\lim_{h\to 0,\Delta t\to 0} \|\varphi_{h,\Delta t} - I_{h,\Delta t}(\varphi)\| = 0,$$

provided φ *is smooth enough.*

Theorem 3.6. *Denote by* $\|\cdot\|$ *a family (depending on* M *and* N*) of norms on* $\mathbb{R}^{(N+2)\times(M+1)}$*, such that* $\|\cdot\| \le C\|\cdot\|_\infty$ *for a constant* C *independent of* M *and* N*. If the scheme* $\mathcal{L}_{h,\Delta t}(\varphi_{h,\Delta t}) = \tilde{I}_{h,\Delta t}(\mathcal{G})$ *is consistent and stable with respect to the norms* $\|\cdot\|$*, then it is convergent. If the scheme* $\mathcal{L}_{h,\Delta t}(\varphi_{h,\Delta t}) = \tilde{I}_{h,\Delta t}(\mathcal{G})$ *is of order* k_t *(resp.,* k_x*) with respect to* t *(resp.,* x*) and stable with respect to the norms* $\|\cdot\|$*, and if* φ *is smooth enough, then for a constant* $C(\varphi)$ *independent of* h *and* Δt*,*

$$\|\varphi_{h,\Delta t} - I_{h,\Delta t}(\varphi)\| \le C(\varphi)(h^{k_x} + \Delta t^{k_t}).$$

Proof. Calling $\mathcal{E}_{h,\Delta t}$ the consistency error for φ, we have $\varphi_{h,\Delta t} - I_{h,\Delta t}(\varphi) = \mathcal{L}_{h,\Delta t}^{-1}\mathcal{E}_{h,\Delta t}$. The stability of the scheme implies that $\|\varphi_{h,\Delta t} - I_{h,\Delta t}(\varphi)\| \le \|\mathcal{E}_{h,\Delta t}\|$ for a constant C independent of h and Δt. The consistency of the scheme tells us that for φ smooth enough, $\lim_{h\to 0,\Delta t\to 0}\|\mathcal{E}_{h,\Delta t}\| = 0$. The previous two observations yield the convergence.

The proof of the second assertion is done in the same manner. □

Consistency of the Euler Explicit Scheme. Calling ϵ_j^m the entries of the consistency error $\mathcal{E}_{h,\Delta t}$, we have

$$\epsilon_j^m = \begin{pmatrix} \dfrac{\varphi(x_j,t_{m+1}) - \varphi(x_j,t_m)}{\Delta t} - \dfrac{\partial \varphi}{\partial t}(x_j,t_m) \\ -\dfrac{\sigma^2(x_j,t_m)}{2}\left(\dfrac{\varphi(x_{j+1},t_m) - 2\varphi(x_j,t_m) + \varphi(x_{j-1},t_m)}{h^2} - \dfrac{\partial^2\varphi}{\partial x^2}(x_j,t_m)\right) \\ -\beta(x_j,t_m)\left(\dfrac{\varphi(x_{j+1},t_m) - \varphi(x_{j-1},t_m)}{2h} - \dfrac{\partial\varphi}{\partial x}(x_j,t_m)\right) \end{pmatrix},$$

$$1 \le j \le N, \quad 1 \le m \le M,$$

$$\epsilon_0^m = \epsilon_{N+1}^m = \epsilon_j^0 = 0, \quad 1 \le j \le N, \quad 0 \le m \le M.$$

From Lemmas 3.1 and 3.2, we know that if $\varphi \in C^0([0,T]; C^4[-\bar{x},\bar{x}]) \cup C^2([0,T]; C^0[-\bar{x},\bar{x}])$, then

$$|\epsilon_j^m| \le C(\varphi)(h^2 + \Delta t). \tag{3.9}$$

Therefore, the Euler explicit scheme is of order one with respect to t and of order two with respect to x.

Stability of the Euler Explicit Scheme. For simplicity, we focus on the case when the coefficients σ, β, and r are constant. The general case will be treated completely in §3.3. Let us consider a family of norms $\|\cdot\|$ on $\mathbb{R}^{N+2}$, such that, for a constant C independent of

Δt, $\|\cdot\| \le C\|\cdot\|_\infty$. For this norm, we define the norm $\|\cdot\|$ on $\mathbb{R}^{(N+2)\times(M+1)}$:

$$\|V\| = \max_{m=0,\dots,M} \|(v_0^m, \dots, v_{N+1}^m)^T\|. \tag{3.10}$$

From the proof of Theorem 3.6, we have to estimate $\|\mathcal{L}_{h,\Delta t}^{-1}\mathcal{E}_{h,\Delta t}\|$, and since $\epsilon_0^m = \epsilon_{N+1}^m = \epsilon_{j=0}^0$, $0 \le m \le M$, $1 \le j \le N$, it is enough to find a condition on Δt and h such that

$$\sup_{V \in V_{h,\Delta t}^0 \backslash \{0\}} \frac{\|\mathcal{L}_{h,\Delta t}^{-1} V\|}{\|V\|} < C', \tag{3.11}$$

with $V_{h,\Delta t}^0 = \{V \in V_{h,\Delta t};\ v_0^m = v_{N+1}^m = v_{j=0}^0,\ 0 \le m \le M,\ 1 \le j \le N\}$, holds for a constant C' independent of Δt and h.

Notation. For $V \in V_{h,\Delta t}^0 \backslash \{0\}$, let us call

$$U = \mathcal{L}_{h,\Delta t}^{-1} V \quad \text{and} \quad U^m = (u_1^m, \dots, u_N^m)^T, \quad V^m = (v_1^m, \dots, v_N^m)^T.$$

We have

$$U^m = \Delta t V^m + (I - \Delta t A) U^{m-1},$$

where $A \in \mathbb{R}^{N\times N}$ is the tridiagonal matrix

$$A_{j,j} = \frac{\sigma^2}{h^2} + r, \quad j = 1, \dots, N,$$

$$A_{j,j+1} = -\frac{\beta}{2h} - \frac{\sigma^2}{2h^2}, \quad j = 1, \dots, N-1,$$

$$A_{j,j-1} = \frac{\beta}{2h} - \frac{\sigma^2}{2h^2}, \quad j = 2, \dots, N.$$

With the notation defined above,

$$\sup_{V \in V_{h,\Delta t}^0 \backslash \{0\}} \frac{\|\mathcal{L}_{h,\Delta t}^{-1} V\|}{\|V\|} = \sup_{V \in V_{h,\Delta t}^0 \backslash \{0\}} \frac{\max_{m=1,\dots,M} \|U^m\|}{\max_{m=1,\dots,M} \|V^m\|}.$$

Lemma 3.7. *For all $V \in V_{h,\Delta t}^0 \backslash \{0\}$, we have, for $U = \mathcal{L}_{h,\Delta t}^{-1} V$,*

$$U^m = \Delta t \sum_{l=0}^{m-1} (I - \Delta t A)^{m-l} V^{l+1} \quad \forall m = 1, \dots, M. \tag{3.12}$$

Recall that we have, for any matrix $B \in \mathbb{R}^{N\times N}$,

$$\rho(B) \le \|B\|, \tag{3.13}$$

where $\rho(B)$ is the spectral radius of B, i.e.,

$$\rho(B) = \max_{\lambda \in \text{spectrum}(B)} |\lambda|,$$

and where $\|B\| = \sup_{V \in \mathbb{R}^N} \frac{\|BV\|}{\|V\|}$. From the previous lemma, we see that a necessary condition for (3.11) is that $\rho(I - \Delta t A)$ is less than or equal to 1.

It is possible to find the spectrum of the matrix A as follows.

Lemma 3.8. *If $\beta \neq -\frac{\sigma^2}{h}$, we have*

$$A\chi_l = \lambda_l \chi_l, \quad \lambda_l = r + \frac{\sigma^2}{h^2} - \sqrt{\frac{\sigma^4}{h^4} - \frac{\beta^2}{h^2}} \cos\left(\frac{l\pi}{N+1}\right), \quad l = 1, \dots, N,$$

$$\chi_l = \left(\left(\frac{\frac{\sigma^2}{h} - \beta}{\frac{\sigma^2}{h} + \beta}\right)^{\frac{1}{2}} \sin\left(\frac{l\pi}{N+1}\right), \dots, \left(\frac{\frac{\sigma^2}{h} - \beta}{\frac{\sigma^2}{h} + \beta}\right)^{\frac{N}{2}} \sin\left(\frac{Nl\pi}{N+1}\right)\right)^T. \tag{3.14}$$

Remark 3.3. *As a consequence of Lemma* 3.8, *we see that if $\sigma^2 = 2$, $\beta = 0$, $r = 0$, then the matrix A is symmetric and we have the discrete Poincaré inequality*

$$V^T A V = \frac{1}{h^2} \sum_{i=0}^{N} (v_i - v_{i+1})^2 \geq \lambda_{\min} \|V\|_2^2 = \frac{4}{h^2} \sin^2\left(\frac{\pi}{2(N+1)}\right) \|V\|_2^2 \approx \frac{\pi^2}{4\bar{x}^2} \|V\|_2^2. \tag{3.15}$$

Theorem 3.9. *For the scheme to be stable in norm $\|\cdot\|$, a necessary condition is that*

$$\frac{\sigma^2}{h} > |\beta| \quad \textit{and} \quad \Delta t \left(\frac{r}{2} + \frac{\sigma^2}{2h^2} + \frac{1}{2}\sqrt{\frac{\sigma^4}{h^4} - \frac{\beta^2}{h^2}} \cos\left(\frac{\pi}{N+1}\right)\right) \leq 1, \tag{3.16}$$

or

$$\frac{\sigma^2}{h} \leq |\beta| \quad \textit{and} \quad \Delta t \leq \frac{2(r + \frac{\sigma^2}{h^2})}{(r + \frac{\sigma^2}{h^2})^2 + (\frac{\beta^2}{h^2} - \frac{\sigma^4}{h^4}) \cos^2(\frac{\pi}{N+1})}. \tag{3.17}$$

Proof. From (3.13) and Lemma 3.8, we see that, if $\frac{\sigma^2}{h} > \beta$,

$$\begin{aligned} \|I - \Delta t A\| \geq \rho(1 - \Delta t A) &= \max_{l=1,\dots,N} |1 - \Delta t \lambda_l| \\ &= \left|1 - \Delta t \left(r + \frac{\sigma^2}{h^2} + \sqrt{\frac{\sigma^4}{h^4} - \frac{\beta^2}{h^2}} \cos\left(\frac{\pi}{N+1}\right)\right)\right|. \end{aligned}$$

Therefore, from Lemma 3.7, a necessary condition for (3.11) is that $\|I - \Delta t A\| \leq 1$, and we obtain (3.16) in the case $\frac{\sigma^2}{h} > |\beta|$.

If $\frac{\sigma^2}{h} < |\beta|$,

$$|1 - \Delta t \lambda_l|^2 = \left(1 - \Delta t \left(r + \frac{\sigma^2}{h^2}\right)\right)^2 + \Delta t^2 \left(\frac{\beta^2}{h^2} - \frac{\sigma^4}{h^4}\right) \cos^2\left(\frac{\pi l}{N+1}\right).$$

Thus,

$$\rho^2(I - \Delta t A) = \left(1 - \Delta t \left(r + \frac{\sigma^2}{h^2}\right)\right)^2 + \Delta t^2 \left(\frac{\beta^2}{h^2} - \frac{\sigma^4}{h^4}\right) \cos^2\left(\frac{\pi}{N+1}\right).$$

Therefore, a necessary condition for (3.11) is (3.17).

If $\frac{\sigma^2}{h} = |\beta|$, the matrix A has one eigenvalue, $r + \frac{\sigma^2}{h^2}$, so the claim holds in this case. □

Remark 3.4. *The condition* (3.16) *was first found by Courant, Friedrichs, and Lewy* [35] *for similar problems. For that reason, it is classically called a CFL condition.*

The CFL condition (3.16) is quite restrictive. Indeed, it says that for small values of h, Δt must scale like h^2, which means that the number of time steps must scale like the square of the number of steps in the x-direction. In practice, the CFL condition is responsible for very long CPU times when refining the grid in the x-variable. For this reason, explicit schemes are seldom used for parabolic problems, except when the dynamic of the solution is very fast and justifies a very fine time discretization.

We have given a necessary condition for stability. To find a sufficient condition, we have to compute or estimate the norm $\|I - \Delta t A\|$. For example, if we choose for the norm in $\mathbb{R}^{N+2}$: $\frac{1}{\sqrt{N}}\|\cdot\|_2$, then we have to estimate $\|I - \Delta t A\|_2$: we see that

$$\begin{aligned}\|(I - \Delta t A)U\|_2^2 &= \|U\|_2^2 - \Delta t U^T(A + A^T)U + \Delta t^2 U^T A^T A U\\ &= (1 - 2r\Delta t)\|U\|_2^2 - \frac{\sigma^2 \Delta t}{h^2}\sum_{i=0}^{N} |u_{i+1} - u_i|^2\\ &+ \Delta t^2 \sum_{i=1}^{N}\left(\frac{\sigma^2}{2h^2}(2u_i + u_{i-1} + u_{i+1}) - \frac{\beta}{2h}(u_{i+1} - u_{i-1}) + ru_i\right)^2\end{aligned}$$

and we see that if Δt is small enough compared to h^2, then $\|I - \Delta t A\|_2 < 1$, so the CFL condition gives the correct scaling for a sufficient stability condition. Note that it is possible (but really tedious) to compute exactly $\|I - \Delta t A\|_2$, since $\|I - \Delta t A\|_2^2 = \rho((I - \Delta t A)^T(I - \Delta t A))$, and $(I - \Delta t A)^T(I - \Delta t A)$ is a pentadiagonal matrix with constant coefficients on the diagonals (except for the first and last diagonal coefficients), so a necessary and sufficient stability condition in the norm $\|\cdot\|_2$ can be derived.

Remark 3.5. *Note that if $\beta = 0$, then the matrix A is symmetric, and we have $\|I - \Delta t A\|_2 = \rho(I - \Delta t A)$, so the necessary condition* (3.16) *is also sufficient for the stability in the 2-norm.*

Remark 3.6. *Replacing $r^m\varphi_j^m$ by $r^{m+1}\varphi_j^{m+1}$ in* (3.7) *still enables us to compute φ_j^{m+1} in an explicit manner. This choice improves (often only slightly) the stability of the scheme.*

Convergence. From (3.9) and Theorem 3.9, we see that if the stability condition (3.11) holds and if φ is smooth enough, i.e., if $\varphi \in C^0([0, T]; C^4([-\bar{x}, \bar{x}])) \cap C^2([0, T]; C^0([-\bar{x}, \bar{x}]))$, then

$$\|\varphi_{h,\Delta t} - I_{h,\Delta t}(\varphi)\| \le C(h^2 + \Delta t).$$

Note that, for vanilla options, for example, φ cannot be so regular, because φ_0 does not even belong to $C^2([-\bar{x}, \bar{x}])$. In order to obtain the convergence, one must first approximate φ_0 by a smooth function $\tilde{\varphi}_0$ (such that $\tilde{\varphi}_0(\pm\bar{x}) = 0$), then solve the boundary value problem

$$\begin{aligned}&\frac{\partial\tilde{\varphi}}{\partial t} - \frac{1}{2}\sigma^2(x,t)\frac{\partial^2\tilde{\varphi}}{\partial x^2} - \beta(x,t)\frac{\partial\tilde{\varphi}}{\partial x} + r(t)\tilde{\varphi} = f(x,t), \quad x \in (-\bar{x}, \bar{x}),\ t \in (0, T],\\ &\tilde{\varphi}(t, -\bar{x}) = \tilde{\varphi}(t, \bar{x}) = 0, \quad t \in (0, T],\\ &\tilde{\varphi}(0, x) = \tilde{\varphi}_0(x), \quad x \in (-\bar{x}, \bar{x}).\end{aligned} \tag{3.18}$$

For each $\epsilon > 0$, one can choose $\tilde{\varphi}_0$ in order to have

$$\|\varphi - \tilde{\varphi}\|_{L^\infty((0,T)\times(-\bar{x},\bar{x}))} \le \epsilon. \tag{3.19}$$

Then one approximates $\tilde{\varphi}$ by the Euler explicit scheme and obtains $\tilde{\varphi}_{h,\Delta t}$. If the CFL condition (3.11) is satisfied, then it follows from the stability analysis that

$$\|\tilde{\varphi}_{h,\Delta t} - \varphi_{h,\Delta t}\| \le C\epsilon. \tag{3.20}$$

Finally, we have

$$\|\tilde{\varphi}_{h,\Delta t} - I_{h,\Delta t}(\tilde{\varphi})\| \le C(\tilde{\varphi})(h^2 + \Delta t), \tag{3.21}$$

since $\tilde{\varphi}$ is smooth enough, and h and Δt can be chosen small enough so that $\|\tilde{\varphi}_{h,\Delta t} - I_{h,\Delta t}(\tilde{\varphi})\| \le \epsilon$. From this, (3.19), and (3.20), we see that $\lim_{h,\Delta t\to 0} \|\varphi_{h,\Delta t} - I_{h,\Delta t}(\varphi)\| = 0$. It is also possible to find rates of convergence in weaker norms.

Exercise 3.1. *Modify the scheme* (3.7), (3.8) *to discretize the new boundary value problem obtained by replacing the second line of* (3.6) *by* $\varphi(-\bar{x}, t) = 0$; $\frac{\partial \varphi}{\partial x}(\bar{x}, t) = g(t)$. *To discretize* $\frac{\partial \varphi}{\partial x}(\bar{x}, t)$, *one can use Lemma* 3.2. *Analyze the stability and the consistency of the scheme.*

Exercise 3.2. *Prove the following assertion: let* u *be a* C^3 *function defined on a closed bounded interval* I *of* $\mathbb{R}$. *Then, for* z *and* h *such that* $z, z-h, z-2h \in I$,

$$\left|\frac{1}{2h}(3u(z) - 4u(z-h) + u(z-2h)) - u'(z)\right| \le Ch^2\|u^{(3)}\|_{L^\infty(I)}. \tag{3.22}$$

Exercise 3.3. *Deduce from Exercise* 3.2 *a new second order scheme for the boundary value problem in Exercise* 3.1.

Exercise 3.4. *We consider the variable coefficients boundary value problem* "in divergence form" *(we assume that* $0 < \sigma_* \le \sigma(x) \le \sigma^* < +\infty$ *and* $0 < a_* \le a(x) \le a^* < +\infty$*)*

$$\begin{aligned} & a(x)\frac{\partial u}{\partial t} - \frac{\partial}{\partial x}\left(\sigma(x)\frac{\partial u}{\partial x}\right) = 0, \quad x \in (0,1), t > 0, \\ & u(0,t) = u(1,t) = 0, \qquad t > 0, \\ & u(x,0) = u_0(x). \end{aligned} \tag{3.23}$$

Propose a scheme for (3.23) *based on approximating*

$$\frac{\partial}{\partial x}\left(\sigma\frac{\partial u}{\partial x}\right)(x_j)$$

by

$$\frac{1}{h}\left(\left(\sigma\frac{\partial u}{\partial x}\right)(x_{j+\frac{1}{2}}) - \left(\sigma\frac{\partial u}{\partial x}\right)(x_{j-\frac{1}{2}})\right), \quad \textit{where } x_{j\pm\frac{1}{2}} = x_j \pm \frac{h}{2}$$

using a centered finite difference method to approximate $\frac{\partial u}{\partial x}(x_{j\pm\frac{1}{2}})$.

Prove that if σ *is smooth enough, then this scheme is second order accurate with respect to* x.

ALGORITHM 3.1. Euler explicit scheme.

```
#include <iostream>
#include <math.h>
#include <stdlib.h>
#include <fstream.h>
#include<gsl/gsl_sf_erf.h>
using namespace std;

const int NT=10000;                              //   number of time steps
const int NX=201;                               //   nb of space intervals
const int L=200;                                        //    val max of x
const int K = 100;                                         //     strike
double sigmap=0.2, r=0.1;
double * u=new double[NX];
double phi_T(double s){ if(s>K) return s-K; else return 0;}
void PDEfiniteDiff()
{
    double dx= (double)L/(double)(NX-1), dt = 1./NT, t=0;
    for(int i=0;i<NX;i++)
      u[i] = phi_T(i*dx);                             //   final condition

    for(int j=0;j<NT;j++)
      {
        t+=dt;
        for(int i=1;i<NX-1;i++)
          {
            double x=i*dx;
            u[i] += (0.5*sigmap*x*sigmap*x*(u[i+1]-2*u[i]+u[i-1])/dx/dx
                      +r*x*(u[i+1]-u[i])/dx-r*u[i]) * dt;
          }
        u[NX-1]=L-K*exp(-r*t);
        u[0]=0;
      }
}

int main()
{
  ofstream ff("fd.dat");
  PDEfiniteDiff();
  double dx= (double)L/(double)(NX-1);
  for(int i=0;i<NX;i++)
    {
      double x=i*dx;
      ff<< x<< "\t" << u[i]<<endl;
    }
}
```

We have plotted in Figure 3.1 the price of the vanilla call option computed with Algorithm 3.1.

3.1.3 The Euler Implicit Scheme

We have seen above that the Euler explicit scheme becomes much too slow when the mesh is fine because stability requires $\Delta t = O(h^2)$.

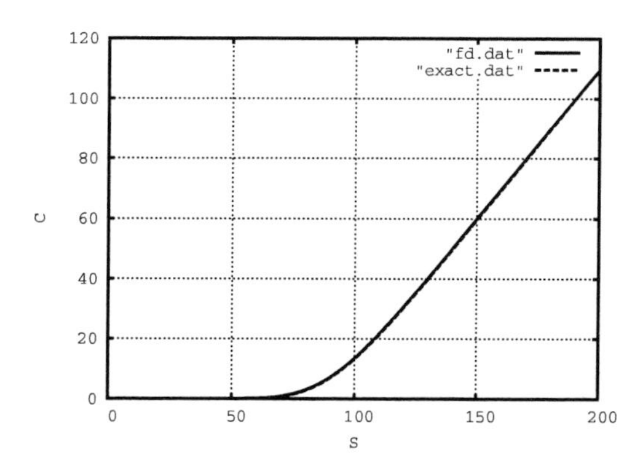

Figure 3.1. *The price of the European vanilla call option with $K = 100, \sigma = 0.2, r = 0.1$ one year to maturity computed by Euler's explicit scheme and by the Black–Scholes formula.*

One idea is to modify the scheme in the following way:

Find φ_j^m, $m \in \{0, \dots, M\}$, $j \in \{0, \dots, N+1\}$, satisfying (3.8), and, for $1 \le m \le M$, $1 \le j \le N$,

$$\frac{1}{\Delta t}(\varphi_j^m - \varphi_j^{m-1}) - \frac{1}{2h^2}(\sigma_j^m)^2(\varphi_{j+1}^m - 2\varphi_j^m + \varphi_{j-1}^m) - \frac{\beta_j^m}{2h}(\varphi_{j+1}^m - \varphi_{j-1}^m) + r^m\varphi_j^m = f_j^m. \quad (3.24)$$

In order to obtain $(\varphi_1^m, \dots, \varphi_N^m)^T$ from $(\varphi_1^{m-1}, \dots, \varphi_N^{m-1})^T$, one has to solve a nontrivial system of linear equations with N unknowns. This is why the method is called implicit.

The Euler implicit scheme (3.24), (3.8) can be recast in the abstract form as

$$\begin{aligned}
\mathcal{L}_{h,\Delta t}(\varphi_{h,\Delta t}) = \mathcal{G}_{h,\Delta t}, \text{ where } \mathcal{G}_{h,\Delta t} &= \tilde{I}_{h,\Delta t}(\mathcal{G}) = (g_j^m)_{0\le j\le N+1, 0\le m\le M},\\
g_j^m &= f(x_j, t_m) \quad \text{for } 1 \le m \le M, 1 \le j \le N,\\
g_0^m &= g_{N+1}^m = 0 \quad \text{for } 1 \le m \le M,\\
g_j^0 &= \varphi_0(x_j) \quad \text{for } 0 \le j \le N+1.
\end{aligned}$$

Consistency of the Euler Implicit Scheme. Calling ϵ_j^m the entries of the consistency error $\mathcal{E}_{h,\Delta t}$, we have

$$\epsilon_j^m = \left(\begin{array}{l}
\dfrac{\varphi(x_j, t_m) - \varphi(x_j, t_{m-1})}{\Delta t} - \dfrac{\partial \varphi}{\partial t}(x_j, t_m) \\
-\dfrac{\sigma^2(x_j, t_m)}{2}\left(\dfrac{\varphi(x_{j+1}, t_m) - 2\varphi(x_j, t_m) + \varphi(x_{j-1}, t_m)}{h^2} - \dfrac{\partial^2 \varphi}{\partial x^2}(x_j, t_m) \right) \\
-\beta(x_j, t_m)\left(\dfrac{\varphi(x_{j+1}, t_m) - \varphi(x_{j-1}, t_m)}{2h} - \dfrac{\partial \varphi}{\partial x}(x_j, t_m) \right)
\end{array} \right),$$

$$1 \le j \le N, \quad 1 \le m \le M,$$

$$\epsilon_0^m = \epsilon_{N+1}^m = \epsilon_j^0 = 0, \quad 1 \le j \le N, 0 \le m \le M.$$

From Lemmas 3.1 and 3.2, we know that (3.9) holds if

$$\varphi \in \mathcal{C}^0([0,T];\mathcal{C}^4[-\bar{x},\bar{x}]) \cup \mathcal{C}^2([0,T];\mathcal{C}^0[-\bar{x},\bar{x}]).$$

Therefore, the Euler implicit scheme is of order one with respect to t and of order two with respect to x.

Stability of the Euler Implicit Scheme. For simplicity, we assume here that the coefficients are constant. We take the norm

$$\|V\| = \max_{m=0,\dots,M} \left(\frac{1}{N} \sum_{j=0}^{N+1} (v_j^m)^2 \right)^{\frac{1}{2}}. \tag{3.25}$$

It is enough to show that (3.11) is satisfied. For $V \in V_{h,\Delta t}^0 \backslash \{0\}$, let us call

$$U = \mathcal{L}_{h,\Delta t}^{-1} V \quad \text{and} \quad U^m = (u_1^m, \dots, u_N^m)^T, \quad V^m = (v_1^m, \dots, v_N^m)^T.$$

At each time step, one has to solve the system of linear equations

$$(I + \Delta t A) U^m = \Delta t V^m + U^{m-1},$$

where $A \in \mathbb{R}^{N \times N}$ is the tridiagonal matrix introduced in §3.1.2.

Lemma 3.10.

- *The matrix $I + \Delta t A$ is invertible.*
- $$\|(I + \Delta t A)^{-1}\|_2 \le 1, \tag{3.26}$$

and the Euler implicit scheme (3.24) *is stable in the norm* $\|\cdot\|$.

Proof. Let W be a vector in $\mathbb{R}^N$: we have from Remark 3.3

$$\begin{aligned} W^T A W &= \frac{\sigma^2}{2h^2} \sum_{i=0}^{N} (w_i - w_{i+1})^2 + r\|W\|_2^2 \\ &\ge \left(r + \frac{2\sigma^2}{h^2} \sin^2\left(\frac{\pi}{2(N+1)} \right) \right) \|W\|_2^2 \approx \left(r + \frac{\pi^2 \sigma^2}{8\bar{x}^2} \right) \|W\|_2^2. \end{aligned} \tag{3.27}$$

Therefore $W^T(I + \Delta t A)W \ge \|W\|_2^2$, which implies that $I + \Delta t A$ is invertible. Furthermore, if $(I + \Delta t A)W = F$, then

$$\left(1 + \Delta t \left(r + \frac{2\sigma^2}{h^2} \sin^2\left(\frac{\pi}{2(N+1)} \right) \right) \right) \|W\|_2^2 \le \|F\|_2 \|W\|_2,$$

which implies estimate (3.26). For all $V \in V_{h,\Delta t}^0 \backslash \{0\}$, we have

$$\frac{1}{\sqrt{N}} \|U^m\|_2 = \frac{\Delta t}{\sqrt{N}} \sum_{l=0}^{m-1} \|(I + \Delta t A)^{-1}\|_2^{m-l} \|V^{l+1}\|_2, \quad 1 \le m \le M; \tag{3.28}$$

thus

$$\|U\| \le M \Delta t \|V\| = T\|V\|. \qquad \square$$

Remark 3.7. *The Euler implicit scheme* (3.24) *is unconditionally stable in norm* $\|\cdot\|$ *with respect to* Δt.

Convergence of the Euler Implicit Scheme. It is possible to replicate the arguments used for Euler's explicit scheme, and prove that the Euler implicit scheme is convergent in norm $\|\cdot\|$.

Exercise 3.5. *Using Exercise* 3.2, *prove that the Gear scheme*

$$\begin{aligned}&\frac{1}{2\Delta t}(3\varphi_j^m - 4\varphi_j^{m-1} + \varphi_j^{m-2}) - \frac{1}{2h^2}(\sigma_j^m)^2(\varphi_{j+1}^m - 2\varphi_j^m + \varphi_{j-1}^m)\\ &-\frac{\beta_j^m}{2h}(\varphi_{j+1}^m - \varphi_{j-1}^m) + r^m\varphi_j^m = f_j^m\end{aligned} \tag{3.29}$$

is second order with respect to both t and x. This is a two-step scheme, so it is not possible to use it for $j = 1$. *Analyze the stability of this scheme.*

3.1.4 The Crank–Nicolson Scheme

One of the best one-step implicit schemes is the Crank–Nicolson scheme:

Find φ_j^m, $m \in \{0, \dots, M\}$, $j \in \{0, \dots, N+1\}$, satisfying (3.8), and, for $1 \le m \le M$, $1 \le j \le N$,

$$\begin{aligned}&\frac{1}{\Delta t}(\varphi_j^m - \varphi_j^{m-1}) - \frac{(\sigma_j^m)^2}{4h^2}(\varphi_{j+1}^m - 2\varphi_j^m + \varphi_{j-1}^m)\\ &-\frac{(\sigma_j^{m-1})^2}{4h^2}(\varphi_j^{m-1} - 2\varphi_j^{m-1} + \varphi_{j-1}^{m-1}) - \frac{\beta_j^m}{4h}(\varphi_{j+1}^m - \varphi_{j-1}^m)\\ &-\frac{\beta_j^{m-1}}{4h}(\varphi_{j+1}^{m-1} - \varphi_{j-1}^{m-1}) + \frac{r^m}{2}\varphi_j^m + \frac{r^{m-1}}{2}\varphi_j^{m-1} = f\left(x_j, \frac{1}{2}(t_m + t_{m+1})\right).\end{aligned} \tag{3.30}$$

As we shall see, it is more accurate than Euler's schemes and has the same kind of stability as the Euler implicit scheme. Indeed, the Crank–Nicolson scheme (3.30), (3.8) can be recast in the abstract form

$$\mathcal{L}_{h,\Delta t}(\varphi_{h,\Delta t}) = \mathcal{G}_{h,\Delta t},$$

where

$$\mathcal{G}_{h,\Delta t} = \tilde{I}_{h,\Delta t}(\mathcal{G}) = (g_j^m)_{0\le j\le N+1, 0\le m\le M},$$

with

$$\begin{aligned}&g_j^m = f(x_j, \tfrac{1}{2}(t_m + t_{m-1})) \quad \text{for } 1 \le m \le M, 1 \le j \le N,\\ &g_0^m = g_{N+1}^m = 0 \quad \text{for } 1 \le m \le M,\\ &g_j^0 = \varphi_0(x_j) \quad \text{for } 0 \le j \le N+1,\end{aligned}$$

and $[\mathcal{L}_{h,\Delta t}(\phi)]_j^m = \phi_j^m$ if $j = 0$, $j = N+1$, or $m = 0$,

$$[\mathcal{L}_{h,\Delta t}(\phi)]_j^m$$
$$= \frac{1}{\Delta t}(\phi_j^m - \phi_j^{m-1}) - \frac{1}{4h^2}(\sigma_j^{m-1})^2(\phi_{j+1}^{m-1} - 2\phi_j^{m-1} + \phi_{j-1}^{m-1})$$
$$- \frac{1}{4h^2}(\sigma_j^m)^2(\phi_{j+1}^m - 2\phi_j^m + \phi_{j-1}^m) - \frac{\beta_j^{m-1}}{4h}(\phi_{j+1}^{m-1} - \phi_{j-1}^{m-1}) - \frac{\beta_j^m}{4h}(\phi_{j+1}^m - \phi_{j-1}^m)$$
$$+ \frac{r^{m-1}}{2}\phi_j^{m-1} + \frac{r^m}{2}\phi_j^m$$

for $m = 1, \ldots, M$, $1 \le j \le N$.

Consistency of the Crank–Nicolson Scheme. From Lemmas 3.1 and 3.2, we know that if the solution to (3.6) is smooth enough, then the consistency error scales like $\Delta t^2 + h^2$. Therefore, the Crank–Nicolson is of order two with respect to both x and t.

Stability of the Crank–Nicolson Scheme. We assume that the coefficients σ, β, and r are constant, and we keep the notation as for the Euler schemes. It is enough to show that (3.11) is satisfied. For $V \in V_{h,\Delta t}^0\backslash\{0\}$, let us call $U = \mathcal{L}_{h,\Delta t}^{-1}V$ and $U^m = (u_1^m, \ldots, u_N^m)^T$, $V^m = (v_1^m, \ldots, v_N^m)^T$. At each time step, one has to solve the system of linear equations

$$\left(I + \frac{\Delta t}{2}A\right)U^m = \Delta t V^m + \left(I - \frac{\Delta t}{2}A\right)U^{m-1}, \tag{3.31}$$

where A has been introduced in §3.1.2.

Lemma 3.11. *The matrix $I + \frac{\Delta t}{2}A$ is invertible, and*

$$\left\|\left(I + \frac{\Delta t}{2}A\right)^{-1}\right\|_2 \le 1, \qquad \left\|\left(I + \frac{\Delta t}{2}A\right)^{-1}\left(I - \frac{\Delta t}{2}A\right)\right\|_2 \le 1.$$

Proof. We leave the first point to the reader.

For the second point, we have that

$$\left(I + \frac{\Delta t}{2}A\right)^{-1}\left(I - \frac{\Delta t}{2}A\right) = -I + 2\left(I + \frac{\Delta t}{2}A\right)^{-1}. \tag{3.32}$$

Therefore, for $W \in \mathbb{R}^N$,

$$\left\|\left(I + \frac{\Delta t}{2}A\right)^{-1}\left(I - \frac{\Delta t}{2}A\right)W\right\|_2^2$$
$$= \|W\|_2^2 - 2W^T\left(I + \frac{\Delta t}{2}A\right)^{-1}W - 2W^T\left(I + \frac{\Delta t}{2}A\right)^{-T}W + 4\left\|\left(I + \frac{\Delta t}{2}A\right)^{-1}W\right\|_2^2$$
$$= \|W\|_2^2 - \Delta t\left(\left(I + \frac{\Delta t}{2}A\right)^{-1}W\right)^T(A + A^T)\left(I + \frac{\Delta t}{2}A\right)^{-1}W$$
$$\le \|W\|_2^2,$$

which proves the desired result, because $A + A^T$ is positive definite; see (3.27). □

Thanks to Lemma 3.11, we have

$$U^m = \Delta t \left(I + \frac{\Delta t}{2} A\right)^{-1} V^m + \left(I + \frac{\Delta t}{2} A\right)^{-1} \left(I - \frac{\Delta t}{2} A\right) U^{m-1}, \tag{3.33}$$

which implies that

$$\|U^m\|_2 \le \Delta t \sum_{l=0}^{m-1} \left\| \left(I + \frac{\Delta t}{2} A\right)^{-1} \left(I - \frac{\Delta t}{2} A\right) \right\|_2^{m-l-1} \left\| \left(I + \frac{\Delta t}{2} A\right)^{-1} \right\|_2 \|V^{l+1}\|_2, \tag{3.34}$$

and Lemma 3.11 implies that

$$\|U\| \le T \|V\|. \tag{3.35}$$

The Crank–Nicolson scheme is stable with respect to $\|\cdot\|$, unconditionally with respect to h and Δt, and convergence is proved as in §3.1.2.

Exercise 3.6. *For θ, $0 \le \theta \le 1$, consider the scheme*

$$\begin{aligned} &\frac{1}{\Delta t}(\varphi_j^m - \varphi_j^{m-1}) \\ &+ \theta \left(-\frac{(\sigma_j^m)^2}{2h^2}(\varphi_{j+1}^m - 2\varphi_j^m + \varphi_{j-1}^m) - \frac{\beta_j^m}{2h}(\varphi_{j+1}^m - \varphi_{j-1}^m) + r^m \varphi_j^m \right) \\ &+ (1-\theta) \left(-\frac{(\sigma_j^{m-1})^2}{2h^2}(\varphi_j^{m-1} - 2\varphi_j^{m-1} + \varphi_{j-1}^{m-1}) - \frac{\beta_j^{m-1}}{2h}(\varphi_{j+1}^{m-1} - \varphi_{j-1}^{m-1}) + r^{m-1}\varphi_j^{m-1} \right) \\ &= f(x_j, \theta t_m + (1-\theta) t_{m+1}). \end{aligned} \tag{3.36}$$

One recovers Euler's implicit scheme for $\theta = 1$, Euler's explicit scheme for $\theta = 0$, and the Crank–Nicolson scheme for $\theta = \frac{1}{2}$.

Studying the stability of this scheme consists of considering the sequence given by

$$(I + \theta \Delta t A) U^m = \Delta t V^m + (I - (1-\theta)\Delta t A) U^{m-1}, \tag{3.37}$$

where A has been introduced in §3.1.2. Prove that the scheme is unconditionally stable as soon as $\theta \ge \frac{1}{2}$ (the proof is similar to that for the Crank–Nicolson scheme and relies on the fact that $A + A^T$ is positive definite; see (3.27)*).*

3.2 Upwinding

Let us assume that the coefficients σ, β, and r are constant and discuss the stability of the Euler implicit scheme:

Find φ_j^m, $m \in \{0, \dots, M\}$, $j \in \{0, \dots, N+1\}$, satisfying $\varphi_j^0 = \varphi_0(x_j)$, $0 \le j \le N+1$, $\varphi_0^m = \varphi_{N+1}^m = 0$, $0 \le m \le M$, and, for $1 \le m \le M$, $1 \le j \le N$,

$$\frac{1}{\Delta t}(\varphi_j^m - \varphi_j^{m-1}) - \frac{\sigma^2}{2h^2}(\varphi_{j+1}^m - 2\varphi_j^m + \varphi_{j-1}^m) - \frac{\beta}{2h}(\varphi_{j+1}^m - \varphi_{j-1}^m) + r\varphi_j^m = 0, \tag{3.38}$$

with respect to $\|\cdot\|_\infty$: $\|V\|_\infty = \max_{0 \le m \le M} \max_{1 \le j \le N} |v_j^m|$. For that, we need a few additional notions on matrix analysis; for a vector $V \in \mathbb{R}^N$, the notation $V \ge 0$ (resp., $V \le 0$) means that all the components of V are nonnegative (resp., nonpositive).

Definition 3.12. *A matrix $B \in \mathbb{R}^{N\times N}$ is an M-matrix if there exists a diagonal matrix D with positive diagonal entries such that*

$$\begin{aligned} &B_{i,i} > 0, \quad 1 \le i \le N, \\ &B_{i,j} \le 0, \quad 1 \le i, j \le N, i \ne j, \\ &\sum_{j=1}^{N} B_{i,j} D_j > 0, \quad 1 \le i \le N. \end{aligned} \tag{3.39}$$

An M-matrix is the right product of a strictly diagonal dominant matrix by an invertible diagonal matrix, so it is invertible. The M-matrices are an important class of matrices, because they have monotonicity properties.

Lemma 3.13 (discrete maximum principle). *Let $F \in \mathbb{R}^N$ be a vector such that $F \ge 0$. Let $B \in \mathbb{R}^{N\times N}$ be an M-matrix. Then $V = B^{-1}F \ge 0$.*

Proof. By easy algebraic manipulations, it is enough to prove the result when the matrix B is diagonal dominant, i.e., $D = I_d$. Let i_0 be the index such that $v_{i_0} = \min_j v_j$. We have

$$B_{i_0,i_0} v_{i_0} = F_{i_0} - \sum_{j \ne i_0} B_{i_0,j} v_j \ge F_{i_0} - \sum_{j \ne i_0} B_{i_0,j} v_{i_0}$$

because the off-diagonal coefficients of B are nonpositive, which implies

$$\left(\sum_j B_{i_0,j} \right) v_{i_0} \ge F_{i_0} \ge 0.$$

Therefore $v_{i_0} \ge 0$, from the third property in (3.39). □

Corollary 3.14. *The entries of the inverse of an M-matrix are all nonnegative.*

Proof. Let B be an M-matrix and $K_j = ((B^{-1})_{1,j}, \dots, (B^{-1})_{N\,j})^T$ the jth column of B^{-1}. We have $BK_j = F_j$, where $F_j = (\delta_{i,j})^T_{1\le i\le N}$. Lemma 3.13 implies that $K \ge 0$. □

Exercise 3.7. *Let a matrix $B \in \mathbb{R}^{N\times N}$ be invertible and such that*

$$\begin{aligned} &B_{i,i} > 0, \quad 1 \le i \le N, \\ &B_{i,j} \le 0, \quad 1 \le i, j \le N, i \ne j, \\ &\sum_{j=1}^{N} B_{i,j} \ge 0, \quad 1 \le i \le N. \end{aligned}$$

Prove that Lemma 3.13 *holds for B.*

Let us consider the scheme (3.38). If $|\beta| \le \frac{\sigma^2}{h}$, then the matrices A and $I + \Delta t A$ are M-matrices and we have the following stability estimate.

Lemma 3.15. *If* $|\beta| \leq \frac{\sigma^2}{h}$, *then* $\|(I + \Delta t A)^{-1}\|_\infty \leq 1$.

Proof. Consider the system of linear equations $(I + \Delta t A)U = V$. Let $V_{\max}$, (resp., $V_{\min}$) be the vector whose components are all equal to $\max_{j=1,\dots,N} v_j$ (resp., $\min_{j=1,\dots,N} v_j$). It can be easily checked that

$$(I + \Delta t A)\left(U - \frac{1}{1 + r\Delta t} V_{\max}\right) = V - V_{\max} + E,$$

where E is a vector whose components are all nonpositive. Therefore $U - \frac{1}{1+r\Delta t} V_{\max} \leq 0$. Similarly, one can prove that $U - \frac{1}{1+r\Delta t} V_{\min} \geq 0$. Therefore,

$$\|U\|_\infty \leq \frac{1}{1 + r\Delta t} \max\left(\left|\max_{j=1,\dots,N} v_j\right|, \left|\min_{j=1,\dots,N} v_j\right|\right) = \frac{1}{1 + r\Delta t}\|V\|_\infty,$$

which ends the proof. □

In fact, we have proved the following slightly stronger result.

Lemma 3.16. *Assume that* $|\beta| \leq \frac{\sigma^2}{h}$ *and that* $\min_{j=0,\dots,N+1} \varphi_j^0 = 0$. *Then* $(\varphi_j^m)_{j,m}$ *given by* (3.38) *satisfy*

$$0 \leq \varphi_j^m \leq \frac{1}{1 + r\Delta t} \max_{i=0,\dots,N+1} \varphi_i^{m-1} \leq \frac{1}{(1 + r\Delta t)^m} \max_{i=0,\dots,N+1} \varphi_i^0.$$

If the condition $|\beta| \leq \frac{\sigma^2}{h}$ is not fulfilled, then the scheme may not be monotone; i.e., oscillations in the x-variable may appear (see the numerical example below), and $\|(I + \Delta t A)^{-1}\|_\infty$ may be larger than 1. In this case, the scheme becomes unstable in the $\|\cdot\|_\infty$ norm. So, if σ^2 is much smaller than β, then step h needs to be very small.

Remark 3.8. *Note that it is quite important to obtain nonnegative values of* φ *since* φ *stands for a price!*

Remark 3.9. *For European options, the volatility is often not small compared with* r, *so the scheme* (3.38) *is most often stable. For Asian options or options with stochastic volatility, we shall see that some diffusion coefficients in the partial differential equation may vanish completely, so the scheme* (3.38) *will be unstable in the* $\|\cdot\|_\infty$ *norm.*

One way to cure these instability phenomena is to use an alternative discretization for the term $-\beta\frac{\partial\varphi}{\partial x}(x_j, t_m)$, i.e., assuming that $\beta > 0$, $-\beta\frac{\varphi(x_{j+1},t_m)-\varphi(x_j,t_m)}{h}$, so the scheme becomes

$$\frac{1}{\Delta t}(\varphi_j^m - \varphi_j^{m-1}) - \frac{\sigma^2}{2h^2}(\varphi_{j+1}^m - 2\varphi_j^m + \varphi_{j-1}^m) - \frac{\beta}{h}(\varphi_{j+1}^m - \varphi_j^m) + r\varphi_j^m = 0. \tag{3.40}$$

Such a discretization of $-\beta\frac{\partial\varphi}{\partial x}(x_j, t_m)$ is called an *upwind* discretization, because $-\beta$ can be interpreted as a velocity, and the new discretization takes information upstream and not downstream. On the contrary, the scheme (3.38) is called *centered*.

With the upwind discretization, the Euler implicit scheme reads

$$(I + \Delta t B)\varphi^m = \varphi^{m-1}, \text{ where } B \text{ is the tridiagonal matrix,}$$

$$\begin{aligned} B_{j,j} &= \frac{\sigma^2}{h^2} + \frac{\beta}{h} + r, \quad j = 1, \dots, N, \\ B_{j,j+1} &= -\frac{\beta}{h} - \frac{\sigma^2}{2h^2}, \quad j = 1, \dots, N-1, \\ B_{j,j-1} &= -\frac{\sigma^2}{2h^2}, \quad j = 2, \dots, N. \end{aligned} \tag{3.41}$$

The matrices B and $I + \Delta t B$ are M-matrices, independent of Δt and h, and the Euler implicit scheme (3.40) is always stable in the norm $\|\cdot\|_\infty$, because of the following lemma.

Lemma 3.17. *The matrix B satisfies $\|(I + \Delta t B)^{-1}\|_\infty \leq 1$.*

Proof. The proof is left to the reader. □

Such a gain in stability has a cost: indeed, thanks to Lemmas 3.1 and 3.2, one can prove that the scheme (3.40) is first order in the variables t and x, so it is less accurate than the scheme (3.38).

Remark 3.10. *The idea of using upwind schemes originates from computational fluid dynamics, where convective effects usually dominate diffusion.*

An Example. We consider the steady state boundary value problem

$$\begin{aligned} &-\sigma^2 u''(x) + \beta u'(x) + r u(x) = 0, \quad x \in (0,1), \\ &u(0) = 0, \quad u(1) = 1, \end{aligned} \tag{3.42}$$

with $\beta > 0$. The solution to (3.42) is

$$u(x) = e^{\frac{\beta}{2\sigma^2}(x-1)} \frac{\sinh\left(\frac{\sqrt{\beta^2+4r\sigma^2}}{2\sigma^2}x\right)}{\sinh\left(\frac{\sqrt{\beta^2+4r\sigma^2}}{2\sigma^2}\right)}. \tag{3.43}$$

If $\frac{\beta}{2\sigma^2} \ll 1$, $u(x)$ is very close to 0, except in a small neighborhood of $x = 1$ of diameter $O(\frac{\sigma^2}{\beta})$ (this zone is called a boundary layer); see Figure 3.2. Let us consider the two discretizations (3.38) and (3.40) of (3.42), with $\Delta t = \infty$. It is possible to compute explicitly the discrete solutions given by these two schemes (the calculations are a bit long) and to see that the centered scheme produces oscillations, whereas the upwind scheme gives a qualitatively correct solution. In Figure 3.3, we have plotted the solutions given by three schemes: the centered scheme, the upwind scheme, and the downwind scheme (i.e., upwinding has been used in the wrong direction): we observe that the centered scheme is

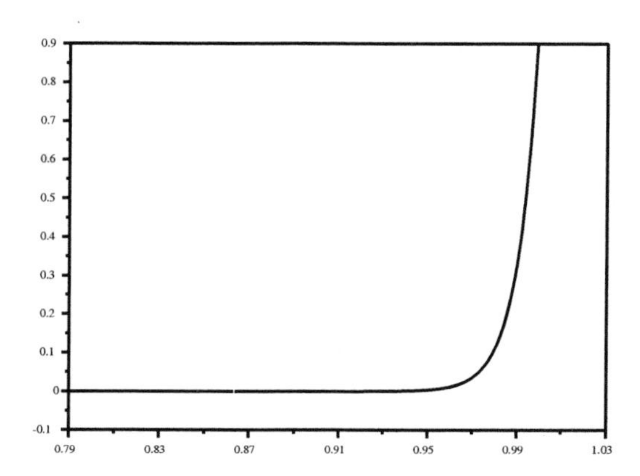

Figure 3.2. *The solution of* (3.42) *for* $\sigma^2 = 0.09,\ r = 1,\ \beta = 10$*: the solution vanishes outside a boundary layer.*

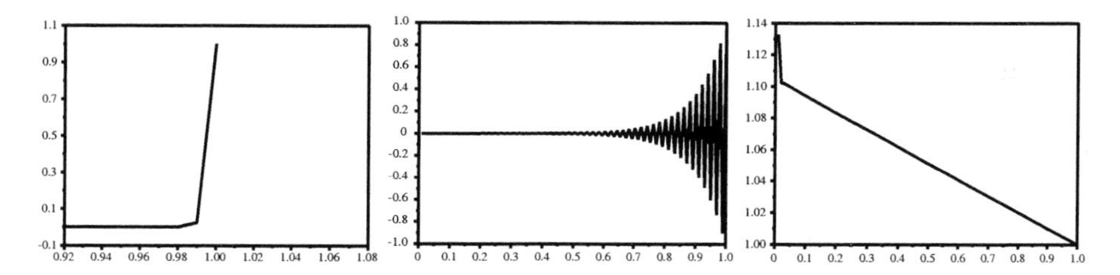

Figure 3.3. *Solutions given by three schemes for* $\sigma^2 = 0.0025,\ r = 1,\ \beta = 10,$ *and* $N = 100$. *Left: upwind scheme (zoom); center: centered scheme; right: downwind scheme. Only the upwind scheme gives a qualitatively correct result.*

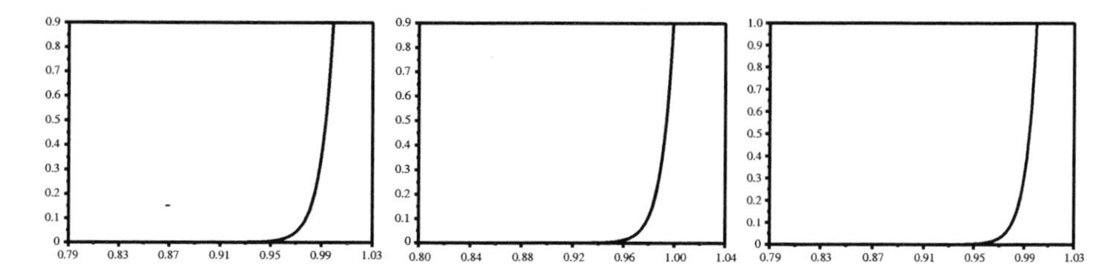

Figure 3.4. *Solutions given by three schemes for* $\sigma^2 = 0.09,\ r = 1,\ \beta = 10,$ *and* $N = 1000$. *Left: upwind scheme (zoom); center: centered scheme; right: downwind scheme.*

unstable in the norm $\|\cdot\|_\infty$ when the grid step is larger than the boundary layer, and that upwinding in the wrong direction produces a disaster.

Remark 3.11. *If the grid is sufficiently fine, i.e., if* $\frac{\beta h}{\sigma^2} \ll 1$, *then all the schemes will end up giving the correct result. This is the case for the results plotted in Figure* 3.4. *In the limit when* $h \to 0$, *the centered scheme is more accurate since it is of order two.*

One has to remember that upwind schemes are safer but less accurate than centered schemes, and that they should be used when transport phenomena dominate diffusion.

It is possible to use second order upwinding, by making use of the second order finite difference formula in Exercise 3.2: assuming that $\beta > 0$, this leads to discretizing the term $-\beta \frac{\partial \varphi}{\partial x}(x_j, t_m)$ by $-\frac{\beta}{2h}[-\varphi(x_{j+2}, t_m) + 4\varphi(x_{j+1}, t_m) - 3\varphi(x_j, t_m)]$, so the Euler implicit scheme becomes

$$\begin{aligned} &\frac{1}{\Delta t}\left(\varphi_j^m - \varphi_j^{m-1}\right) - \frac{\sigma^2}{2h^2}\left(\varphi_{j+1}^m - 2\varphi_j^m + \varphi_{j-1}^m\right) \\ &-\frac{\beta}{2h}\left(-\varphi_{j+2}^m + 4\varphi_{j+1}^m - 3\varphi_j^m\right) + r\varphi_j^m = 0 \end{aligned} \tag{3.44}$$

for $j < N-1$. This formula cannot be applied at $j = N-1$, and one can use, for instance, a first order scheme there, or something else. Although the matrix obtained with (3.44) cannot be an M-matrix (note that it has four nontrivial diagonals instead of three before), we see that the diagonal term is increased compared to a center scheme, so the stability is improved. Of course, it is possible to use the Crank–Nicolson scheme for time stepping, and obtain a second order scheme in both t and x.

3.3 Finite Differences in the Primitive Variables

We consider a European put whose price satisfies the Black–Scholes equation

$$\frac{\partial}{\partial t}P(S,t) - \frac{\sigma^2(S,t)S^2}{2}\frac{\partial^2}{\partial S^2}P(S,t) - r(t)S\frac{\partial}{\partial S}P(S,t) + r(t)P(S,t) = 0, \tag{3.45}$$

with the Cauchy data

$$P(S,0) = P_0(S), \tag{3.46}$$

where P_0 is the payoff function. We truncate the domain in the S variable, so (3.45) holds in $(0,T] \times (0,\bar{S})$, and we must add a boundary condition at the artificial boundary $S = \bar{S}$: for example, we impose the Dirichlet condition

$$P(\bar{S},t) = 0. \tag{3.47}$$

We suppose that the coefficients σ and r are smooth enough and we make the assumptions (2.40) and (2.41).

Consider two integers M and N and call $\Delta t = \frac{T}{M}$, $h = \frac{\bar{S}}{N+1}$, $S_i = ih$ for $i = 0, \dots, N+1$, and $t_m = mh$ for $m = 0, \dots, M$. The nodes (S_i, t_m) form a uniform grid of the rectangle $[0,\bar{S}] \times [0,T]$.

3.3.1 Euler Implicit Scheme

The Euler implicit scheme for discretizing (3.45), (3.46), and (3.47) reads

$$
\begin{aligned}
&\frac{p_j^{m+1}-p_j^m}{\Delta t}+\frac{(\sigma_j^{m+1})^2 S_j^2}{2}\frac{2p_j^{m+1}-p_{j-1}^{m+1}-p_{j+1}^{m+1}}{h^2}-r^{m+1}S_j\frac{p_{j+1}^{m+1}-p_{j-1}^{m+1}}{2h}\\
&+r^{m+1}p_j^{m+1}=0, \qquad m=0,\dots M-1, \quad j=1,\dots,N,\\
&\frac{p_0^{m+1}-p_0^m}{\Delta t}+r^{m+1}p_0^{m+1}=0, \quad m=0,\dots M-1,\\
&p_{N+1}^m=0, \quad m=1,\dots,M,\\
&p_j^0=P_0(S_j), \quad j=0,\dots N+1,
\end{aligned}
\tag{3.48}
$$

where $\sigma_j^m=\sigma(S_j,t_m)$ and $r^m=r(t_m)$.

We call A^m the tridiagonal matrix of $\mathbb{R}^{(N+1)\times(N+1)}$:

$$
\begin{aligned}
A_{j,j}^m &= \frac{(\sigma_j^m)^2 S_j^2}{h^2}+r^m=(j\sigma_j^m)^2+r^m, \quad j=0,\dots,N,\\
A_{j,j+1}^m &= -\frac{(\sigma_j^m)^2 S_j^2}{2h^2}-r^m S_j 2h=-\frac{1}{2}\left((j\sigma_j^m)^2+r^m j\right), \quad j=0,\dots,N-1,\\
A_{j,j-1}^m &= -\frac{(\sigma_j^m)^2 S_j^2}{2h^2}+r^m S_j 2h=-\frac{1}{2}\left((j\sigma_j^m)^2-r^m j\right), \quad j=1,\dots,N,
\end{aligned}
\tag{3.49}
$$

and $P^m\in\mathbb{R}^{N+1}$ the vector $P^m=(p_0^m,\dots,p_N^m)^T$. The Euler implicit scheme reads

$$(I+\Delta t A^{m+1})P^{m+1}=P^m. \tag{3.50}$$

We denote by $\|\cdot\|_2$ the norm in $\mathbb{R}^{N+1}$: $\|Q\|_2^2=\sum_{i=0}^N q_i^2$, and by $|\cdot|$ the norm

$$|Q|^2=\sum_{j=1}^{N+1} S_j^2(q_j-q_{j-1})^2 h^2=\sum_{j=1}^{N+1} j^2(q_j-q_{j-1})^2, \text{ posing } q_{N+1}=0.$$

Lemma 3.18. *Under assumptions* (2.40) *and* (2.41), *there exists a nonnegative constant* C_2 *such that, for all* m, $m=1,\dots,M$,

$$Q^T A^m Q\geq\frac{\underline{\sigma}^2}{4}|Q|^2-C_2\|Q\|_2^2 \quad \forall Q\in\mathbb{R}^{N+1}. \tag{3.51}$$

Proof. We have

$$Q^T A^m Q = \mathcal{S}_1 + \mathcal{S}_2 + \mathcal{S}_3,$$

where

$$\mathcal{S}_1 = \frac{1}{2}\sum_{j=1}^{N}(j\sigma_j^m)^2 q_j(2q_j - q_{j-1} - q_{j+1}),$$

$$\mathcal{S}_2 = -\frac{r^m}{2}\sum_{j=1}^{N} jq_j(q_{j+1} - q_{j-1}), \quad \mathcal{S}_3 = r^m \|Q\|_2^2.$$

Posing $q_{N+1} = 0$, we have that, for any $\eta > 0$,

$$\begin{aligned}
\mathcal{S}_1 &= \frac{1}{2}\sum_{j=1}^{N+1}(j\sigma_j^m)^2(q_j - q_{j-1})^2 + \frac{1}{2}\sum_{j=1}^{N+1}\left(((j-1)\sigma_{j-1}^m)^2 - (j\sigma_j^m)^2\right) q_{j-1}(q_{j-1} - q_j) \\
&= \frac{1}{2}\sum_{j=1}^{N+1}(j\sigma_j^m)^2(q_j - q_{j-1})^2 \\
&\quad + \frac{1}{2}\sum_{j=1}^{N+1}\left((j-1)\sigma_{j-1}^m + j\sigma_j^m\right)\left((j-1)\sigma_{j-1}^m - j\sigma_j^m\right) q_{j-1}(q_{j-1} - q_j) \\
&\geq \frac{1}{2}\sum_{j=1}^{N+1}(j\sigma_j^m)^2(q_j - q_{j-1})^2 - \frac{\eta}{4}\sum_{j=1}^{N+1}\left((j-1)\sigma_{j-1}^m - j\sigma_j^m\right)^2 q_{j-1}^2 \\
&\quad - \frac{1}{4\eta}\sum_{j=1}^{N+1}\left((j-1)\sigma_{j-1}^m + j\sigma_j^m\right)^2 (q_{j-1} - q_j)^2.
\end{aligned}$$

From (2.40), $\sum_{j=1}^{N+1}(j\sigma_j^m)^2(q_j - q_{j-1})^2 \geq \underline{\sigma}^2|Q|^2$.
From (2.41), $\sum_{j=1}^{N+1}\left((j-1)\sigma_{j-1}^m - j\sigma_j^m\right)^2 q_{j-1}^2 \leq C_\sigma^2\|Q\|_2^2$.
From (2.40), $\sum_{j=1}^{N+1}\left((j-1)\sigma_{j-1}^m + j\sigma_j^m\right)^2 (q_{j-1} - q_j)^2 \leq 4\bar{\sigma}^2|Q|^2$.
Therefore,

$$\mathcal{S}_1 \geq \left(\frac{\underline{\sigma}^2}{2} - \frac{\bar{\sigma}^2}{\eta}\right)|Q|^2 - \frac{\eta C_\sigma^2}{4}\|Q\|_2^2. \tag{3.52}$$

For the term $\mathcal{S}_2$, we have that, for any $\mu > 0$,

$$\begin{aligned}
|\mathcal{S}_2| &= \frac{r^m}{2}\left|\sum_{j=1}^{N+1} jq_j(q_j - q_{j-1}) + \sum_{j=0}^{N}(j-1)q_{j-1}(q_j - q_{j-1})\right| \\
&\leq r^m\|Q\|_2|Q| \\
&\leq \frac{r^m}{2\mu}|Q|^2 + \frac{\mu r^m}{2}\|Q\|_2^2.
\end{aligned} \tag{3.53}$$

From (3.52) and (3.53), for any $\eta > 0$ and $\mu > 0$,

$$Q^T A^m Q \geq \left(\frac{\underline{\sigma}^2}{2} - \frac{\bar{\sigma}^2}{\eta} - \frac{r^m}{2\mu}\right)|Q|^2 + \left(r^m - \frac{\eta C_\sigma^2}{4} - \frac{\mu r^m}{2}\right)\|Q\|_2^2,$$

and (3.51) follows by taking η and μ large enough. $\square$

Corollary 3.19. *If $\Delta t \leq \frac{1}{C_2}$, then, for $m = 1, \ldots, M$, $I + \Delta t A^m$ is invertible, and it is possible to use Euler's implicit scheme.*

Proof. The equality $(I + \Delta t A)Q = 0$ implies

$$(1 - \Delta t C_2)\|Q\|_2 + \frac{\Delta t \underline{\sigma}^2}{4}|Q|^2 \leq Q^T(I + \Delta t A)Q = 0,$$

and therefore $Q = 0$. $\square$

Consistency. By using Lemmas 3.1 and 3.2, one can easily check that if the solution P of (3.45), (3.46), and (3.47) is smooth enough, then, for $j = 1, \ldots, N$, the consistency error

$$\begin{aligned}\epsilon_j^{m+1} = &\frac{P(S_j, t_{m+1}) - P(S_j, t_m)}{\Delta t} \\ &+ \frac{(\sigma_j^{m+1})^2 S_j^2}{2} \frac{2P(S_j, t_{m+1}) - P(S_{j-1}, t_{m+1}) - P(S_{j+1}, t_{m+1})}{h^2} \\ &- r^{m+1} S_j \frac{P(S_{j+1}, t_{m+1}) - P(S_{j-1}, t_{m+1})}{2h} + r^{m+1} P(S_j, t_{m+1})\end{aligned}$$

is bounded by $C(P)(h^2 + \Delta t)$, whereas ϵ_0^{m+1} is bounded by $C(P)\Delta t$. Therefore, the Euler implicit scheme is of order one with respect to t and two with respect to S.

Stability: The Energy Method. Let $\|\cdot\|$ be the norm in $\mathbb{R}^{(N+1)\times M}$: $\|Q\| = \max_{m=1,\ldots,M} \frac{1}{\sqrt{N+1}}\|Q^m\|_2$, where $Q = (Q^1, \ldots, Q^m)$ and $Q^m \in \mathbb{R}^{N+1}$.

Lemma 3.20. *There exists a constant $C_3 > 0$ (independent of N) such that, if $\Delta t \leq \frac{1}{C_3}$, then the Euler implicit scheme is stable in the norm $\|\cdot\|$.*

Proof. Proving the stability of the method in the norm $\|\cdot\|$ consists of showing that if, for $m = 1, \ldots, M$,

$$(I + \Delta t A^m)U^m = U^{m-1} + \Delta t V_m, \tag{3.54}$$

then, for a positive constant independent of V, N, and M,

$$\|U\| \leq C\left(\|V\| + \frac{1}{\sqrt{N+1}}\|U^0\|_2\right). \tag{3.55}$$

To do it, we take the scalar product of (3.54) with U^m:

$$(U^m)^T(U^m - U^{m-1}) + \Delta t(U^m)^T A^m U^m = \Delta t(U^m)^T V_m,$$

and we use the well-known identity $U^T(U - V) = \frac{1}{2}(\|U\|_2^2 + \|U - V\|_2^2 - \|V\|_2^2)$. We obtain that

$$\|U^m\|_2^2 + \|U^m - U^{m-1}\|_2^2 + 2\Delta t(U^m)^T A^m U^m = \|U^{m-1}\|_2^2 + 2\Delta t(U^m)^T V_m,$$

which implies, thanks to (3.51),

$$(1 - C_3\Delta t)\|U^m\|_2^2 + \|U^m - U^{m-1}\|_2^2 + \frac{\Delta t \underline{\sigma}^2}{2}|U^m|^2 \le \|U^{m-1}\|_2^2 + \Delta t\|V_m\|_2^2, \quad (3.56)$$

where $C_3 = 2C_2 + 1$. Assuming that $\Delta t < \frac{1}{C_3}$, multiplying (3.56) by $(1 - C_3\Delta t)^{m-M}$, and summing over m, we obtain that

$$\begin{aligned}
&\|U^M\|_2^2 + \sum_{m=1}^{M}(1 - C_3\Delta t)^{m-M-1}\|U^m - U^{m-1}\|_2^2 \\
&+ \frac{\Delta t \underline{\sigma}^2}{2}\sum_{m=1}^{M}(1 - C_3\Delta t)^{m-M-1}|U^m|^2 \\
&\le (1 - C_3\Delta t)^{-M}\|U^0\|_2^2 + \Delta t\sum_{m=1}^{M}(1 - C_3\Delta t)^{m-M-1}\|V^m\|_2^2.
\end{aligned}$$

Let C_4 be a positive constant independent of Δt such that $\log(1 - C_3\Delta t) \ge -C_4\Delta t$. We have that $(1 - C_3\Delta t)^{-M} \le e^{C_4 M\Delta t} = e^{C_4 T}$, and

$$\|U^M\|_2^2 \le e^{C_4 T}\left(\|U^0\|_2^2 + \Delta t\sum_{l=1}^{M}\|V^l\|_2^2\right) \le e^{C_4 T}(\|U^0\|_2^2 + T(N+1)\|V\|^2). \quad (3.57)$$

Since (3.57) holds when M is replaced by $m \le M$, we have proved (3.55), with $C = \max(1, \sqrt{T})e^{\frac{1}{2}C_4 T}$. □

Remark 3.12. *Note that here, the stability is more difficult to analyze than in the previous cases, because the coefficients of the partial differential equation are not constant. The method used here to prove the stability is called an energy method.*

Convergence. It is possible to replicate the arguments used for Euler's explicit scheme in §3.1.2 and prove that the Euler implicit scheme in the primitive variables is convergent in the norm $\|\cdot\|$.

Exercise 3.8. *Write down the Crank–Nicolson scheme for the Black–Scholes equation in the primitive variables and analyze its stability by the energy method.*

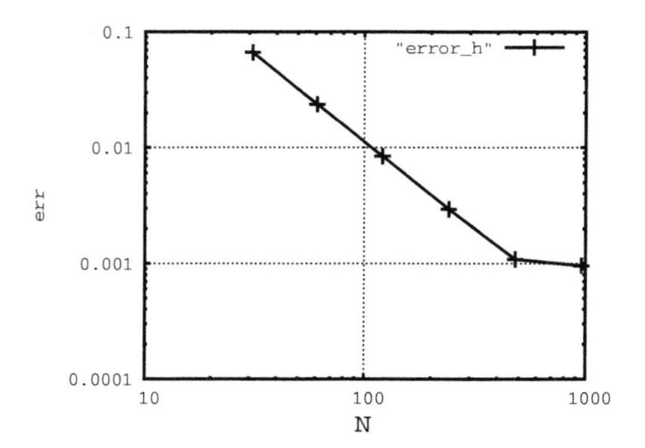

Figure 3.5. *The error produced by the Crank–Nicolson scheme with a centered discretization in the variable S, as a function of N (in log-scales).*

3.4 Numerical Results

We consider a vanilla European put with strike $K = 100$ and maturity 1 and we assume that the volatility and interest rates are constant: $\sigma = 0.2$ and $r = 0.04$. In this case, it is possible to use Black–Scholes formula (1.18), and to compute the error produced by the same finite difference schemes as in §3.3, except that we use the Crank–Nicolson time scheme instead of the Euler implicit scheme. We compute the error in the norm $\|\cdot\|$ introduced above.

In Figure 3.5, we take a very small time step, i.e., $M = 4000$, so that the consistency error due to time discretization is very small, and we plot the error in the norm $\|\cdot\|$ with respect to N. As a function of N, the error is decreasing and limited from below by the error produced by the time discretization. When this last error is negligible compared to the error due to the discretization in S, we see that the convergence order (the slope of the curve in log-scales) is less than 2: this is due to the fact that the payoff is singular at $S = K$. In fact, with more careful theoretical considerations, it could be seen that the error decays faster than $h^{\frac{3}{2}-\epsilon}$ for all $\epsilon > 0$, and slower than $h^{\frac{3}{2}}$. This is indeed observed in Figure 3.5.

In Figure 3.6, we take a small step in the S variable, i.e., $N = 120$, so that the consistency error due to discretization with respect to S is small, and we plot the error in the norm $\|\cdot\|$ with respect to M, for both the Crank–Nicolson and the Euler implicit scheme. As a function of M, the two errors are decreasing and limited from below by the error produced by the discretization in S. When this last error is negligible compared to the error due to the time discretization, we see that the convergence order (the slope of the curve in log-scales) is less than 2: this is due to the fact that the payoff is singular at $S = K$. In fact, with more careful considerations, it could be checked that the error decays faster than $\Delta t^{\frac{3}{4}-\epsilon}$ for all $\epsilon > 0$, and slower than $h^{\frac{3}{4}}$. This is indeed observed in Figure 3.6. Due to the choice of the norm $\|\cdot\|$, the two curves have the same behavior, because the error is driven by the singularity at $t = 0$, and it does not help to take an accurate scheme near $t = 0$.

At N and M fixed ($M = 230$, $N = 120$), plotting the pointwise error for the Crank–Nicolson and Euler implicit schemes (see Figure 3.7) shows that the error is concentrated around the singularity, i.e., $t = 0$ and $S = K$. We see also that the error decays faster in

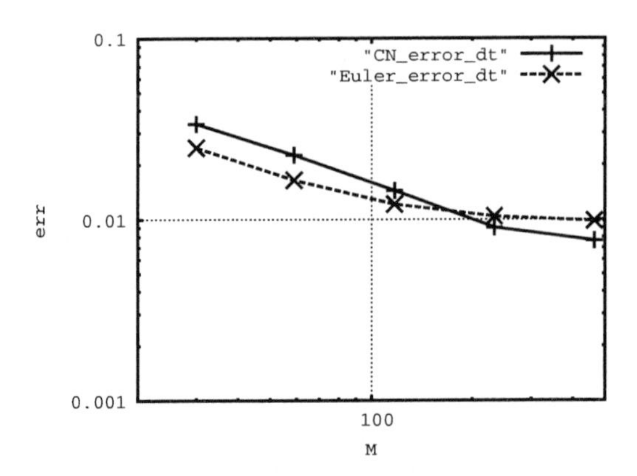

Figure 3.6. *The error produced by the Crank–Nicolson and Euler implicit schemes with a centered discretization in the variable S, as a function of M (in log-scales).*

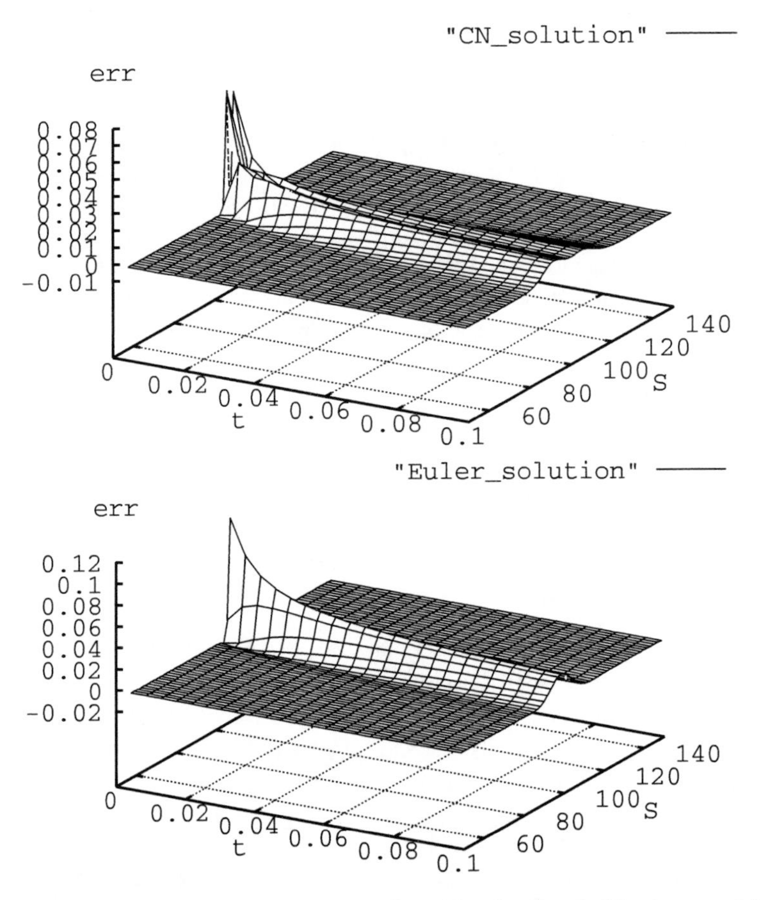

Figure 3.7. *The pointwise error produced by the Crank–Nicolson and Euler implicit schemes with a centered discretization in the variable S, as functions of t and S.*

time with the Crank–Nicolson scheme: indeed, away from $t = 0$, the solution is smooth and the Crank–Nicolson behaves better than the Euler schemes.

Remark 3.13. *The previous observations tell us that in order to improve the accuracy, choosing a uniform grid is not a good idea, and that the grid should be highly refined near $t = 0$ and $S = K$. However, the schemes presented so far rely on uniform grids in the variable S. It is possible to design second order schemes that work on nonuniform grids, but in our opinion, their construction is easier understood with the formalism of finite elements. Therefore, at this point, we keep in mind that nonuniform grids should definitely be used, and we postpone the description of schemes for such grids to the next chapter.*

Project 3.1. *On the example of European call, compare the computing times of a Monte-Carlo method and of a Crank–Nicolson finite difference scheme. For the Monte-Carlo method, apply one of the variance reduction techniques given in Chapter* 1.

3.5 Which Variable Is Better?

We are ready now to answer the question, Should one prefer the primitive variables or the logarithmic price? The answer is, it does not matter; what is important is the grid. Indeed, a uniform grid in the logarithmic price, when converted in the primitive variable, has a large density of nodes near $S = 0$, which is completely unnecessary for a vanilla option, because the price is almost affine at $S = 0$. Conversely, we have seen that a uniform grid in the primitive variable may use too many nodes for large values of S. Therefore, for both choices of variables, what really matters is to use a suitable grid, properly refined near the singularity of the payoff function. The schemes discussed so far are designed for uniform grids in S or x only: to obtain good schemes for nonuniform grids, one should be more careful, and things are better understood with the concepts of finite elements; this will be the topic of the next two chapters.

3.6 Options on a Basket of Two Assets

Consider an option on a basket of two assets: its price is given by the multidimensional Black–Scholes equation

$$\begin{aligned} &\frac{\partial P}{\partial t} - \frac{1}{2}\sum_{k,l=1}^{2} \Xi_{k,l} S_k S_l \frac{\partial^2 P}{\partial S_k \partial S_l} - \sum_{k=1}^{2} r S_k \frac{\partial P}{\partial S_k} + rP = 0, \qquad t \in (0, T],\ S_1, S_2 > 0, \\ &P(S_1, S_2, 0) = P_0(S_1, S_2), \quad S_1, S_2 > 0, \end{aligned} \tag{3.58}$$

where t is the time to maturity, P_0 is the payoff function, and

$$\Xi = \begin{pmatrix} \sigma_{11}^2 & \frac{2\rho}{1+\rho^2}\sigma_{11}\sigma_{22} \\ \frac{2\rho}{1+\rho^2}\sigma_{11}\sigma_{22} & \sigma_{22}^2 \end{pmatrix}$$

is positive definite.

For computing, we truncate the domain and consider (3.58) for $0 < S_1 < \bar{S}_1$ and $0 < S_2 < \bar{S}_2$, where $\bar{S}_1$, $\bar{S}_2$ are large enough. Additional boundary conditions have to be imposed on the artificial boundaries $S_1 = \bar{S}_1$ and $S_2 = \bar{S}_2$. For puts and many reasonable payoff functions like in (2.64), (2.65), it is sensible to impose that $P(\bar{S}_1, S_2) = P(S_1, \bar{S}_2) = 0$ for all $0 < S_1 < \bar{S}_1, 0 < S_2 < \bar{S}_2$.

Consider three positive integers N_1, N_2, and M and pose $h_k = \frac{\bar{S}_k}{N_k+1}$, $k = 1, 2$, and $\Delta t = \frac{T}{N}$. We consider the real numbers $S_{k,i} = ih_k$ for $k = 1, 2$ and $0 \le i \le N_k + 1$ and $t_m = m\Delta t$ for $0 \le m \le M$. The points $(t_m, S_{1,i}, S_{2,j})$ are the nodes of a uniform grid of $[0, T] \times [0, \bar{S}_1] \times [0, \bar{S}_2]$.

The value of

$$-\sum_{k,l=1}^{2} \Xi_{k,l} \frac{S_k S_l}{2} \frac{\partial^2 P}{\partial S_k \partial S_l} - \sum_{k=1}^{2} -r S_k \frac{\partial P}{\partial S_k} P + rP$$

at $(S_{1,i}, S_{2,j}) \in (0, \bar{S}_1) \times (0, \bar{S}_2)$ is approximated by the centered scheme

$$-\sum_{k,l=1}^{2} \Xi_{k,l} \frac{S_k S_l}{2} \mathcal{D}^{kl}_{i,j} - \sum_{k=1}^{2} -r S_k \mathcal{D}^{k}_{i,j} + rP(S_{1,i}, S_{2,j}),$$

where

$$\begin{aligned}
\mathcal{D}^1_{i,j} &= \frac{1}{2h_1}\big(P(S_{1,i+1}, S_{2,j}) - P(S_{1,i-1}, S_{2,j})\big),\\
\mathcal{D}^2_{i,j} &= \frac{1}{2h_2}\big(P(S_{1,i}, S_{2,j+1}) - P(S_{1,i}, S_{2,j-1})\big),\\
\mathcal{D}^{11}_{i,j} &= \frac{1}{h_1^2}\big(P(S_{1,i+1}, S_{2,j}) - 2P(S_{1,i}, S_{2,j}) + P(S_{1,i-1}, S_{2,j})\big),\\
\mathcal{D}^{22}_{i,j} &= \frac{1}{h_2^2}\big(P(S_{1,i}, S_{2,j+1}) - 2P(S_{1,i}, S_{2,j}) + P(S_{1,i}, S_{2,j-1})\big),\\
\mathcal{D}^{12}_{i,j} &= \mathcal{D}^{21}_{i,j} = \frac{1}{4h_1h_2}(P(S_{1,i+1}, S_{2,j+1}) - P(S_{1,i+1}, S_{2,j-1})\\
&\quad - P(S_{1,i-1}, S_{2,j+1}) + P(S_{1,i-1}, S_{2,j-1})).
\end{aligned} \tag{3.59}$$

For $i = 0$ or $j = 0$, we take the same scheme, but some terms vanish, so, on the two boundaries $S_1 = 0$ and S_2, we obtain schemes for the one-dimensional Black–Scholes equation.

Finally, the Euler implicit scheme consists of finding $P^m_{i,j}$, $0 \le m \le M, 0 \le i \le N_1 + 1, 0 \le j \le N_2 + 1$, such that

$$\begin{aligned}
&\frac{1}{\Delta t}(P^m_{i,j} - P^{m-1}_{i,j}) - \sum_{k,l=1}^{2} \Xi_{k,l} \frac{S_k S_l}{2} D^{kl}_{i,j} - \sum_{k=1}^{2} -r S_k D^{k}_{i,j} + rP^m_{i,j} = 0,\\
&1 \le m \le M, 0 \le i \le N_1, 0 \le j \le N_2,\\
&P^m_{N_1+1,j} = P^m_{i,N_2+1} = 0, \quad 1 \le m \le M, 0 \le i \le N_1 + 1, 0 \le j \le N_2 + 1,\\
&P^0_{i,j} = P_0(S_{1,i}, S_{2,j}),
\end{aligned} \tag{3.60}$$

with the convention that $P^m_{-1,j} = P^m_{i,-1} = 0$ and with

$$
\begin{aligned}
D^1_{i,j} &= \frac{1}{2h_1}\left(P^m_{i+1,j} - P^m_{i-1,j}\right), \quad D^2_{i,j} = \frac{1}{2h_2}\left(P^m_{i,j+1} - P^m_{i,j-1}\right),\\
D^{11}_{i,j} &= \frac{1}{h_1^2}\left(P^m_{i+1,j} - 2P^m_{i,j} + P^m_{i-1,j}\right),\\
D^{22}_{i,j} &= \frac{1}{h_2^2}\left(P^m_{i,j+1} - 2P^m_{i,j} + P^m_{i,j-1}\right),
\end{aligned}
\tag{3.61}
$$

and

$$
\begin{aligned}
D^{21}_{i,j} &= \frac{1}{4h_1h_2}\left(P^m_{i+1,j+1} - P^m_{i+1,j-1} - P^m_{i-1,j+1} + P^m_{i-1,j-1}\right),\\
D^{12}_{i,j} &= D^{21}_{i,j}.
\end{aligned}
\tag{3.62}
$$

It can be checked that the scheme (3.60) is first order in time, and second order in the variables S_1 and S_2. The stability and convergence analyses follow the same line as in §3.3.

Exercise 3.9. *Write the Euler implicit scheme with an upwind discretization of the first order derivatives.*

Project 3.2. *Compare various implicit schemes for a vanilla European call: Euler and Crank–Nicolson with/without upwinding and Crank–Nicolson with Gear's formula* (3.22) *for the first order derivative. Plot the errors; plot also a finite difference approximation of* $\partial_{SS}C$.

3.7 An Asian Put with Fixed Strike

3.7.1 An Upwind Scheme with Respect to A

For the financial modeling of Asian options, we refer the reader to [117, 116] and the references therein.

We consider an Asian put with fixed strike whose payoff is $P_0(S, A) = (A - K)_+$: calling A the average value of the asset in time, $A = \frac{1}{t}\int_0^t S(\tau)d\tau$, the price of the option is found by solving the Cauchy problem (hereafter, t denotes the time to maturity, so $A = \frac{1}{T-t}\int_0^{T-t} S(\tau)d\tau$)

$$
\begin{aligned}
&\frac{\partial P}{\partial t} - \frac{\sigma^2 S^2}{2}\frac{\partial^2 P}{\partial S^2} - rS\frac{\partial P}{\partial S} - \frac{S-A}{T-t}\frac{\partial P}{\partial A} + rP = 0, \quad t \in]0, T],\ S > 0, A > 0,\\
&P(0, S, A) = P_0(S, A), \quad S > 0, A > 0.
\end{aligned}
\tag{3.63}
$$

Note that when $t \to T$, then $A \to S$, so the price of the option at $t = T$ (today) is given by $P(t, S, S)$, and for pricing, we are interested only in the value of P on the diagonal $S = A$. Nevertheless, we have to solve (3.63).

On $S = 0$, the coefficient $\frac{\sigma^2 S^2}{2}$ vanishes, so the equation degenerates into a partial differential equation with respect to t and A:

$$\frac{\partial P}{\partial t} + \frac{A}{T-t}\frac{\partial P}{\partial A} + rP = 0.$$

Near $A = 0$, we have

$$-\frac{S-A}{T-t}\frac{\partial P}{\partial A} \simeq -\frac{S}{T-t}\frac{\partial P}{\partial A}.$$

This term is analogous to a transport term with a velocity pointing outward the domain in fluid mechanics, and for that reason no boundary condition has to be imposed on the boundary $A = 0$ (in fluid mechanics, this is called an outflow boundary).

Remark 3.14. *Note that there is no diffusion term with respect to A (i.e., $\frac{\partial^2 P}{\partial A^2}$), so, in view of §3.2, upwinding in the variable A will be necessary.*

Exactly as for the Black–Scholes equation, we truncate the domain in the variables S and A, i.e., we consider (3.63) in $(0, \bar{S}) \times (0, \bar{A})$, and we have to supply additional boundary conditions on the boundaries $S = \bar{S}$ and $A = \bar{A}$. An argument similar to that for the European put shows that $\lim_{A\to\infty} P(t, S, A) = 0$, so we can choose to impose the Dirichlet condition $P(t, S, \bar{A}) = 0$ in the zone where $S > \bar{A}$. No condition has to be imposed for $S < \bar{A}$ for the same reasons as above (if $S < \bar{A}$, the term $-\frac{S-\bar{A}}{T-t}\frac{\partial P}{\partial A}$ is a transport term with an outgoing velocity).

Consider now the boundary $S = \bar{S}$: on this boundary, we impose the Neumann condition $\frac{\partial P}{\partial S} = 0$, because P_0 does not depend on S, so P should not depend on S for large values of S.

Take three integers N_S, N_A, and M, pose $h_S = \frac{\bar{S}}{N_S+1}$, $h_A = \frac{\bar{A}}{N_A+1}$, and $\Delta t = \frac{T}{M}$, and consider the real numbers $S_i = ih_S$, $A_j = jh_A$, $t_m = m\Delta t$ for $0 \le i \le N_S + 1$, $0 \le j \le N_A + 1$, $0 \le m \le M$. The nodes (t_m, S_i, A_j) form a uniform grid of $[0, T] \times [0, \bar{S}] \times [0, \bar{A}]$. At a node (S_i, A_j) in $(0, \bar{S}) \times (0, \bar{A})$,

$$-\frac{\sigma^2 S^2}{2}\frac{\partial^2 P}{\partial S^2} - rS\frac{\partial P}{\partial S} - \frac{S-A}{T-t}\frac{\partial P}{\partial A} + rP$$

is approximated by the upwind scheme

$$-\frac{S_i^2}{2}D_{S,S} - rS_i D_S - \frac{S_i - A_j}{T-t}D_A + rP(S_i, A_j),$$

where

$$\begin{aligned}
D_{S,S} &= \frac{P(S_{i+1}, A_j) - 2P(S_i, A_j) + P(S_{i-1}, A_j)}{h_S^2}, \\
D_S &= \frac{P(S_{i+1}, A_j) - P(S_{i-1}, A_j)}{2h_S}, \\
D_A &= \begin{cases} \dfrac{P(S_i, A_{j+1}) - P(S_i, A_j)}{h_A} & \text{if } S_i > A_j, \\[2ex] \dfrac{P(S_i, A_j) - P(S_i, A_{j-1})}{h_A} & \text{if } S_i \le A_j. \end{cases}
\end{aligned} \tag{3.64}$$

It is possible to use a second order scheme by making use of Exercise 3.3 for discretizing $-\frac{S-A}{T-t}\frac{\partial P}{\partial A}$ at (S_i, A_j):

$$
\begin{aligned}
&-\frac{S-A}{T-t}\frac{\partial P}{\partial A}(S_i, A_j)\\
&\simeq \begin{cases} -\dfrac{S_i - A_j}{T-t}\dfrac{3P(S_i, A_j) - 4P(S_i, A_{j-1}) + P(S_i, A_{j-2})}{2h} & \text{if } S_i \le A_j, \\[2ex] -\dfrac{S_i - A_j}{T-t}\dfrac{-3P(S_i, A_j) + 4P(S_i, A_{j+1}) - P(S_i, A_{j+2})}{2h} & \text{if } S_i > A_j. \end{cases}
\end{aligned}
\tag{3.65}
$$

Of course, this scheme cannot be applied near the boundaries $A = 0$ and $A = \bar{A}$.

3.7.2 A Program in C++

The following program computes the price of an Asian put with a Crank–Nicolson scheme and the finite difference schemes described in §3.7.1. For simplicity, we have used only band matrices (the bandwidth is equal to the number of steps in the S variable) and the systems of linear equations are solved by LU factorization. It would have been better to use sparse matrix and a good preconditioned iterative solver like GMRES (see [107, 106]) with an incomplete LU factorization (see [106]).

The time grid may have variable steps, whereas the grid in S and A does not vary with time. We use the vector + sparse matrix class called RNM of Danaila, Hecht, and Pironneau [38]. It is not necessary to study these classes in order to use them. One need only know how to call them and how to use the overloaded operators. It may seem like killing a fly with a gun, but it will allow us to hide low-level implementations using the `blas` library for speed and adaptation to the computer architecture.

The RNM class is templated over a basic type: the real numbers (either `double` or `float`). We will later use `ddouble` for automatic differentiation as a fast way to compute derivatives and sensitivities (Greeks).

To create an RNM vector v of size m and a band matrix A of size $m \times m$ and bandwidth $d = 1$, one writes

```
const int N=10;
typedef double Real;
typedef KN<Real> vec;
typedef SkyLineMatrix<Real> mat;
vec v(N);
mat a(N,1);
```

It is indeed convenient to define the types `vec` and `mat` for readability, as these will be used often.

The RNM classes implement efficiently the operations of vector calculus such as

```
vec c=0,b=2 * c + 1;
a.LU(); a.Solve(v,c);
cout<<v<<endl;
```

which means that, with $c_i = 1, \ i = 1 \ldots, N$, and a a tridiagonal band matrix, the system of linear equations $av = c$ is solved by the Gauss LU factorization method, and the result is displayed.

A "CN-scheme" class is defined for the Crank–Nicolson scheme.

ALGORITHM 3.2. Crank–Nicolson.

```
class CN_Scheme_Asian
{
private:
  int order;
  double T, S_max, A_max;                      //   bounds of the comp.  domain
  int NS,NA;                                   //    number of nodes in S and A
  vector<double> grid_t;                      //   time grid, can be nonuniform
  MatriceProfile<double> AA_S;               //   auxiliary matrix computed once
                                                                //   and for all
  MatriceProfile<double> AA_A;               //   auxiliary matrix computed once
                                                                //   and for all
  MatriceProfile<double> BB;                //   the matrix of the linear system
  double rate ;                                               //   interest rate
  double sigma;                                                   //   volatility
public:
  CN_Scheme_Asian( const int orderg,
                   const double Tg,
                   const double S_maxg,
                   const double A_maxg,
                   const int NSg,
                   const int NAg, const vector<double> &grid_tg,
                   const double rateg,const double sigmag)
: order(orderg),T(Tg),S_max(S_maxg),A_max(A_maxg), NS(NSg),NA(NAg),
 grid_t(grid_tg), AA_S(NSg*NAg,2*NAg),AA_A(NSg*NAg,2*NAg),BB(NSg*NAg,2*NAg),
  rate(rateg), sigma(sigmag)
  {
    Assemble_Matrices();
  };                                                            //   constructor
  void Assemble_Matrices( );        //   assembles time independent matrices
  void Time_Step(int it,  vector<KN<double> >  &P);          //   one time step
  void build_rhs(KN<double> &u, const KN<double> &u_p,
                 const double t,const double dt);      //    construct RHS of
                                                              //   linear system
  void build_matrix( const double t,  const double dt); //   builds BB from
AA_S and AA_A
}
;
```

A time step of the method is implemented in the function `Time_Step`: it consists of building the matrix $\mathbf{B} \simeq \mathbf{M} + \frac{\Delta t_m}{2}\mathbf{A}^m$ (not equal because of the Neumann boundary condition at $S = \bar{S}$), constructing the right-hand side of the system, and solving the system.

ALGORITHM 3.3. Time step.

```
void CN_Scheme_Asian::Time_Step(int it,  vector< KN<double> >& P)
{
  double t=grid_t[it];                                    //    current time
  double dt=t-grid_t[it-1];                            //   current time step
  build_rhs(P[it],P[it-1],T-t+dt,dt);                    //   computes the RHS
  build_matrix(T-t,dt);                //   computes the matrix B = (M+dt/2* A)
  BB.LU();                                         //   LU factorization of B
  BB.Solve(P[it],P[it]);                                //   solves the system
}
```

The matrix has to be recomputed at each time step, since it depends on t; however, it is possible to write $\mathbf{A}^m = \mathbf{A}_S + \frac{1}{T-t_m}\mathbf{A}_A$, where the matrices $\mathbf{A}_S$ and $\mathbf{A}_A$ do not vary in time and can be computed once and for all. The matrices $\mathbf{A}_S$ and $\mathbf{A}_A$ are computed by the function `void Assemble_Matrices()`.

ALGORITHM 3.4. Matrix assembly 1.

```
void  CN_Scheme_Asian::Assemble_Matrices()
{
  assert(S_max<=A_max);                         //    checks that S_max<=A_max
  double hS=S_max/(NS-1);                                       //    step in S
  double hA=A_max/(NA-1);                                       //    step in A
  vector<double> S(NS);                                           //    S grid
  vector<double> A(NA);                                           //    A grid
  for(int i=0;i<NS;i++)
    S[i]=(S_max/(NS-1))*i;
  for(int i=0;i<NA;i++)
    A[i]=(A_max/(NA-1))*i;
  int k=-1;
  AA_S=0;                    //    stiffness matrix except derivatives wrt A
  AA_A=0;                                             //    derivative wrt A
  double aux;

  for(int i=0;i<NS-1;i++)
    for (int j=0;j<NA;j++)
      {
        k=i*NA+j;
        aux=pow(sigma*i,2);
        AA_S(k,k)=rate+aux;
        AA_S(k,k+NA)=   -(aux +rate*i)/2;
      }
  if (order==1)
    for(int i=0;i<NS-1;i++)
      for (int j=0;j<NA;j++)
        {
          aux= fabs(-S[i]+A[j])/hA;
          k=i*NA+j;
          AA_A(k,k)= aux;
          if (S[i]>=A[j])
            AA_A(k,k+1)=-aux;
          else
```

```
            AA_A(k,k-1)=-aux;
        }
    else  if (order==2)
      {
        for(int i=0;i<NS-1;i++)
          for (int j=0;j<2;j++)
            {
              aux= fabs(-S[i]+A[j])/hA;
              k=i*NA+j;
              AA_A(k,k)= aux;
              if (S[i]>=A[j])
                AA_A(k,k+1)=-aux;
              else
                AA_A(k,k-1)=-aux;
            }
        for(int i=0;i<NS-1;i++)
          for (int j=NA-2;j<NA;j++)
            {
              aux= fabs(-S[i]+A[j])/hA;
              k=i*NA+j;
              AA_A(k,k)= aux;
              if (S[i]>=A[j])
                AA_A(k,k+1)=-aux;
              else
                AA_A(k,k-1)=-aux;
            }
        for(int i=0;i<NS-1;i++)
          for (int j=2;j<NA-2;j++)
            {
              aux= fabs(-S[i]+A[j])/hA;
              k=i*NA+j;
              if (S[i]>=A[j])
                {
                  AA_A(k,k)= 3*aux/2;
                  AA_A(k,k+1)=-4*aux/2;
                  AA_A(k,k+2)=aux/2;
                }
              else
                {
                  AA_A(k,k)= 3*aux/2;
                  AA_A(k,k-1)=-4*aux/2;
                  AA_A(k,k-2)=aux/2;
                }
            }
      }
    for (int j=0;j<NA;j++)
      for(int i=1;i<NS-1;i++)
        {
          k=i*NA+j;
          AA_S(k,k-NA)=  -(pow(sigma*i,2) -rate*i)/2;
        }
              //  note that the Neumann condition at S=S_max is not done yet
}
```

The function `:build_matrix` is as follows.

ALGORITHM 3.5. Matrix assembly 2.

```
void  CN_Scheme_Asian::build_matrix(const double t,const double dt)
{
  int i;
  BB=0;

  for(int k=0;k<NS-1;k++)
    for (int j=0;j<NA;j++)
      {
        i=k*NA+j;
        BB(i,i)=1+dt/2*(AA_S(i,i)+AA_A(i,i)/t);
        BB(i,i+NA)=dt/2*AA_S(i,i+NA);
      }

  for(int k=1;k<NS-1;k++)
    for (int j=0;j<NA;j++)
      {
        i=k*NA+j;
        BB(i,i-NA)=dt/2*AA_S(i,i-NA);
      }

  for(int k=0;k<NS-1;k++)
    for (int j=1;j<NA;j++)
      {
        i=k*NA+j;
        BB(i,i-1)=dt/2*AA_A(i,i-1)/t;
      }

  for(int k=0;k<NS-1;k++)
    for (int j=0;j<NA-1;j++)
      {
        i=k*NA+j;
        BB(i,i+1)=dt/2*AA_A(i,i+1)/t;
      }
  if (order==2)
    {

      for(int k=0;k<NS-1;k++)
        for (int j=0;j<NA-2;j++)
          {
            i=k*NA+j;
            BB(i,i+2)=dt/2*AA_A(i,i+2)/t;
          }
      for(int k=0;k<NS-1;k++)
        for (int j=2;j<NA;j++)
          {
            i=k*NA+j;
            BB(i,i-2)=dt/2*(AA_A(i,i-2)/t);
          }
    }

  for (int j=0;j<NA;j++)                    //   the Neumann condition at S=S_max
    {
      i=j+(NS-1)*NA;
```

```
        BB(i,i)=BB(i-NA,i-NA);
        BB(i,i-NA)=-BB(i,i);
    }
}
```

Finally, the right-hand side of the system of linear equations is constructed by the following program.

ALGORITHM 3.6. Right-hand side.

```
void CN_Scheme_Asian::build_rhs(KN<double> &u, const KN<double> &u_p,
                                const double t,const double dt)
{
  u=0;
  AA_A.addMatMul(u_p, u);
  u/=t;
  AA_S.addMatMul(u_p, u);
  u*=-dt/2;

  for(int k=0;k<NS-1;k++)
    for (int j=0;j<NA;j++)
      {
        int i=j+NA*k;
        u(i)+=u_p(i);
      }
  for (int j=0;j<NA;j++)                  //  the Neumann condition at S=S_max
    {
      int i=j+(NS-1)*NA;
      u(i)=0;
    }
}
```

We use this program for computing an Asian put with strike $K = 100$. We have chosen $\bar{S} = \bar{A} = 200$. The maturity is 1 year, $r = 0.05$, and $\sigma = 0.2$. In Figure 3.8, we have plotted the surface $P(S, A)$ at $t \sim 0$, $t \sim T/2$, and $t \sim T$. The grids are uniform and we have used the first order in A upwind scheme with 100 nodes in S and A, and 50 nodes in t.

In Figure 3.9, we plot a zoom of the price of the put at $t \sim \frac{T}{2}$, when we use an upwind second order and a centered scheme instead of the first upwind scheme for discretizing $-\frac{S-A}{T-t}\frac{\partial P}{\partial A}$: we see that the second scheme is not monotone and that the centered scheme causes spurious oscillations. Here the grid in $(S < A)$ has 60×60 nodes.

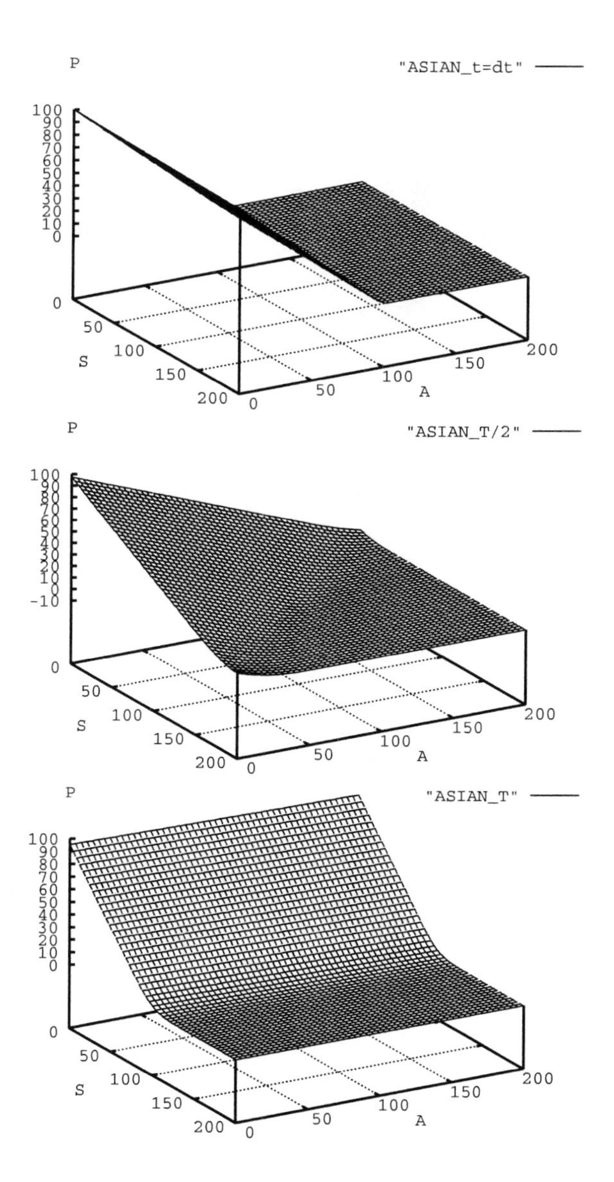

Figure 3.8. *The Asian put computed by the Crank–Nicolson scheme at $t \sim 0$, $t \sim T/2$, and $t \sim T$.*

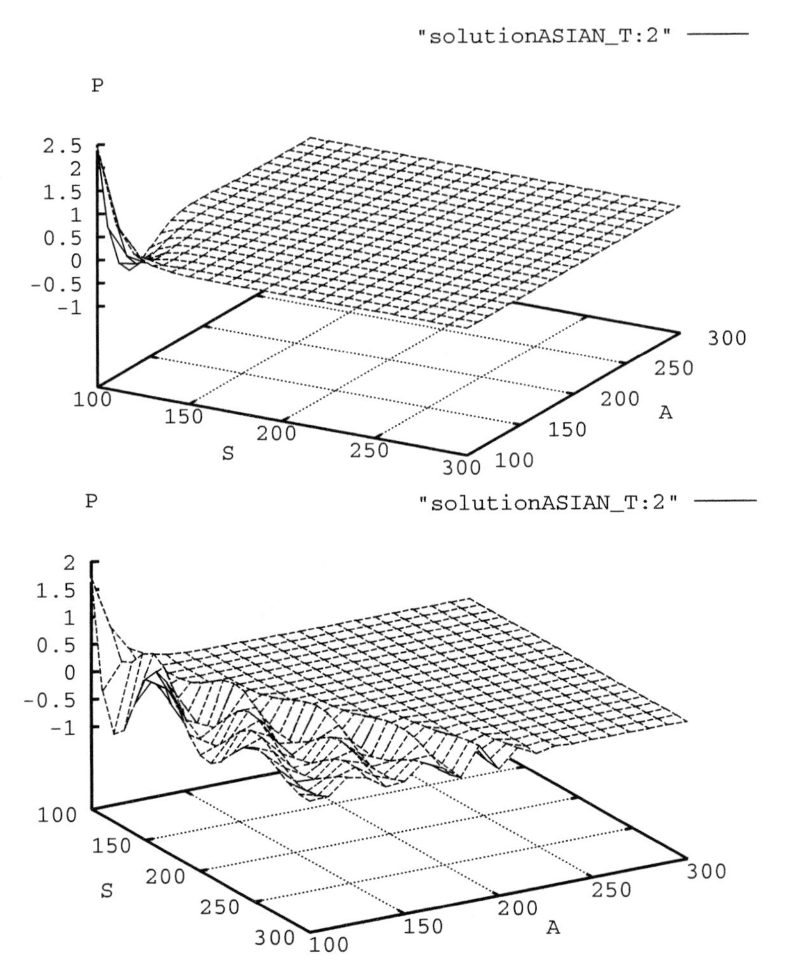

Figure 3.9. *The Asian put computed by the Crank–Nicolson scheme with a second order upwind scheme in A and with a centered scheme at $t \sim T/2$.*

Chapter 4

The Finite Element Method

4.1 Orientation

Conforming finite element methods are numerical approximations closely linked to the theory of variational or weak formulations presented in Chapter 2. The first finite element method can be attributed to Courant [34].

The framework is the same in any dimension of space d: for a weak formulation posed in an infinite-dimensional function space V, for instance,

$$V = H^1(\Omega) = \{w \in L^2(\Omega) \; : \; \nabla w \in L^2(\Omega)^d\},$$

it consists of choosing a finite-dimensional subspace V_h of V, for instance, the space of continuous piecewise affine functions on a triangulation of Ω, and of solving the problem with test and trial functions in V_h instead of V. In the simpler finite element methods, the construction of the space V_h is done as follows:

- The domain is partitioned into nonoverlapping cells (elements) whose shapes are simple and fixed: for example, intervals in one dimension, triangles or quadrilaterals in two dimensions, tetrahedra, prisms, or hexahedra in three dimensions. The set of the elements is in general an unstructured mesh called a *triangulation.*

- The maximal degree k of the polynomial approximation in the elements is chosen (mostly degree one in this book).

- V_h is made of functions of V whose restriction to the elements are polynomial of degree less than k.

Programming the method is also somewhat similar in any dimension, but mesh generation is very much dimension-dependent.

There is a very well understood theory on error estimates for finite elements. It is possible to distinguish a priori and a posteriori error estimates: in a priori estimates, the error is bounded by some quantity depending on the solution of the continuous problem (which is unknown, but for which estimates are available), whereas, in a posteriori estimates, the error is bounded by some quantity depending on the solution of the discrete problem which is available.

For a priori error estimates, one can see the books of Raviart and Thomas [103], Strang and Fix [111], Braess [18], Brenner and Scott [20], Ciarlet [27, 28], and Thomée [112] on parabolic problems. By and large, deriving error estimates for finite element methods consists of the following:

1. establishing the stability of the discretization with respect to some norms related to $\|\cdot\|_V$, as we did for finite difference methods;

2. once this is done, one sees that in simple cases the error depends on some distance of the solution of the continuous problem to the space V_h. This quantity cannot be computed exactly since the solution is unknown. However, it can be estimated from a priori knowledge on the regularity of the solution.

When sharp results on the solution of the continuous problem are available, the a priori estimates give very valuable information on how to choose the discretization a priori; see the nice papers by Schötzau and Schwab [109] and Werder et al. [115], in the case of homogeneous parabolic problems with smooth coefficients.

A posteriori error estimates are a precious tool, since they give practical information that can be used to refine the mesh when needed. In Chapter 5, we consider a posteriori error estimates for a finite element method applied to the Black–Scholes equation.

In this chapter, we insist on implementation rather than error estimates. The chapter is organized as follows: We first describe the finite element method on a generic parabolic boundary value problem in two dimensions. Then we focus on the Black–Scholes equation in one and two dimensions.

4.2 A Generic Problem

4.2.1 Variational Formulation

Consider the following:

- let Ω be a polygonal domain of $\mathbb{R}^2$ (Ω is open and bounded);
- let Γ be the boundary of Ω;
- we assume that $\Gamma = \Gamma_d \cup \Gamma_n$, where the one-dimensional measure of $\Gamma_d \cap \Gamma_n$ is 0; for $x \in \Gamma$,
- we denote by n the unit normal vector to Γ at x, pointing outward;
- we consider smooth enough functions:

$$\kappa : \Omega \mapsto \mathbb{R}^{2\times 2}, \qquad \alpha : \Omega \mapsto \mathbb{R}^2, \qquad \beta : \Omega \mapsto \mathbb{R}, \qquad b : \Gamma_n \mapsto \mathbb{R}.$$

Remark 4.1. *It is possible to consider more general domains whose boundaries are locally the graph of a Lipschitz continuous function.*

For suitable functions

$$u_0 : \Omega \mapsto \mathbb{R}, \quad \phi : \Omega \times (0,T) \mapsto \mathbb{R}, \quad g : \Gamma_d \times (0,T] \mapsto \mathbb{R}, \quad f : \Gamma_n \times (0,T] \mapsto \mathbb{R},$$

we are interested in finding $u(x,t)$ solving the parabolic boundary value problem

$$\begin{aligned} \frac{\partial u}{\partial t} - \nabla \cdot (\kappa \nabla u) - \nabla \cdot (\alpha u) + \beta u = \phi \qquad & \text{in } \Omega \times (0,T), \\ u|_{t=0} = u_0(x) \qquad & \text{in } \Omega, \\ u = g \qquad & \text{on } \Gamma_d \times (0,T), \\ -bu - (\kappa \nabla u) \cdot n = f \qquad & \text{on } \Gamma_n \times (0,T), \end{aligned} \tag{4.1}$$

where

$$\begin{aligned} \nabla u = \left(\frac{\partial}{\partial x_1}, \frac{\partial}{\partial x_2} \right)^T, \qquad & \nabla \cdot (\kappa \nabla u) = \sum_{i,j=1}^{2} \frac{\partial}{\partial x_i} \left(\kappa_{i,j} \frac{\partial u}{\partial x_j} \right), \\ \nabla \cdot (\alpha u) = \sum_{i=1}^{2} \frac{\partial}{\partial x_i} (\alpha_i u), \qquad & (\kappa \nabla u) \cdot n = \sum_{i,j=1}^{2} n_i \left(\kappa_{i,j} \frac{\partial u}{\partial x_j} \right). \end{aligned} \tag{4.2}$$

We recall Green's formula,

$$\int_\Omega [(\kappa \nabla u) \cdot \nabla v + v \nabla \cdot (\kappa \nabla u)] = \int_\Gamma (\kappa \nabla u) \cdot n v, \tag{4.3}$$

which holds whenever the integrals are defined.

The variational formulation of (4.1) involves the Sobolev space

$$W = H^1(\Omega) = \{ v \in L^2(\Omega); \nabla v \in (L^2(\Omega))^2 \},$$

which is a Hilbert space endowed with the norm

$$\|v\|_W = \left(\|v\|^2_{L^2(\Omega)} + \sum_{i=1}^{2} \left\| \frac{\partial v}{\partial x_i} \right\|^2_{L^2(\Omega)} \right)^{\frac{1}{2}}.$$

Calling

$$\mathcal{D}(\bar{\Omega}) \text{ the space of the restrictions of the functions of } \mathcal{D}(\mathbb{R}^2) \text{ to } \Omega,$$

we recall that $\mathcal{D}(\bar{\Omega})$ is dense in W.

The linear operator $u \mapsto u|_{\Gamma_d}$, which maps a function to its restriction to Γ_d, is bounded from $\mathcal{D}(\bar{\Omega})$ with the norm $\|\cdot\|_W$ to $L^2(\Gamma_d)$. Therefore, we can define a continuous extension γ_d of this operator on W, called the trace operator on Γ_d. We define V as the kernel of γ_d. The space V is a closed subspace of W.

Note that it is also possible to define a trace operator from W to $L^2(\Gamma_n)$.

For simplicity, we assume that the coefficients κ, α, and β and the function g are smooth enough so that there exists a function

$$u_g \in L^2((0,T); W) \cap C^0([0,T]; L^2(\Omega)), \quad \frac{\partial u_g}{\partial t} \in L^2((0,T); V'),$$

with

- $\gamma_d(u_g) = g$ a.e. with respect to t;
- $\frac{\partial u_g}{\partial t} - \nabla \cdot (\kappa \nabla u_g) - \nabla \cdot (\alpha u_g) + \beta u_g \in L^2(\Omega \times (0,T))$;
- the normal trace $(bu_g + (\kappa \nabla u_g) \cdot n)|_{\Gamma_n}$ can be defined and belongs to $L^2(\Gamma_n \times (0,T))$.

Reasoning as in §2.6, we introduce the bilinear form on W:

$$a(w, v) = \int_{\Omega} ((\kappa \nabla w) \cdot \nabla v - \nabla \cdot (\alpha w) v + \beta w v) + \int_{\Gamma_n} b w v. \tag{4.4}$$

Assuming that

- for a.e. $x \in \Omega$, $\kappa(x)$ is a symmetric tensor and there exist two positive constants $0 < \underline{\kappa} \leq \bar{\kappa}$ such that, for all $\xi \in \mathbb{R}^2$,
$$\underline{\kappa}|\xi|^2 \leq \kappa(x)\xi \cdot \xi \leq \bar{\kappa}|\xi|^2 \quad \text{for a.e. } x \in \Omega;$$
- $a \in (L^\infty(\Omega))^2$, $\beta \in L^\infty(\Omega)$, $b \in L^\infty(\Gamma_n)$,

it can be proved that there exist two positive constants $\underline{c} \leq \bar{c}$ and a nonnegative constant λ such that for all $v, w \in W$,

$$a(v, w) \leq \bar{c}|v|_W |w|_W \tag{4.5}$$

and

$$a(v, v) \geq \underline{c}|v|_W^2 - \lambda \|v\|_{L^2(\Omega)}^2. \tag{4.6}$$

The *variational formulation* of (4.1) is as follows: Find u: $u - u_g \in L^2((0, T); V)$, $u \in C^0([0, T]; L^2(\Omega))$, and $\frac{\partial u}{\partial t} \in L^2((0, T); V')$, with $u|_{t=0} = u_0$, and, for a.e. $t \in (0, T)$,

$$\forall v \in V, \quad {}_{V'}\left(\frac{\partial u}{\partial t}(t), v\right)_V + a(u(t), v) = \int_{\Omega} \phi(t) v + \int_{\Gamma_n} f(t) v. \tag{4.7}$$

Thanks to the bound (4.5) and to Gårding's inequality (4.6), it can be proved that, if there exists u_g satisfying the condition above, then the variational formulation has a unique solution, which satisfies the first line of (4.1) in the sense of distributions.

4.2.2 The Time Semidiscrete Problem

We introduce a partition of the interval $[0, T]$ into subintervals $[t_{m-1}, t_m]$, $1 \leq m \leq M$, such that $0 = t_0 < t_1 < \cdots < t_m = T$.

We denote by Δt_m the length $t_m - t_{m-1}$, and by Δt the maximum of Δt_m, $1 \leq m \leq M$. For simplicity, we assume that $u_0 \in W$.

We discretize (4.7) by means of a Crank–Nicolson scheme; i.e., we look for $u^m \in W$, $m = 0, \ldots, M$, such that $u^0 = u_0$ and for all $m = 1, \ldots, M$, $u^m - u_g(t_m) \in V$, and for all $v \in V$,

$$\left(\frac{u^m - u^{m-1}}{\Delta t_m}, v\right) + \frac{1}{2}\left(a(u^m, v) + a(u^{m-1}, v)\right) = \int_{\Omega} \phi^{m-1/2} v + \int_{\Gamma_n} f^{m-1/2} v, \tag{4.8}$$

where $\phi^{m-1/2}$ means $\phi(\frac{t_{m-1}+t_m}{2})$, and similarly for f. This scheme is second order.

Remark 4.2. *If u_0 does not belong to W, then we have to approximate first u_0 by a function in W, at the cost of an additional error.*

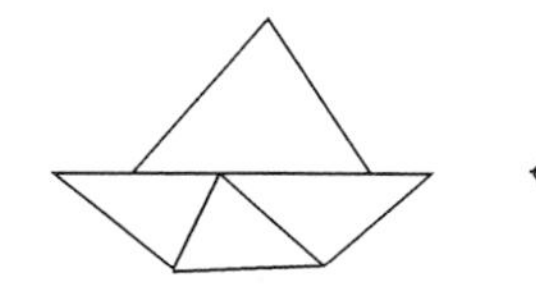

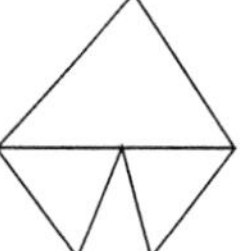

Figure 4.1. *Left: a finite element mesh made of triangles. This mesh has been obtained by pricing adaptively an American basket option. Right: these cases are ruled out.*

4.2.3 The Full Discretization: Lagrange Finite Elements

Discretization with respect to x is obtained by replacing W (resp., V) by a subspace of finite dimension $W_h \subset W$ (resp., $V_h \subset V$). For example, one may choose for V_h a space of continuous piecewise polynomial functions on a triangulation of Ω vanishing on Γ_d. For a positive real number h, consider a partition $\mathcal{T}_h$ of Ω into nonoverlapping closed triangles ($\mathcal{T}_h$ is the set of all the triangles forming the partition) such that

- $\bar{\Omega} = \cup_{K \in \mathcal{T}_h} K$;
- for all $K \neq K'$, two triangles of $\mathcal{T}_h$, $K \cap K'$ is either empty, a vertex of both K and K', or a whole edge of both K and K';
- for all $K \in \mathcal{T}_h$, the one-dimensional measure of $K \cap \Gamma_d$ (resp., $K \cap \Gamma_n$) is either 0 or $K \cap \Gamma_d$ (resp., $K \cap \Gamma_n$) is a whole edge of K;
- $\max_{K \in \mathcal{T}_h} \text{diameter}(K) = h$.

For these conditions to hold, Ω must be polygonal (to be covered exactly by a triangulation). However, if Ω is not polygonal but has a smooth boundary, it is possible to find a set $\mathcal{T}_h$ of nonoverlapping triangles of diameters less than h such that the distance between Ω and $\cup_{K \in \mathcal{T}_h} K$ scales like h^2.

In Figure 4.1, we show examples of situations which can or cannot occur with the mesh defined above.

Exercise 4.1. *Call N^T the number of triangles in $\mathcal{T}_h$, N^E the number of edges, N^V the number of vertices, and N_0^V the number of vertices lying in the open domain Ω. Prove that*

$$N^T + N^V = N^E + 1, \qquad 3N^T + N^V = 2N^E + N_0^V.$$

For k a positive integer, we introduce the spaces

$$W_h = \{w_h \in \mathcal{C}^0(\bar{\Omega}) : w_h|_K \in \mathcal{P}^k \ \forall K \in \mathcal{T}_h\}, \qquad V_h = \{v_h \in W_h, v_h|_{\Gamma_d} = 0\}. \tag{4.9}$$

We focus on the case where $k = 1$; i.e., the functions in W_h are piecewise affine. It is clear that W_h is a finite-dimensional subspace of W and that V_h is a finite-dimensional subspace of V.

Assume that for each $m = 1, \dots, M$, there exists a function $u^m_{g,h} \in W_h$ such that the trace of $u^m_{g,h}$ on Γ_d is $g(t_m)$. If it is not the case, $g(t_m)$ must be approximated first by the trace of a function in W_h, at the cost of an additional error. For example, if g is continuous on Γ_d, one can take the Lagrange interpolation of g.

Assuming that $u_0 \in W_h$, the full discretization of the variational formulation consists of finding $u^m_h \in W_h$, $m = 1, \dots, M$, such that $u^m_h - u^m_{g,h} \in V_h$, and, with $u^0_h = u_0$,

$$\begin{aligned} &\forall v_h \in V_h, \\ &\frac{1}{\Delta t_m}(u_h^m - u_h^{m-1}, v_h) + \frac{1}{2}\left(a(u_h^m, v_h) + a(u_h^{m-1}, v_h)\right) \\ &= \int_\Omega \phi^{m-1/2} v_h + \int_{\Gamma_n} f^{m-1/2} v_h. \end{aligned} \tag{4.10}$$

4.2.4 The Discrete Problem in Matrix Form

A basis of V_h is chosen, $(w_i)_{i=1,\dots,N}$. Then, for $1, \dots, M$, u^m_h can be written as

$$u_h^m(x) = u_{g,h}^m(x) + \sum_1^N u_j^m w_j(x), \tag{4.11}$$

and, applying (4.11) to (4.10) with $v_h = w_i$, we obtain a system of linear equations for $\mathbf{u}^m = (u^m_j)^T_{j=1,\dots,N}$:

$$\mathbf{M}(\mathbf{u}^m - \mathbf{u}^{m-1}) + \frac{\Delta t_m}{2}\mathbf{A}(\mathbf{u}^m + \mathbf{u}^{m-1}) = \mathbf{f}, \tag{4.12}$$

where $\mathbf{M}$ and $\mathbf{A}$ are matrices in $\mathbb{R}^{N\times N}$:

$$\begin{aligned} \mathbf{M}_{ij} &= \int_\Omega w_i w_j, \\ \mathbf{A}_{i,j} &= a(w_j, w_i) = \int_\Omega ((\kappa \nabla w_j)\cdot \nabla w_i - \nabla\cdot(\alpha w_j) w_i + \beta w_j w_i) + \int_{\Gamma_n} b w_j w_i, \end{aligned} \tag{4.13}$$

and

$$\mathbf{f}_i = \Delta t_m \left(\int_\Omega \phi^{m-1/2} w_i + \int_{\Gamma_n} f^{m-1/2} w_i - \frac{1}{2} a(u_{g,h}^{m-1} + u_{g,h}^m, w_i) \right) - \int_\Omega (u_{g,h}^m - u_{g,h}^{m-1}) w_i. \tag{4.14}$$

The matrix $\mathbf{M}$ is called the mass matrix and $\mathbf{A}$ the stiffness matrix. It can be proved, thanks to estimates (4.5) and (4.6), that if Δt is small enough, then $\mathbf{M} + \frac{\Delta t_m}{2}\mathbf{A}$ is invertible, so it is possible to solve (4.12).

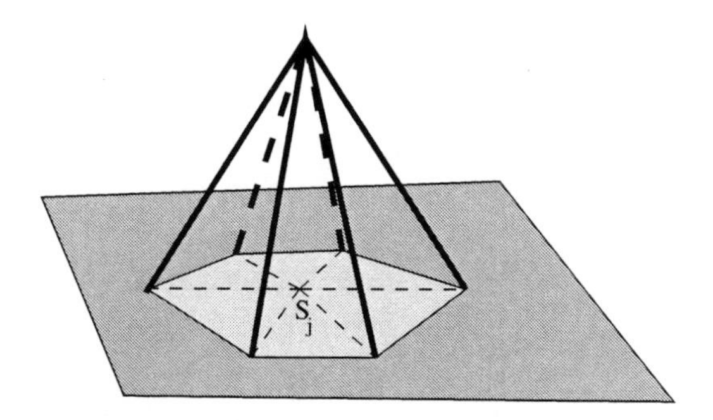

Figure 4.2. *The shape function w^j.*

4.2.5 The Nodal Basis

Hereafter, we take $k = 1$, so we deal with piecewise linear finite elements. On each triangle $K \in \mathcal{T}_h$, denoting by q^i, $i = 1, 2, 3$, the vertices of K, we define for $x \in \mathbb{R}^2$ the barycentric coordinates of x, i.e., the solution of

$$\sum_{i=1,2,3} \lambda_i^K(x) q^i = x, \quad \sum_{i=1,2,3} \lambda_i^K(x) = 1.$$

This 3×3 system of linear equations is never singular because its determinant is twice the area of K. It is obvious that the barycentric coordinates λ_i^K are affine functions of x. Furthermore,

- when $x \in K, \lambda_i^K \geq 0, \;\; i = 1, 2, 3;$
- if $K = [q^{i_1}, q^{i_2}, q^{i_3}]$ and x is aligned with q^{i_1}, q^{i_2}, then $\lambda_{i_3}^K = 0$.

Let v_h be a function in W_h: it is easy to check that, on each triangle $K \in \mathcal{T}_h$,

$$v_h(x) = \sum_{j=1,2,3} v_h(q^{i_j}) \lambda_{i_j}^K(x) \quad \forall x \in K.$$

Therefore, a function in W_h is uniquely defined by its values at the nodes of $\mathcal{T}_h$ and a function in V_h is uniquely defined by its values at the nodes of $\mathcal{T}_h$ not located on Γ_d.

Call $(q^i)_{i=1,\ldots,N}$ the nodes of $\mathcal{T}_h$ not located on Γ_d, and let w^i be the unique function in V_h such that $w^i(q^j) = \delta_{i,j}$ for all $j = 1, \ldots, N$. For a triangle K such that q^i is a vertex of K, it is clear that w^i coincides in K with one of the three barycentric coordinates attached to triangle K. Therefore, we have the identity

$$v_h = \sum_{i=1}^{N} v_h(q^i) w^i, \tag{4.15}$$

which shows that $(w^i)_{i=1,\ldots,N}$ is a basis of V_h. As shown in Figure 4.2, the support of w^i is the union of the triangles of $\mathcal{T}_h$ containing the node q^i, so it is very small when the mesh is fine, and the supports of two basis functions, w^i and w^j, intersect if and only if q^i and q^j are the vertices of a same triangle of $\mathcal{T}_h$. Therefore, the matrices $\mathbf{M}$ and $\mathbf{A}$ constructed with this basis are very sparse. This reduces dramatically the complexity when solving properly (4.12). The basis $(w^i)_{i=1,\dots,N}$ is often called the nodal basis of V_h. The shape functions w^i are sometimes called *hat functions*. For $v_h \in V_h$, the values $v_i = v_h(q^i)$ are called the degrees of freedom of v_h.

If $K = [q^{i_1}, q^{i_2}, q^{i_3}]$, and if b^{i_1} is the point aligned with q^{i_2} and q^{i_3} and such that $\overrightarrow{b^{i_1}q^{i_1}} \perp \overrightarrow{q^{i_2}q^{i_3}}$, then

$$\nabla\lambda_{i_1}^K = \frac{1}{|\overrightarrow{b^{i_1}q^{i_1}}|^2}\overrightarrow{b^{i_1}q^{i_1}}, \tag{4.16}$$

and calling n^{i_1} the unit vector orthogonal to $\overrightarrow{q^{i_2}q^{i_3}}$ and pointing to q^{i_1}, i.e., $n^{i_1} = \frac{1}{|\overrightarrow{b^{i_1}q^{i_1}}|}\overrightarrow{b^{i_1}q^{i_1}}$ and E^{i_1} the length of the edge of K opposite to q^{i_1}, and using the well-known identity $E^{i_1}|\overrightarrow{b^{i_1}q^{i_1}}| = 2|K|$, we obtain

$$\nabla\lambda_{i_1}^K = \frac{E^{i_1}}{2|K|}n^{i_1}. \tag{4.17}$$

This yields in particular

$$\int_K \nabla\lambda_{i_1}^K \cdot \nabla\lambda_{i_2}^K = \frac{1}{2\tan(\alpha_3^K)} = \frac{(q^{i_3} - q^{i_2}) \cdot (q^{i_3} - q^{i_1})}{4|K|}, \tag{4.18}$$

where α_3^K is the angle of K at vertex q^{i_3}.

The following integration formula is very important for the numerical implementation of the finite element method.

Proposition 4.1. *Calling λ_i, $i = 1, 2, 3$, the barycentric coordinates of the triangle K, and ν_1, ν_2, ν_3 three nonnegative integers, and $|K|$ the measure of K,*

$$\int_K (\lambda_1^K)^{\nu_1}(\lambda_2^K)^{\nu_2}(\lambda_3^K)^{\nu_3} = 2|K|\frac{\nu_1!\nu_2!\nu_3!}{(\nu_1 + \nu_2 + \nu_3 + 2)!}. \tag{4.19}$$

Remark 4.3. *It may be useful to use bases other than the nodal basis, for example, bases related to wavelet decompositions, in particular for speeding up the solution of* (4.12)*; see* [95, 114].

Remark 4.4. *The integral of a quadratic function on a triangle K is one-third the sum of the values of the function on the midedges times $|K|$, and therefore* (4.19) *is simpler when $\nu_1 + \nu_2 + \nu_3 = 2$:*

$$\int_K \lambda_i^K\lambda_j^K = \frac{|K|}{12}(1 + \delta_{ij}). \tag{4.20}$$

Remark 4.5 (mass lumping for piecewise linear triangular elements). *Let f be a smooth function and consider the following approximation for the integral of f over $\Omega = \cup_{K\in\mathcal{T}_h} K$, where $\mathcal{T}_h$ is a triangulation of Ω:*

$$\int_\Omega f = \sum_{K\in\mathcal{T}_h}\int_K f \approx \sum_{K\in\mathcal{T}_h}\frac{|K|}{3}\sum_{i=1}^{3} f(q_i^K),$$

where q_1^K, q_2^K, q_3^K are the three vertices of K. If f is affine, this formula is exact; otherwise it computes the integral with an error $O(h^2)$.

This approximation is called mass lumping: for two functions $u_h, w_h \in V_h$, we call U and V the vectors of their coordinates in the nodal basis (see (4.15)): mass lumping permits us to approximate $\int_\Omega u_h v_h$ by $U^T\tilde{\mathbf{M}}V$, where $\tilde{\mathbf{M}}$ is a diagonal matrix with positive diagonal entries.

4.2.6 Stabilization by Least Squares

Exactly as for finite differences, the Galerkin finite element scheme presented above becomes unstable in the maximum norm when the nonsymmetric term in (4.1) becomes dominant (for example, this is the case for some Asian options). One has to stabilize the method: one way is to add a least squares term to (4.8). We consider the sum on the elements of the squared residuals:

$$\begin{aligned}
&\mathcal{J}_m(v, u_h^{m-1})\\
&= \sum_{K\in\mathcal{T}_h}\int_K \left(\begin{array}{c} \dfrac{v-u_h^{m-1}}{\Delta t_m} - \nabla\cdot\left(\kappa\nabla\left(\dfrac{v+u_h^{m-1}}{2}\right)\right) \\ -\nabla\cdot\left(\alpha\dfrac{v+u_h^{m-1}}{2}\right)+\beta\dfrac{v+u_h^{m-1}}{2}-\phi^{m-\frac{1}{2}} \end{array}\right)^2\\
&+\sum_{K\in\mathcal{T}_h}\int_{\partial K\cap\Gamma_n}\left(-b\frac{v+u_h^{m-1}}{2}-\left(\kappa\nabla\left(\frac{v+u_h^{m-1}}{2}\right)\right)\cdot n - f^{m-\frac{1}{2}}\right)^2.
\end{aligned}$$

Finding u_h^m as the minimizer of $\mathcal{J}_m(v, u_h^{m-1})$ over $u_{g,h}^m + V_h$ amounts to solving a least squares approximation to (4.1). It consists of solving the Euler equations

$$\mathcal{R}(u_h^m, v_h) = \mathcal{F}(v_h) \quad \forall v_h \in V_h, \tag{4.21}$$

where

$$\mathcal{R}(u_h, v_h) = \left(\begin{array}{c} \displaystyle\sum_{K\in\mathcal{T}_h}\int_K \mathcal{L}_1(u_h)(x)\mathcal{L}_1(v_h)(x)dx \\ \displaystyle+\sum_{K\in\mathcal{T}_h}\int_{\partial K\cap\Gamma_n}\mathcal{L}_2(u_h)(x)\mathcal{L}_2(v_h)(x) \end{array}\right),$$

with

$$\begin{aligned}
\mathcal{L}_1(v) &= \frac{v}{\Delta t_m} - \nabla\cdot\left(\kappa\nabla\frac{v}{2}\right) - \nabla\cdot\left(\alpha\frac{v}{2}\right)+\beta\frac{v}{2},\\
\mathcal{L}_2(v) &= -b\frac{v}{2} - \left(\kappa\nabla\left(\frac{v}{2}\right)\right)\cdot n,
\end{aligned}$$

and where $\mathcal{F}(v_h)$ is a linear form depending on u_h^{m-1}, $\phi^{m-\frac{1}{2}}$ and $f^{m-\frac{1}{2}}$.

This method has serious drawbacks: the condition number of the matrix in the system of linear equations (4.21) scales as the square of the condition number of the matrix in the Galerkin method. Therefore, the solution is much harder to compute by iterative methods and more sensitive to roundoff errors. Also, this method is less accurate in the regions where the solution is smooth.

Therefore, it is much better to mix together the Galerkin and the least squares approximations, and the resulting method is called the least squares Galerkin method: for a well-chosen parameter δ, the new discrete problem is to find $u_h^m \in u_{g,h}^m + V_h, m = 1, \dots, M$, with $u_h^0 = u_0$, and

$$\begin{aligned} &\forall v_h \in V_h, \\ &\frac{1}{\Delta t_m}(u_h^m - u_h^{m-1}, v_h) + \frac{1}{2}\left(a(u_h^m, v_h) + a(u_h^{m-1}, v_h)\right) + \delta \mathcal{R}_m(u_h^m, v_h) \\ &= \int_\Omega \phi^{m-1/2} v_h + \int_{\Gamma_n} f^{m-1/2} v_h + \delta \mathcal{F}(v_h). \end{aligned} \tag{4.22}$$

Of course, this problem amounts to solving a system of linear equations for the values of u_h^m at the vertices, with a new matrix $\tilde{\mathbf{M}} + \frac{\Delta t_m}{2}\tilde{\mathbf{A}}$. The stability is increased because the diagonal coefficients of the matrix are larger now. It is possible to study this procedure thoroughly, including a priori error estimates, and to choose δ in an optimal way. There are many references on this topic; see, for example, [54].

4.3 The Black–Scholes Equation with Local Volatility

We are interested in discretizing the Black–Scholes equation for a vanilla put, i.e., (3.45), (3.46), (3.47), with a finite element method. The variational formulation that we start from has been introduced in §2.3.2 and is given in (2.23), (2.24).

We introduce a partition of the interval $[0, \bar{S}]$ into subintervals $\kappa_i = [S_{i-1}, S_i]$, $1 \le i \le N+1$, such that $0 = S_0 < S_1 < \cdots < S_N < S_{N+1} = \bar{S}$. The size of the interval T_i is called h_i and we set $h = \max_{i=1,\dots,N+1} h_i$. We define the mesh $\mathcal{T}_h$ of $[0, \bar{S}]$ as the set $\{\kappa_1, \dots, \kappa_{N+1}\}$. In what follows, we will assume that the strike K coincides with some node of $\mathcal{T}_h$: there exists k_0, $0 < k_0 < N+1$, such that $S_{k_0} = K$. We define the discrete space V_h by

$$V_h = \left\{v_h \in \mathcal{C}^0([0, \bar{S}]),\ v_h(\bar{S}) = 0\ \forall \kappa \in \mathcal{T}_h, v_{h|\kappa} \text{ is affine}\right\}. \tag{4.23}$$

The assumption on the mesh ensures that $P_0 \in V_h$. The discrete problem obtained by applying the Crank–Nicolson scheme in time reads as follows:

Find $(P_h^m)_{0 \le m \le M}$, $P_h^m \in V_h$ satisfying

$$P_h^0 = P_0, \tag{4.24}$$

and for all m, $1 \le m \le M$,

$$\forall v_h \in V_h, \quad \left(P_h^m - P_h^{m-1}, v_h\right) + \frac{\Delta t_m}{2}(a_m(P_h^m, v_h) + a_{m-1}(P_h^{m-1}, v_h)) = 0, \tag{4.25}$$

where $a_m = a_{t_m}$ and

$$a_t(v,w) = \int_0^{\bar S} \frac{S^2\sigma^2(S,t)}{2}\frac{\partial v}{\partial S}\frac{\partial w}{\partial S} + \int_0^{\bar S}\left(-r(t)+\sigma^2(S,t)+S\sigma(S,t)\frac{\partial \sigma}{\partial S}(S,t)\right)S\frac{\partial v}{\partial S}w + r(t)\int_0^{\bar S} vw. \tag{4.26}$$

Note that, for $v, w \in V_h$, we have a simpler expression for $a_t(v,w)$ when σ is continuous with respect to S:

$$a_t(v,w) = -\sum_{i=1}^{N}\frac{S_i^2\sigma^2(S_i,t)}{2}\left[\frac{\partial v}{\partial S}\right](S_i)w(S_i) - r(t)\int_0^{\bar S} S\frac{\partial v}{\partial S}w + r(t)\int_0^{\bar S} vw, \tag{4.27}$$

where $[\frac{\partial v}{\partial S}](S_i)$ is the jump of $\frac{\partial v}{\partial S}$ at S_i:

$$\left[\frac{\partial v}{\partial S}\right](S_i) = \frac{\partial v}{\partial S}(S_i^+) - \frac{\partial v}{\partial S}(S_i^-). \tag{4.28}$$

Let $(w^i)_{i=0,\dots N}$ be the nodal basis of V_h, and let $\mathbf{M}$ and $\mathbf{A}^m$ in $\mathbb{R}^{(N+1)\times(N+1)}$ be the mass and stiffness matrices defined by $\mathbf{M}_{i,j} = (w^i, w^j)$, $\mathbf{A}^m_{i,j} = a_{t_m}(w^j, w^i)$, $0 \le i, j \le N$. Calling $P^m = (P_h^m(S_0), \dots, P_h^m(S_N))^T$ and $P^0 = (P_0(S_0), \dots, P_0(S_N))^T$, (4.25) is equivalent to

$$\mathbf{M}\left(P^m - P^{m-1}\right) + \frac{\Delta t_m}{2}\left(\mathbf{A}^m P^m + \mathbf{A}^{m-1}P^{m-1}\right) = 0. \tag{4.29}$$

The shape functions w^i corresponding to vertex S_i are supported in $[S_{i-1}, S_{i+1}]$. This implies that the matrices $\mathbf{M}$ and $\mathbf{A}^m$ are tridiagonal because when $|i-j| > 1$, the intersection of the supports of w^i and w^j has measure 0. Furthermore,

$$\begin{aligned} w^i(S) &= \frac{S - S_{i-1}}{h_i}, & \frac{\partial w^i}{\partial S} &= \frac{1}{h_i} & \forall S \in (S_{i-1}, S_i),\\ w^i(S) &= \frac{S_{i+1} - S}{h_{i+1}}, & \frac{\partial w^i}{\partial S} &= -\frac{1}{h_{i+1}} & \forall S \in (S_i, S_{i+1}), \end{aligned} \tag{4.30}$$

giving

$$\begin{aligned} &\int_0^{\bar S} w^{i-1}w^i = \frac{h_i}{6}, & &\int_0^{\bar S} Sw^i\frac{\partial w^{i-1}}{\partial S} = -\frac{S_{i-1}}{6} - \frac{S_i}{3},\\ &\int_0^{\bar S} w^{i2} = \frac{h_i + h_{i+1}}{3}, & &\int_0^{\bar S} Sw^i\frac{\partial w^i}{\partial S} = -\frac{1}{2}\int_0^{\bar S}(w^i)^2 = -\frac{h_i+h_{i+1}}{6} \quad \text{if } i > 0,\\ &\int_0^{\bar S} w^{02} = \frac{h_1}{3}, & &\int_0^{\bar S} Sw^0\frac{\partial w^0}{\partial S} = -\frac{1}{2}\int_0^{\bar S}(w^0)^2 = -\frac{h_1}{6},\\ &\int_0^{\bar S} w^{i+1}w^i = \frac{h_{i+1}}{6}, & &\int_0^{\bar S} Sw^i\frac{\partial w^{i+1}}{\partial S} = \frac{S_{i+1}}{6} + \frac{S_i}{3}. \end{aligned}$$

From this, a few calculations show that the entries of $\mathbf{A}^m$ are

$$\mathbf{A}^m_{i,i-1} = -\frac{S_i^2\sigma^2(t_m, S_i)}{2h_i} + \frac{r(t_m)S_i}{2}, \quad 1 \le i \le N,$$

$$\mathbf{A}^m_{i,i} = \frac{S_i^2\sigma^2(t_m, S_i)}{2}\left(\frac{1}{h_i} + \frac{1}{h_{i+1}}\right) + \frac{r(t_m)}{2}(h_{i+1} + h_i), \quad 1 \le i \le N,$$

$$\mathbf{A}^m_{0,0} = \frac{r(t_m)}{2}h_1,$$

$$\mathbf{A}^m_{i,i+1} = -\frac{S_i^2\sigma^2(t_m, S_i)}{2h_{i+1}} - \frac{r(t_m)S_i}{2}, \quad 0 \le i \le N-1.$$

Note that when the mesh is uniform, we recover the matrix A (up to a scaling by h for the rows $i > 1$, and $\frac{h}{2}$ for the row $i = 0$) found when using the finite difference method in §3.3; see (3.49). The entries of $\mathbf{M}$ are

$$\mathbf{M}_{i,i-1} = \frac{h_i}{6}, \quad 1 \le i \le N,$$

$$\mathbf{M}_{i,i} = \frac{h_i + h_{i+1}}{3}, \quad 1 \le i \le N,$$

$$\mathbf{M}_{0,0} = \frac{h_1}{3},$$

$$\mathbf{M}_{i,i+1} = \frac{h_{i+1}}{6}, \quad 0 \le i \le N-1.$$

Therefore, when the mesh is uniform, the scheme (4.29) is not completely equivalent to the finite difference scheme because $\mathbf{M}$ is not diagonal. However, in this case, it is possible to use a quadrature formula which makes of $\mathbf{M}$ a diagonal matrix $\tilde{\mathbf{M}}$: $\tilde{\mathbf{M}}_{i,i} = \sum_j \mathbf{M}_{i,j}$ (this process is called mass lumping). Doing so, one obtains a scheme completely equivalent to the finite difference scheme. So, when the mesh is uniform, the finite element method with mass lumping is equivalent to the finite difference centered scheme, although it has not been obtained in the same manner.

When the mesh is not uniform, the scheme (4.29) can also be seen as a finite difference scheme of second order with respect to S. Indeed, calling $S_{i\pm\frac{1}{2}} = \frac{1}{2}(S_i + S_{i\pm1})$, it corresponds to using the following finite difference approximations for the second order derivative:

$$\frac{\partial^2 P}{\partial S^2}(S_i) \approx \frac{2}{h_i + h_{i+1}}\left(\frac{\partial P}{\partial S}(S_{i+\frac{1}{2}}) - \frac{\partial P}{\partial S}(S_{i-\frac{1}{2}})\right),$$

with $$\frac{\partial P}{\partial S}(S_{i-\frac{1}{2}}) \approx \frac{P(S_i) - P(S_{i-1})}{h_i},$$

and $$\frac{\partial P}{\partial S}(S_{i+\frac{1}{2}}) \approx \frac{P(S_{i+1}) - P(S_i)}{h_{i+1}};$$

and for the first order derivative,

$$\frac{\partial P}{\partial S}(S_i) \approx 2\frac{P(S_{i+\frac{1}{2}}) - P(S_{i-\frac{1}{2}})}{h_i + h_{i+1}},$$

$$\text{with} \qquad P(S_{i\pm\frac{1}{2}}) \approx \frac{P(S_{i\pm 1}) + P(S_i)}{2}.$$

Exercise 4.2. *Prove that the finite difference scheme obtained by the two sets of formulas above is second order accurate.*

4.4 A Black–Scholes Equation Solver in C++

The following program solves the one-dimensional Black–Scholes equation with variable σ with a Crank–Nicolson scheme and piecewise affine finite element in S. The systems of linear equations are solved by LU factorization. The time grid may have variable steps, whereas the mesh in S does not vary with time. A program where the mesh in S can vary in time is given in the next chapter.

As in §3.7.2, we use the RNM vector + sparse matrix class of Danaila, Hecht, and Pironneau [38]. It is not necessary to study these classes in order to use them. One need only know how to call them and how to use the overloaded operators. It may seem like overkill, but it will allow us to hide low-level implementations using the `blas` library for speed and adaptation to the computer architecture.

A `CN_Scheme` class is defined for the Crank–Nicolson scheme as follows.

Algorithm 4.1. CN-scheme.

```
class CN_Scheme{
private:
  vector<double> S_nodes,S_steps;                  //  node and element sizes
  vector<double> grid_t;                                  //    the time grid
  double rate (double);            //    interest rate is a function of time
  double vol(double,double);         //   vol.  is a function of time and S
  MatriceProfile<double> B;                             //     matrix M+dt/2 A
public:
  CN_Scheme(const vector<double>  &g_grid_S,
            const  vector<double>  &g_S_steps,
            const vector<double> & g_grid_t) :S_nodes(g_grid_S),
                S_steps(g_S_steps), grid_t(g_grid_t),
                B(g_grid_S.size()-1,2) {};                   //   constructor

  void Time_Step(int i,  vector<KN<double> >  &P);
  void build_rhs(KN<double> &u, const KN<double> &u_p,
                       const double t,const double dt);
  void build_matrix(MatriceProfile<double> &B , const double t,
                       const double dt);
};
```

A time step of the method is implemented in the function `Time_Step`: it consists of building the matrix $B = \mathbf{M} + \frac{\Delta t_m}{2}\mathbf{A}^m$ (we have assumed that local volatility is a function of time and S, so the matrix has to be recomputed at each time step), computing its LU factorization, constructing the right-hand side of the system of linear equations, and solving this system.

ALGORITHM 4.2. Time step.

```
void CN_Scheme::Time_Step(int it,  vector< KN<double> >& P)
{
  double t=grid_t[it];                                    //    current time
  double dt=t-grid_t[it-1];                            //   current time step
  build_matrix(B,t,dt);                          //   computes B = (M+dt/2* A)
  B.LU();                                          //   LU factorization of B
  build_rhs(P[it],P[it-1],t-dt,dt);                       //   computes the RHS
  B.Solve(P[it],P[it]);                                //   solves the system
}
```

The matrix $B = \mathbf{M} + \frac{\Delta t_m}{2}\mathbf{A}^m$ is assembled as follows.

ALGORITHM 4.3. Matrix assembly.

```
void     CN_Scheme::build_matrix(MatriceProfile<double> &B,
                            const double t,const double dt)
{  int i;
  double S,h_p,h_n, a,b,c,d;
  double r=rate(t);
                                          //   computes the first row of B
    h_n=S_steps[0];
    B(0,0)=(0.25*dt*r+1./3)*h_n;
    B(0,1)=h_n/6;

  for(i=1;i< S_steps.size()-1;i++)
    {                                        //   computes the i-th row of B
      h_p=h_n;
      S=S_nodes[i];
      h_n=S_steps[i];
      a=pow(S*vol(t,S),2);
      b=a/h_p;
      c=a/h_n;
      d=r*S;

      B(i,i)=0.25*dt*(b+c+r*(h_p+h_n))+ (h_p+h_n)/3 ;
      B(i,i-1)=0.25*dt*(-b+d)+h_p/6;
      B(i,i+1)=0.25*dt*(-c-d)+h_n/6;
    }
  h_p=h_n;                                   //   computes the last row of B
  S=S_nodes[i];
  h_n=S_steps[i];
  a=pow(S*vol(t,S),2);
  b=a/h_p;
  c=a/h_n;
  d=r*S;
  B(i,i)=0.25*dt*(b+c+r*(h_p+h_n))+ (h_p+h_n)/3 ;
  B(i,i-1)=0.25*dt*(-b+d)+h_p/6;
}
```

The right-hand side is computed as follows.

ALGORITHM 4.4. Right-hand side.

```
void CN_Scheme::build_rhs(KN<double> &u,const KN<double> &u_p,
                          const double t,const double dt)
{                         //  u_p is the solution at the previous time step
                     //  u will be the right-hand side of the linear system
                     //   u is computed from u_p by a loop on the elements
  double r,c,d;                                     //  auxiliary variables
  double x_l,x_r;           //  left and right endpoints of current element
  double v_l,v_r;           //  linkS to the values of the vol.  at x_l,x_r
  double u_l,u_r;      //  d.o.f.  of u_p associated to the current element

  r=rate(t);  u=0;         //  all the components of u are initialized to 0
  x_l=0;                   //  some initializations before entering the loop
  u_l=u_p(0);
  v_l=0.5*pow(x_l*vol(t,x_l),2);
  int i;
  for ( i=0; i<S_steps.size()-1;i++)                    //  loop on elements
    {
      //  left val.  of x, u_p and vol at left end of elem.  are known
      x_r=S_nodes[i+1];         //  get x, u_p and vol at right end of elem.
      v_r=0.5*pow(x_r*vol(t,x_r),2);
      u_r=u_p(i+1);

      c=u_r-u_l;
      d=c/S_steps[i];
      u(i)+=S_steps[i]/6.*(2*u_l+u_r)
       +0.5*dt*(d*v_l+(c*(2*x_l+x_r)-(2*u_l+u_r)*S_steps[i])*r/6);
      u(i+1)+=S_steps[i]/6.*(u_l+2*u_r)
       +0.5*dt*(-d*v_r+(c*(2*x_r+x_l)-(2*u_r+u_l)*S_steps[i])*r/6);

                     //  right val.  becomes left val.  for the next elem.
      x_l=x_r;
      u_l=u_r;
      v_l=v_r;
    }
                //  last elem.  is special because of Dirichlet conditions
  c=-u_l;
  d=c/S_steps[i];
  u(i)+=S_steps[i]/3.*(u_p(i))
      +0.5*dt*(d*v_l+(c*(2*x_l+x_r)-2*u_l*S_steps[i])*r/6);
}
```

In Figure 4.3, we compare the error produced by the method programmed above for two grids: a uniform grid with 100 nodes in t and 100 nodes in S, and a nonuniform grid with the same number of nodes and an algebraic refinement near $t = 0$ and $S = K$. We see that the error is much larger with the uniform grid. We see also that when the mesh is nonuniform there is an error at high values of S, which becomes comparable to the error near the money: this error is due to the artificial boundary conditions. For improving on this, one should either take $\bar{S}$ larger or use transparent boundary conditions (see §4.5).

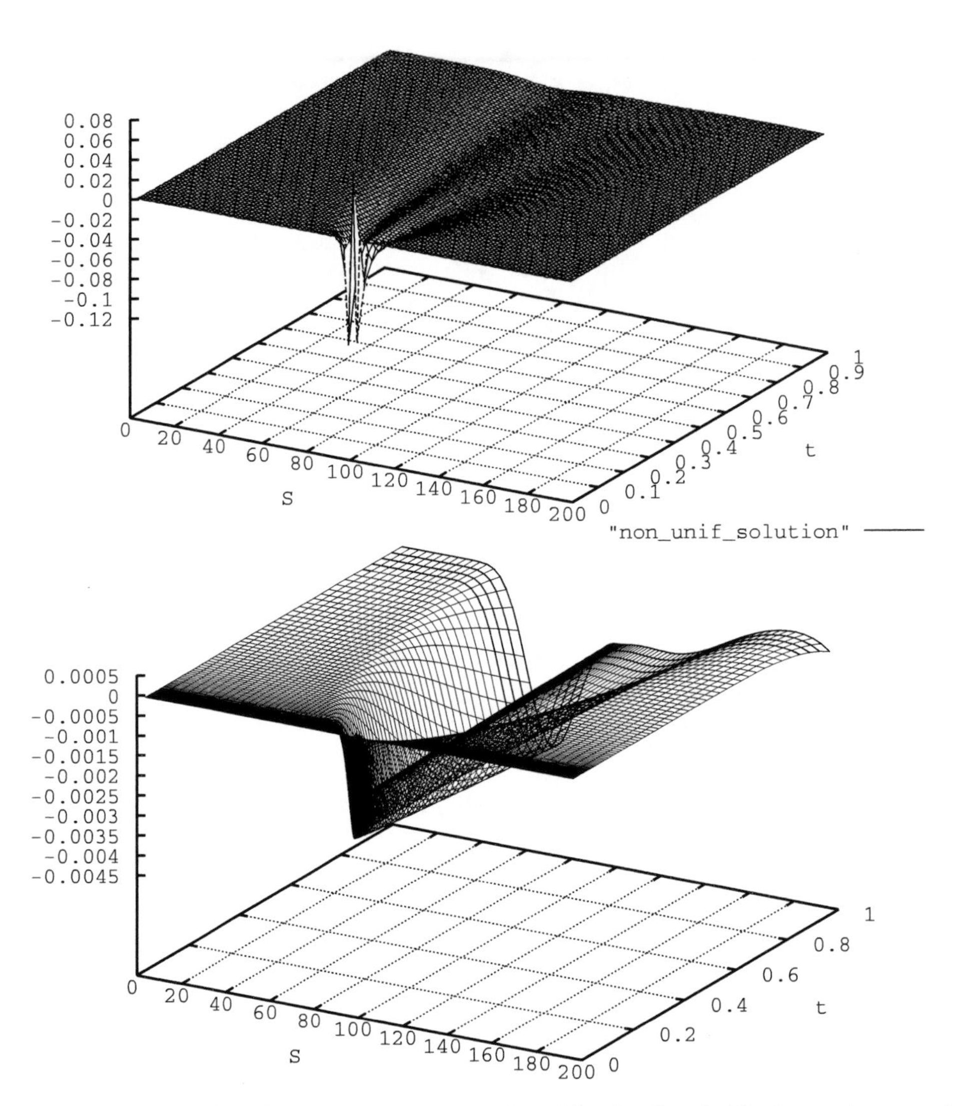

Figure 4.3. *The pointwise error produced by the Crank–Nicolson scheme and piecewise affine finite element methods for uniform and nonuniform grids with* 100×100 *nodes.*

4.5 A Transparent Boundary Condition

While the Black–Scholes equation is set on $(0, +\infty)$, localization at $S \in (0, \bar{S})$ implies an error which decays fast as $\frac{\bar{S}}{K}$ grows. Nevertheless, for saving time, one may need to have $\bar{S}$ as small as possible—for example, $\bar{S} = 1.2K$.

Transparent boundary conditions were proposed in computational physics for linear partial differential equations with constant coefficients, for which the Green's functions are known, and in particular in electromagnetism, where one frequently deals with unbounded domains; see [80]. They permit us to compute the solution in a bounded domain with no errors. We propose adapting the idea to the present situation.

Consider the problem

$$\partial_t u - \partial_{xx} u = 0,\ x > L;\quad u(L,t) = g(t),\ \ u(+\infty,t) = 0,\quad u(x,0) = 0,$$

with the compatibility condition $g(0) = 0$. It corresponds to σ, r constant and is obtained from (2.9) by performing several changes of variables and unknown function; see §2.2.3, in particular the change of variable $x = \log S$. Extending u in $x < L$ by $u(x,t) = g(t)$ and calling $q = -\partial_x u(L,t)$, this problem is equivalent to

$$\partial_t u - \partial_{xx} u = g'(t) I_{x<L} + q(t)\delta(x-L)\ \forall x \in \mathbb{R};\quad u(x,0) = 0.$$

So the solution satisfies

$$\begin{aligned} u(x,t) &= \int_{R\times(0,t)} (g'(\tau) I_{y<L} + q(\tau)\delta(y-L)) G(x-y, t-\tau) \mathrm{d}y\mathrm{d}\tau \\ &= \int_0^t g'(\tau) \int_{-\infty}^a G(x-y,t-\tau)\mathrm{d}y\mathrm{d}\tau + \int_0^t q(\tau) G(x_L, t-\tau)\mathrm{d}\tau, \end{aligned} \tag{4.31}$$

where $G(x,t) = \frac{e^{-\frac{x^2}{4t}}}{\sqrt{4\pi t}}$ is the fundamental solution to the heat equation, which yields at $x = L$

$$g(t) = \int_0^t g'(\tau) \int_0^\infty G(z, t-\tau)\mathrm{d}z\mathrm{d}\tau + \int_0^t \frac{q(\tau)}{\sqrt{4\pi(t-\tau)}}\mathrm{d}\tau,$$

i.e., since $\int_0^\infty G(z, t-\tau)\mathrm{d}z = \frac{1}{2}$,

$$u(L,t) = -\sqrt{\frac{1}{\pi}} \int_0^t \partial_x u(L,\tau) \frac{\mathrm{d}\tau}{\sqrt{t-\tau}}.$$

This, in turn, is approximated numerically by

$$\sqrt{\pi} u + 2\sqrt{\delta t}\,\partial_x u = -\int_0^{t-\delta t} \partial_x u \frac{\mathrm{d}\tau}{\sqrt{t-\tau}}.$$

Returning to the Black–Scholes equation in the variable S, in the special case when $r = \sigma^2/2$, the transparent condition is

$$P(\bar{S},t) = -\sqrt{\frac{1}{2\pi}} \int_0^t \sigma \bar{S} \partial_S P(\bar{S},\tau) \frac{e^{-r(t-\tau)}}{\sqrt{t-\tau}} \mathrm{d}\tau$$

and is approximated by

$$\sqrt{2\pi} P(\bar{S},t) + 2\sqrt{\delta t}\,\sigma \bar{S} \partial_S P(\bar{S},t) = -2\int_0^{t-\delta t} \sigma \bar{S} \partial_S P(\bar{S},\tau) \frac{e^{-r(t-\tau)}}{\sqrt{t-\tau}} \mathrm{d}\tau.$$

When r is a different constant, the new function $v = u(x,t)e^{\frac{(r+\frac{\sigma^2}{2})^2 t}{2\sigma^2}} e^{\frac{(r-\frac{\sigma^2}{2})x}{2\sigma^2}}$ (with $u(x,t) = P(S,t)$, $x = \log(S)$) satisfies

$$v(L,t) = -\sqrt{\frac{1}{2\pi}} \int_0^t \sigma \partial_x v(L,\tau) \frac{d\tau}{\sqrt{t-\tau}},$$

which yields

$$u(L,t) = -\sqrt{\frac{1}{2\pi}} \int_0^t \sigma \left(\frac{r-\frac{\sigma^2}{2}}{2\sigma^2} u(L,\tau) + \partial_x u(L,\tau) \right) \frac{e^{-\frac{(r+\frac{\sigma^2}{2})^2(t-\tau)}{2\sigma^2}}}{\sqrt{t-\tau}} d\tau,$$

and in the S variable,

$$P(\bar{S},t) = -\sqrt{\frac{1}{2\pi}} \int_0^t \left(\frac{r-\frac{\sigma^2}{2}}{2\sigma} P(\bar{S},\tau) + \sigma \bar{S} \partial_S P(\bar{S},\tau) \right) \frac{e^{-\frac{(r+\frac{\sigma^2}{2})^2(t-\tau)}{2\sigma^2}}}{\sqrt{t-\tau}} d\tau. \tag{4.32}$$

As above, this is approximated by

$$\sqrt{2\pi} P(\bar{S},t) + 2\sqrt{\delta t} \left(\frac{r-\frac{\sigma^2}{2}}{2\sigma} P(\bar{S},t) + \sigma \bar{S} \partial_S P(\bar{S},t) \right)$$

$$= -\int_0^{t-\delta t} \left(\frac{r-\frac{\sigma^2}{2}}{2\sigma} P(\bar{S},\tau) + \sigma \bar{S} \partial_S P(\bar{S},\tau) \right) \frac{e^{-\frac{(r+\frac{\sigma^2}{2})^2(t-\tau)}{2\sigma^2}}}{\sqrt{t-\tau}} d\tau.$$

Discretization by finite differences on a uniform grid of step Δt and h gives

$$aP^m_{N+1} - bP^m_N = \sum_{k=0}^{m-1} \frac{e^{-\frac{(r+\frac{\sigma^2}{2})^2(m-k)\Delta t}{2\sigma^2}}}{\sqrt{m-k}} (cP^k_{N+1} - dP^k_N),$$

where

$$a = \sqrt{2\pi} + \sqrt{\Delta t} \left(\frac{r-\frac{\sigma^2}{2}}{\sigma} + \frac{2\sigma \bar{S}}{h} \right), \qquad b = \frac{2\sigma \bar{S} \sqrt{\Delta t}}{h},$$

$$c = -\sqrt{\Delta t} \left(\frac{r-\frac{\sigma^2}{2}}{2\sigma} + \frac{\sigma \bar{S}}{h} \right), \qquad d = -\frac{\sigma \bar{S} \sqrt{\Delta t}}{h}.$$

The program for an Euler scheme with transparent boundary condition is as follows.

ALGORITHM 4.5. Transparent boundary conditions.

```
void Euler_Scheme::Time_Step_Transp_BC(int it,  vector< KN<double> >& P,
                                        int verbose)
{
  int i,n;
  double dt,t,S,h_p,h_n,r;
  double a,b,c,d,e;
  double co_1,co_2;
  double pi=4*atan(1.);

  n=S_steps[it].size();
```

```
  MatriceProfile<double> A(n,2);
  t=grid_t[it];
  dt=t-grid_t[it-1];
  r=rate(t);
  e=0.5*dt;

  h_n=S_steps[it][0];
  A(0,0)=e*r*h_n+ h_n/3;
  A(0,1)=h_n/6;
  for(i=1;i< n-1;i++)
    {
      h_p=h_n;
      S=S_nodes[it][i];
      h_n=S_steps[it][i];
      a=pow(S*vol(t,S),2);
      b=a/h_p;
      c=a/h_n;
      d=r*S;
      A(i,i)=e*(b+c+r*(h_p+h_n))+ (h_p+h_n)/3 ;
      A(i,i-1)=e*(-b+d)+h_p/6;
      A(i,i+1)=e*(-c-d)+h_n/6;
    }
  h_p=h_n;
  S=S_nodes[it][i];
  h_n=S_steps[it][i];
  double Smax=S;
  double eta_infi=vol(t,Smax);
  double eta_infi_sq= pow(eta_infi,2.);
  double aux_tr=r- 0.5*eta_infi_sq;
  A(i,i)=sqrt(2*pi) +  sqrt(dt)*(aux_tr/eta_infi + 2* eta_infi *Smax/ h_n);
  A(i,i-1)=-2* sqrt(dt)*eta_infi*Smax/ h_n;
  if (change_grid[it])
    build_rhs(P[it],P[it-1],S_steps[it-1],S_nodes[it-1],S_steps[it],
              S_nodes[it]);
  else
    build_rhs_same_grid(P[it],P[it-1],S_steps[it-1],S_nodes[it-1],
                        S_steps[it],S_nodes[it]);
  P[it](n-1)=0;
  for (int j=1;j<it;j++)
    {
      double t2=grid_t[j];
      double dt2=t2-grid_t[j-1];
      double r2=rate(t2);
      int siz=  S_steps[j].size()-1;
      double h2= S_steps[j][siz];
      co_1=-dt2*((r2- 0.5*eta_infi_sq)/(2*eta_infi)  + eta_infi*Smax/h2);
      co_2=dt2*eta_infi*Smax/h2;
      P[it](n-1)+=(P[j](siz)*co_1+P[j](siz-1)*co_2)
        * exp(-pow((r2+ 0.5*eta_infi_sq),2.)
              *  (t-t2)/(2*eta_infi_sq))/sqrt(t-t2);
    }
  A.LU();
  A.Solve(P[it],P[it]);
}
```

The sum

$$\sum_{k=0}^{m-1} \frac{e^{-\frac{(r+\frac{\sigma^2}{2})^2(m-k)\Delta t}{2\sigma^2}}}{\sqrt{m-k}}(cP_{N+1}^k - dP_N^k)$$

can be approximated as follows:

$$\sum_{k=0}^{m-1} \frac{e^{-\frac{(r+\frac{\sigma^2}{2})^2(m-k)\Delta t}{2\sigma^2}}}{\sqrt{m-k}}(cP_{N+1}^k - dP_N^k)$$

$$\approx \sum_{k=k_m}^{m-1} \frac{e^{-\frac{(r+\frac{\sigma^2}{2})^2(m-k)\Delta t}{2\sigma^2}}}{\sqrt{m-k}}(cP_{N+1}^k - dP_N^k) + \frac{e^{-\frac{(r+\frac{\sigma^2}{2})^2(m-\frac{k_m}{2})\Delta t}{2\sigma^2}}}{\sqrt{m-\frac{k_m}{2}}} \sum_{k=0}^{k_m-1} cP_{N+1}^k - dP_N^k.$$

This expression can be computed much faster because an induction formula is available for the last sum.

Exercise 4.3. *Write down the transparent boundary condition for* (2.12) *with a constant dividend yield* $q > 0$.

Exercise 4.4. *Write down the transparent boundary condition for Dupire's equation* (2.51) *for constant dividend yield and interest rate.*

Figure 4.4 illustrates the performances of the transparent boundary conditions at $\bar{S} = 1.4K$ and $\bar{S} = 1.2K$, for a put option, with uniform volatility. The option price is well approximated, whereas the Dirichlet condition $P(\bar{S}) = 0$ gives bad results. There remains an error though, due to

- the integration formula for (4.32);
- the fact that the function does not solve exactly the Black–Scholes equation in $(0, \bar{S}) \times (0, T)$.

In Figure 4.5, the price obtained with transparent boundary conditions at $\bar{S} = 1.4K$ and $\bar{S} = 1.2K$ one year to maturity is compared to the exact price.

4.6 Lévy Driven Assets

As in §2.5, one may consider Lévy driven assets. The aim of this section is to describe several possible ways to discretize the partial integrodifferential equation (2.57).

4.6.1 A Scheme with an Explicit Discretization of the Nonlocal Term

In [32], it is shown that when the kernel k is not too singular or when diffusion dominates, an explicit treatment of the nonlocal term is good enough. This means that a minor modification of the program for the usual Black–Scholes equation will suffice: at every time step t_m,

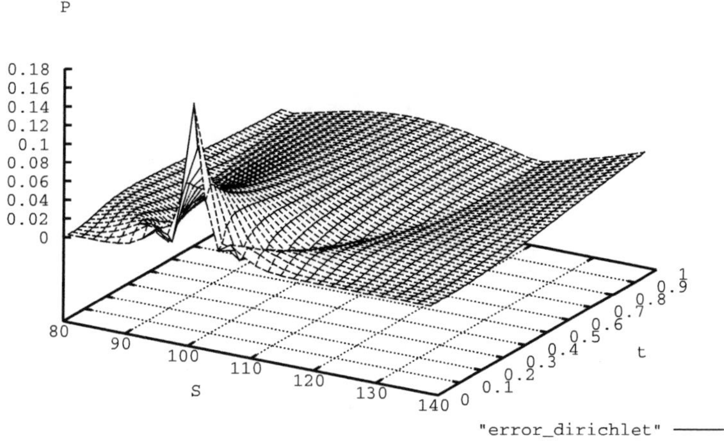

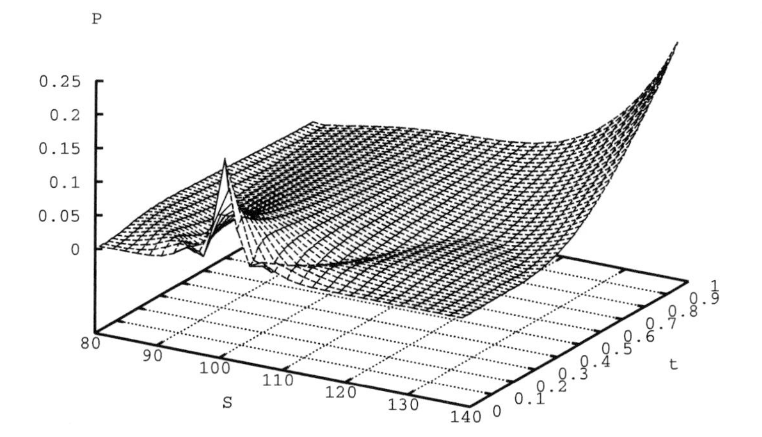

Figure 4.4. *Top: error in the exact price when using an Euler scheme with the new boundary condition on a European put at* $T = 1$ *for* $\sigma = 0.2$, $r = 0.04$, $\bar{S} = 1.4$, $K = 140$. *Bottom: error with the Dirichlet condition* $P(\bar{S}, t) = 0$. *There are* 70 *nodes in the* S *direction and* 50 *time steps.*

before the resolution of the system of linear equations one must change the right-hand side of the linear system from $P^m \Delta t$ to $(P^m + v)\Delta t$ with

$$v(S) = \int_{\mathbb{R}} \left[P(Se^y, t) - P(S, t) - \frac{\partial P}{\partial S}(S, t)S(e^y - 1) \right] k(y)\mathrm{d}y$$

(recall that t is the time to maturity).

Let S_i be the node of a uniform mesh with step size h. Let us denote by P_i^m the approximation of $P(S_i, m\Delta t)$, $i = 0, \ldots, N$. We approximate the integral $v(S)$ by a finite

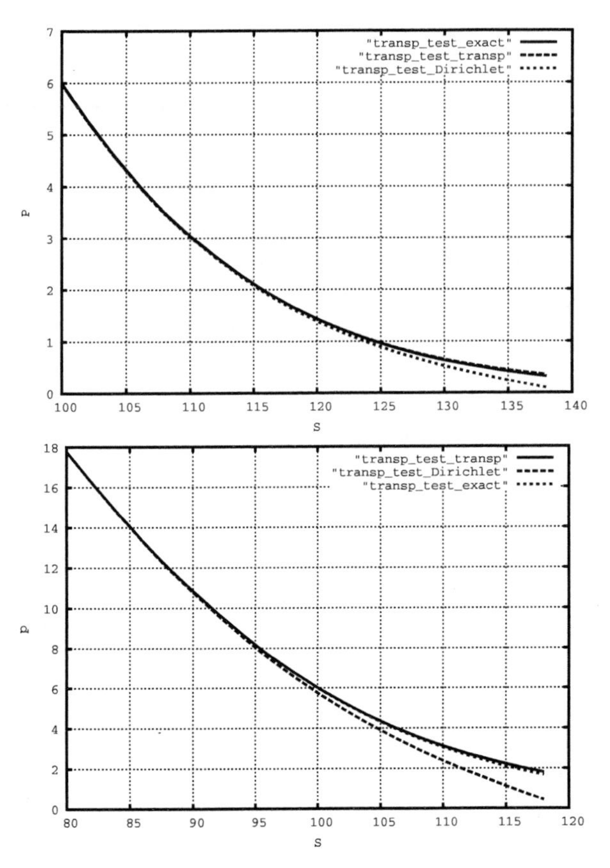

Figure 4.5. *Comparison with the exact solution at $T = 1$ when using transparent and Dirichlet conditions at $\bar{S} = 140$ (top) and $\bar{S} = 120$ (bottom).*

sum over $j = 0, \ldots, N$:

$$v(S_i) \approx \sum_{j=0}^{N} \frac{1}{j+\frac{1}{2}} \left[P_j^m - P_i^m - \frac{P_{i+1}^m - P_i^m}{h}(S_j - S_i) \right] k \left(\log \left(\frac{S_j}{S_i} \right) \right).$$

ALGORITHM 4.6. An elementary program.

```
void getAu() {
    const double c=0.1/sqrt(8*atan(1.0));
    for(int i=1;i<nS-1;i++)
    {
        v[i]=0;
        for(int j=1;j<nS;j++)
        {
          double x = log(double(j)/i);
```

```
            double w = u[j]-u[i]-(u[i+1]-u[i])*(j-i);
            v[i] += w*c*exp(-x*x/2)/(j+0.5);
        }
    }
}
```

In Figure 4.6, we have plotted the price of a vanilla put option computed with the Euler explicit scheme with Algorithm 4.6, when the underlying asset is described by a Lévy process, with $k(z) = e^{-\frac{z^2}{2}}$, as well as the price of the same option given by the standard Black–Scholes model.

4.6.2 A Semi-Implicit Scheme and a Program for Options on CGMY Driven Assets

We are going to propose a scheme where the nonlocal term is discretized partly implicitly; i.e., the nonlocal interaction between nodes close to each other will be treated implicitly. On the contrary, when two nodes are far enough from each other, their interaction will be treated explicitly. The advantage of such a method is twofold:

- the systems of linear equations to be solved at each time step involve banded matrices;
- one can tune the bandwidth of this matrix in order to obtain a stable scheme.

Calling as above w^j the shape functions, we have to compute

$$\int_{\mathbb{R}_+} w^i(S)\left(\int_{\mathbb{R}}\left(w^j(Se^y) - w^j(S) - S(e^y-1)\frac{\partial}{\partial S}w^j(S)\right)k(y)dy\right)dS,$$

with k given by (2.59). The first step is to compute

$$I(S) = \int_{\mathbb{R}}\left(w^j(Se^y) - w^j(S) - S(e^y-1)\frac{\partial}{\partial S}w^j(S)\right)k(y)dy.$$

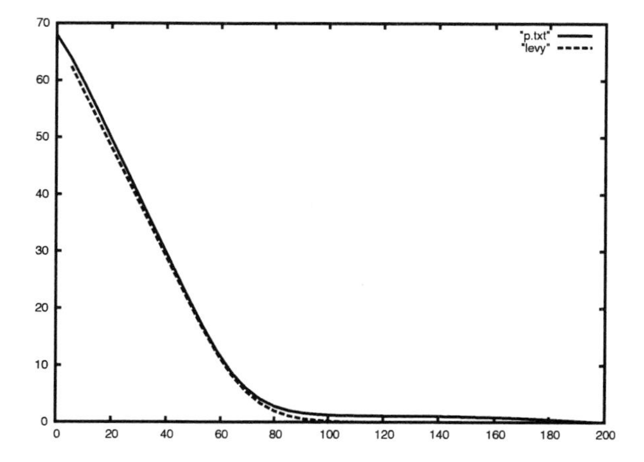

Figure 4.6. *A put option on an underlying asset modeled by a Lévy process and comparison with the same modeled by a Black–Scholes equation.*

To simplify the discussion, we focus on the case where the parameter Y satisfies $0 \le Y \le 1$.

We discuss in detail the case when $S > S_{j+1}$: in this case,

$$
\begin{aligned}
I(S) &= \int_{\log(\frac{S_{j-1}}{S})}^{\log(\frac{S_{j+1}}{S})} w^j(Se^y)k(y)dy \\
&= \int_{\log(\frac{S_{j-1}}{S})}^{\log(\frac{S_j}{S})} \frac{Se^y - S_{j-1}}{h_{j-1}} \frac{C}{|y|^{1+Y}} e^{-G|y|}dy + \int_{\log(\frac{S_j}{S})}^{\log(\frac{S_{j+1}}{S})} \frac{-Se^y + S_{j+1}}{h_j} \frac{C}{|y|^{1+Y}} e^{-G|y|}dy.
\end{aligned}
$$

Calling χ the function $\chi : \mathbb{R}_+ \times \mathbb{R}_+ \to \mathbb{R}_+$,

$$
\chi(x, L) = C \int_x^{\infty} \frac{e^{-Ly}}{|y|^{1+Y}} dy, \tag{4.33}
$$

we see that

$$
I(S) = \left\{
\begin{aligned}
& \frac{S_{j+1}}{h_j}\chi\left(\log\left(\frac{S}{S_{j+1}}\right), G\right) - \left(\frac{S_{j+1}}{h_j} + \frac{S_{j-1}}{h_{j-1}}\right)\chi\left(\log\left(\frac{S}{S_j}\right), G\right) \\
& \qquad + \frac{S_{j-1}}{h_{j-1}}\chi\left(\log\left(\frac{S}{S_{j-1}}\right), G\right) \\
& -S\left(\begin{aligned}
& \frac{1}{h_j}\chi\left(\log\left(\frac{S}{S_{j+1}}\right), 1+G\right) + \frac{1}{h_{j-1}}\chi\left(\log\left(\frac{S}{S_{j-1}}\right), 1+G\right) \\
& -\chi\left(\log\left(\frac{S}{S_j}\right), 1+G\right)\left(\frac{1}{h_j} + \frac{1}{h_{j-1}}\right)
\end{aligned}\right).
\end{aligned}
\right.
$$

Note that χ can be written using special functions:

1. If $Y = 0$,
$$
\chi(x, L) = CE_1(Lx),
$$
where E_1 is the exponential integral $E_1(x) = \int_1^{\infty} \frac{e^{-tx}}{t} dt$.

2. If $Y = 1$,
$$
\chi(x, L) = C\frac{E_2(Lx)}{x},
$$
where E_2 is the second order exponential integral $E_2(x) = \int_1^{\infty} \frac{e^{-tx}}{t^2} dt$.

3. If $0 < Y < 1$,
$$
\chi(x, L) = \frac{C}{Y}\left(e^{-xL}x^{-Y} - L^Y\Gamma(1 - Y, Lx)\right),
$$
where $\Gamma(a, x)$ is the incomplete Γ function: $\Gamma(a, x) = \int_x^{+\infty} t^{a-1}e^{-t} dt$.

After some calculations, one obtains the following in a similar manner:

1. If $S_j < S < S_{j+1}$,

$$I(S) = \begin{cases} -\dfrac{S_{j+1}}{h_j}\chi\left(\log\left(\dfrac{S_{j+1}}{S}\right), M\right) + \dfrac{S}{h_j}\chi\left(\log\left(\dfrac{S_{j+1}}{S}\right), M-1\right) \\ +\dfrac{S_{j-1}}{h_{j-1}}\chi\left(\log\left(\dfrac{S}{S_{j-1}}\right), G\right) - \dfrac{S}{h_{j-1}}\chi\left(\log\left(\dfrac{S}{S_{j-1}}\right), 1+G\right) \\ +\left(\dfrac{S}{h_{j-1}} + \dfrac{S}{h_j}\right)\chi\left(\log\left(\dfrac{S}{S_j}\right), 1+G\right) - \left(\dfrac{S_{j-1}}{h_{j-1}} + \dfrac{S_{j+1}}{h_j}\right)\chi\left(\log\left(\dfrac{S}{S_j}\right), G\right). \end{cases}$$

2. If $S_{j-1} < S < S_j$,

$$I(S) = \begin{cases} \dfrac{S_{j-1}}{h_{j-1}}\chi\left(\log\left(\dfrac{S}{S_{j-1}}\right), G\right) - \dfrac{S}{h_{j-1}}\chi\left(\log\left(\dfrac{S}{S_{j-1}}\right), 1+G\right) \\ -\dfrac{S_{j+1}}{h_j}\chi\left(\log\left(\dfrac{S_{j+1}}{S}\right), M\right) + \dfrac{S}{h_j}\chi\left(\log\left(\dfrac{S_{j+1}}{S}\right), M-1\right) \\ -\left(\dfrac{S}{h_{j-1}} + \dfrac{S}{h_j}\right)\chi\left(\log\left(\dfrac{S_j}{S}\right), M-1\right) + \left(\dfrac{S_{j-1}}{h_{j-1}} + \dfrac{S_{j+1}}{h_j}\right)\chi\left(\log\left(\dfrac{S_j}{S}\right), M\right). \end{cases}$$

3. If $S < S_{j-1}$,

$$I(S) = \begin{cases} -\dfrac{S_{j+1}}{h_j}\chi\left(\log\left(\dfrac{S_{j+1}}{S}\right), M\right) + \left(\dfrac{S_{j+1}}{h_j} + \dfrac{S_{j-1}}{h_{j-1}}\right)\chi\left(\log\left(\dfrac{S_j}{S}\right), M\right) \\ \qquad -\dfrac{S_{j-1}}{h_{j-1}}\chi\left(\log\left(\dfrac{S_{j-1}}{S}\right), M\right) \\ +S\begin{pmatrix} \dfrac{1}{h_j}\chi\left(\log\left(\dfrac{S_{j+1}}{S}\right), M-1\right) + \dfrac{1}{h_{j-1}}\chi\left(\log\left(\dfrac{S_{j-1}}{S}\right), M-1\right) \\ -\chi\left(\log\left(\dfrac{S_j}{S}\right), M-1\right)\left(\dfrac{1}{h_j} + \dfrac{1}{h_{j-1}}\right) \end{pmatrix}. \end{cases}$$

The special functions E_1, E_2, and $\Gamma(\,,\,)$ are programmed in the GSL [59]. Using this library, the code for the function χ is as follows.

ALGORITHM 4.7. χ.

```
#include <gsl/gsl_sf.h>

double chi(double x, double C,double M,double Y)
 {
   if (Y<1e-6)
```

```
      {
        if (x<50)
          return  gsl_sf_expint_E1(M*x)/C;
        else
          return 0;
      }
    else
      if( 1.-Y<1e-6)
        {
          if (x<50)
            return  gsl_sf_expint_E2(M*x)/C/x;
          else
            return 0;
        }
      else
        {
          if (x<50)
            return
              (exp(-x*M)* pow(x, -Y) -pow(M, Y)*
               gsl_sf_gamma(1.-Y)*gsl_sf_gamma_inc_Q(1-Y,M*x))/C/Y;
          else
            return 0;
        }
}
```

Then the program for computing the function $w^i(S)I(S)$ is as follows.

ALGORITHM 4.8. CGMY 1.

```
double fu_I(double x, void * params)
{
  double * alpha= (double*) params;
  double x0=alpha[0];
  double x1=alpha[1];
  double x2=alpha[2];
  double h0=x1-x0;
  double h1=x2-x1;
  double res;
  if (x>x2)
    {
      double log0;

      if(x0>0)
        log0=log(x/x0);
      else
        log0=50;
      double log1=log(x/x1);
      double log2=log(x/x2);
      if ( (1+G)*log2>50)
        res=0;
      else
        {
          res=x2/h1*chi(log2,C,G,Y) - (x2/h1+x0/h0) * chi(log1,C,G,Y)+
```

```
              x0/h0*chi(log0,C,G,Y)
              -x*(chi(log2,C,1+G,Y)/h1+chi(log0,C,1+G,Y)/h0
                  -chi(log1,C,G+1,Y)*(1/h1+1/h0));
           }
       }
    else
      if (x>x1)
        {
          double log0;
          if(x0>0)
            log0=log(x/x0);
          else
            log0=50;
          double log1=log(x/x1);
          double log2=log(x2/x);
          res=-x2/h1*chi(log2,C,M,Y) +x/h1*chi(log2,C,M-1,Y)+
            x0/h0*chi(log0,C,G,Y)-x/h0*chi(log0,C,1+G,Y)
            +(1/h0+1/h1)*x*chi(log1,C,1+G,Y)
            -(x0/h0+x2/h1)*chi(log1,C,G,Y);
        }
      else
        if (x>x0)
          {
            double log0;
            if(x0>0)
              log0=log(x/x0);
            else
              log0=50;
            double log1=log(x1/x);
            double log2=log(x2/x);
            res= x0/h0*chi(log0,C,G,Y)-x/h0*chi(log0,C,1+G,Y)
              -x2/h1*chi(log2,C,M,Y)+x/h1*chi(log2,C,M-1,Y)
              -(1/h0+1/h1)*x*chi(log1,C,M-1,Y)
            +(x0/h0+x2/h1)*chi(log1,C,M,Y);
          }
        else
          {
            double log0=log(x0/x);
            double log1=log(x1/x);
            double log2=log(x2/x);
            if(M*log0>50)
              res= 0;
            else
              {
                res=-x2/h1*chi(log2,C,M,Y) +(x2/h1+x0/h0) * chi(log1,C,M,Y)
                  - x0/h0*chi(log0,C,M,Y)
                  +x*(chi(log2,C,M-1,Y)/h1+chi(log0,C,M-1,Y)/h0
                      -chi(log1,C,M-1,Y)*(1/h1+1/h0));
              }
          }
  res*=alpha[3]+x*alpha[4];
  return res;
}
```

Note that this function has to be modified if $j = 0$.

We propose the following semi-implicit scheme for discretizing (2.57): Calling $\mathbf{B}$ the matrix defined by

$$\mathbf{B}_{i,j} = \int_{\mathbb{R}_+} w^i(S) \left(\int_{\mathbb{R}} \left(w^j(Se^y) - w^j(S) - S(e^y - 1)\frac{\partial}{\partial S} w^j(S) \right) k(y)dy \right) dS, \quad (4.34)$$

we choose an integer b, we call $\tilde{\mathbf{B}}$ the banded matrix such that

$$\begin{aligned} \tilde{\mathbf{B}}_{i,j} &= \mathbf{B}_{i,j} && \text{if} \quad |j-i| < b, \\ \tilde{\mathbf{B}}_{i,j} &= 0 && \text{if} \quad |j-i| \geq b, \end{aligned} \quad (4.35)$$

and we take $\bar{\mathbf{B}} = \mathbf{B} - \tilde{\mathbf{B}}$. With $\mathbf{A}$ the stiffness matrix for the standard Black–Scholes equation and with $P^m = (P_h^m(S_0), \ldots, P_h^M(S_N))^T$ and $P^0 = (P_0(S_0), \ldots, P_0(S_N))^T$, the scheme is

$$\mathbf{M}(P^m - P^{m-1}) + \Delta t_m \mathbf{A}^m P^m + \Delta t_m \tilde{\mathbf{B}} P^m = -\Delta t_m \bar{\mathbf{B}} P^{m-1}. \quad (4.36)$$

Computing the entries of $\tilde{\mathbf{B}}$ amounts to computing an integral of a singular function: for that, we use the GSL function `gsl_integration_qagp`, and a simplified program for a time step is the following.

ALGORITHM 4.9. CGMY 2.

```
#include <gsl/gsl_math.h>
#include <gsl/gsl_integration.h>

void Euler_Scheme_for_Levy::Time_Step(int it,  vector< KN<double> >& P)
{
  int i,n;
  double dt,t,S,h_p,h_n,r;
  double a,b,c,d,e;
  gsl_integration_workspace *w =gsl_integration_workspace_alloc(200);
  double res;
  double err;
  size_t si=200;

  double * param=new double[5];
  gsl_function F;
  n=S_steps[it].size();
  MatriceProfile<double> A(n,max(2,bandwidth));

  t=grid_t[it];
  dt=t-grid_t[it-1];
  r=rate(t);
  e=0.5*dt;
  h_n=S_steps[it][0];
  A(0,0)=e*r*h_n+ h_n/3;
  A(0,1)=h_n/6;
  for(i=1;i< n-1;i++)
    {
      h_p=h_n;
      S=S_nodes[it][i];
      h_n=S_steps[it][i];
      a=pow(S*vol(t,S),2);
      b=a/h_p;
```

```
      c=a/h_n;
      d=r*S;

      A(i,i)=e*(b+c+r*(h_p+h_n))+ (h_p+h_n)/3;
      A(i,i-1)=e*(-b+d)+h_p/6;
      A(i,i+1)=e*(-c-d)+h_n/6;

    }
  h_p=h_n;
  S=S_nodes[it][i];
  h_n=S_steps[it][i];
  a=pow(S*vol(t,S),2);
  b=a/h_p;
  c=a/h_n;
  d=r*S;
  A(i,i)=e*(b+c+r*(h_p+h_n))+ (h_p+h_n)/3;
  A(i,i-1)=e*(-b+d)+h_p/6;

  for(i=0;i< n;i++)
    {
      S=S_nodes[it][i];
      double Sn=S_nodes[it][i+1];
      h_n=S_steps[it][i];
      for(int j=-bandwidth+1;j<bandwidth;j++)
        if ((i+j<n)&&(i+j>=0))
          {
            param[0]=S_nodes[it][i+j-1];
            param[1]=S_nodes[it][i+j];
            param[2]=S_nodes[it][i+j+1];
            param[3]=Sn/h_n;
            param[4]=-1./h_n;
            double * sing=new double[2];
            sing[0]=S;
            sing[1]=Sn;
            if (i+j>0)
              F.function=& fu_I;
            else
              F.function=&fu_I_0;
            F.params=param;
            gsl_integration_qagp(&F,sing,2, 0,0.001,si, w,&res,&err);
            A(i,i+j)-=e*res;
            if (i<n-1)
              {
                param[3]=-S/h_n;
                param[4]=1./h_n;
                F.params=param;
                gsl_integration_qagp(&F,sing,2, 0, 0.001,si, w,&res,&err);
                A(i+1,i+j)-=e*res;
              }
          }

    }
  build_rhs(P[it],P[it-1],S_steps[it],S_nodes[it],it);
  A.LU();
  A.Solve(P[it],P[it]);
}
```

We skip the description of the function `build_rhs`.

Note that the algorithmic complexity for constructing the matrix and the right- hand side is $O(N^2)$, so it is one order larger than for the standard Black–Scholes equation with local volatility. There are ways of improving this: either by using a wavelet basis (see [95, 94] for a sophisticated procedure using nonuniform time meshes and wavelet compression in the price variable— this method can be used for barrier options and American options for Lévy driven assets) or by using the fast Fourier transform [24], with a narrower range of applications.

Numerical Results. We take $\sigma = 0$, $C = 1$, $M = 2.5$, $G = 1.8$, so the Lévy process is a pure jump process. In Figure 4.7, we plot the price of a put with strike $K = 100$ one year to maturity for several values of Y. In Figure 4.8, we plot the price as a function of time to maturity and of S for $Y = 0.5$. We see also that even with such a nonuniform mesh, the scheme is stable. In Figure 4.9, we plot the solution computed with $\sigma = 0$, $C = 1$, $M = 2.5$, $G = 1.8$, and $Y = 0.9$ and by treating the nonlocal term fully explicitly ($b = 0$ in (4.35)), we see that instabilities develop. With the same parameters and mesh, taking $b = 2$ suffices to stabilize the scheme. In Figure 4.10, we set $\sigma = 0.1$ and we compare the price given by the Black–Scholes formula with prices of CGMY driven assets for $Y = 0, 0.5, 0.9$, with $C = 1$, $M = 2.5$, $G = 1.8$.

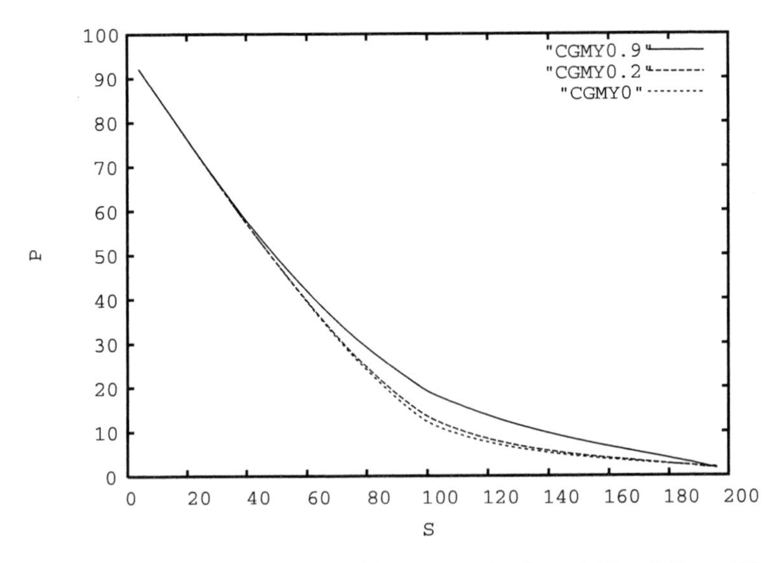

Figure 4.7. *A put option on a CGMY driven asset for $Y = 0$, $Y = 0.2$, and $Y = 0.9$.*

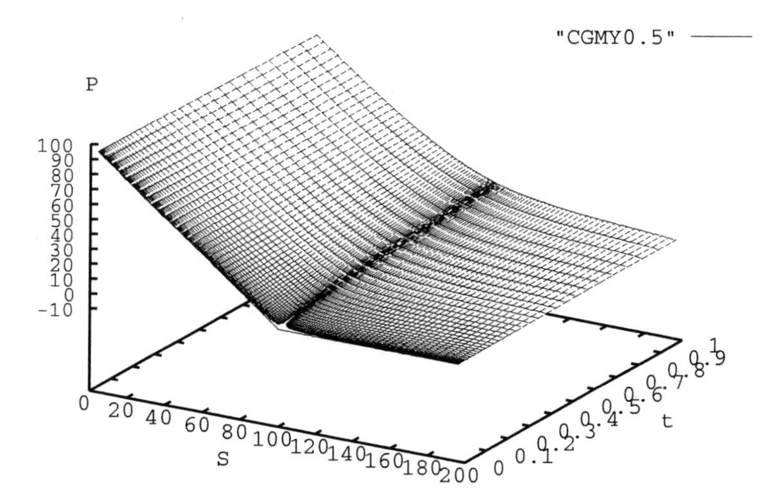

Figure 4.8. *A put option on a CGMY driven asset for $Y = 0.5$ as a function of S and t.*

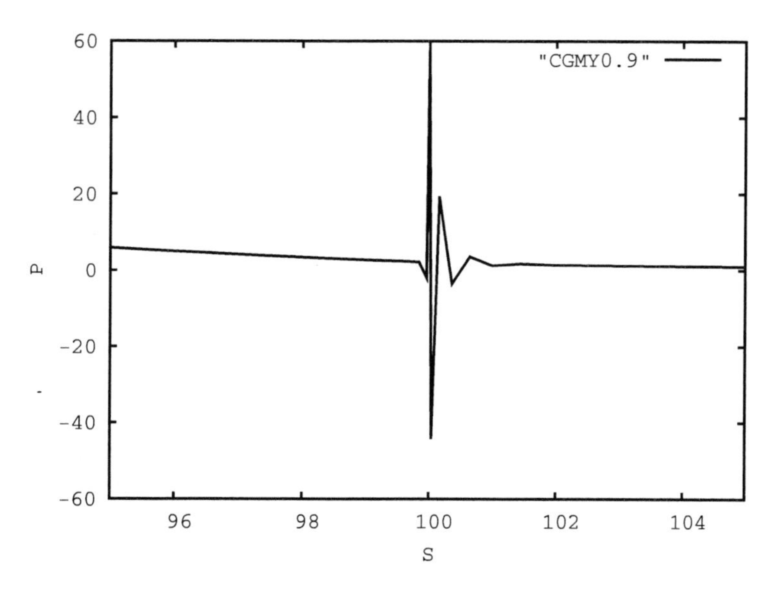

Figure 4.9. *Instabilities caused by a fully explicit treatment of the nonlocal term.*

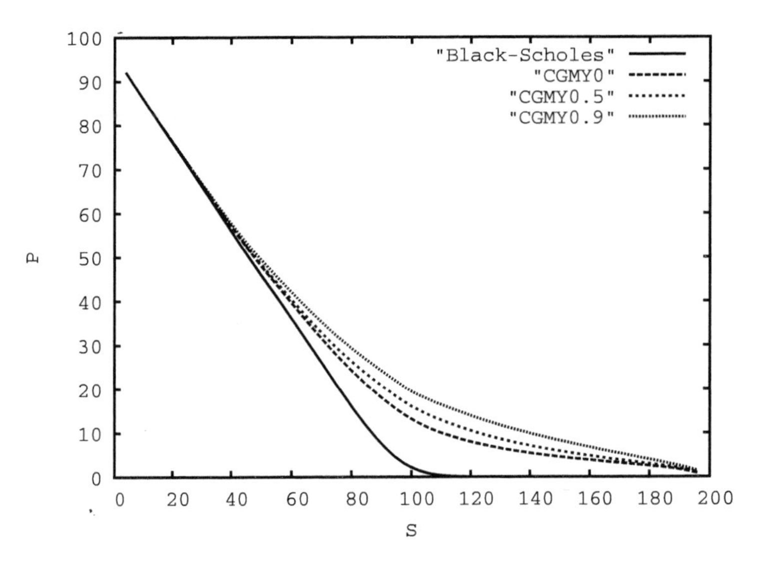

Figure 4.10. *Comparison between the price of a put computed by the Black–Scholes formula with $\sigma = 0.1$ and puts on CGMY driven assets with $\sigma = 0.1$ and $Y = 0, 0.5, 0.9$.*

4.7 Programs for Two-Dimensional Cases

4.7.1 Options on a Basket of Assets

We consider an option on a basket of two assets whose price is obtained by solving the variational problem (2.70), (2.71), where the bilinear form a is given by

$$\begin{aligned} a(v,w) = &\int_Q \frac{1}{2}\sum_{k,l=1}^{2} \Xi_{k,l} S_k S_l \frac{\partial v}{\partial S_k}\frac{\partial w}{\partial S_l} \\ &+ \int_Q \sum_{k=1}^{2}\left(-r + \sum_{l=1}^{2}\frac{1}{2}\Xi_{k,l}\right) S_k \frac{\partial v}{\partial S_k} w + r\int_Q vw, \end{aligned} \tag{4.37}$$

where the domain Q is the rectangle $(0, \bar{S}_1) \times (0, \bar{S}_2)$. Let a triangulation of Q be obtained by three families of parallel lines (like a finite difference mesh where the mesh rectangles are divided into two triangles using always the same diagonal). Using the basis $(w^i)_{i\le N}$ of the shape functions introduced in §4.2.5 (N is the number of vertices in Q), and a lexicographic ordering of the degrees of freedom, the matrix $\mathbf{A}$ has generally nine nonzero diagonals (it may be pentadiagonal in the case when $\Xi_{12} = \Xi_{21} = 0$). The bandwidth, i.e., $\max\{|i-j| : \mathbf{A}_{ij} \neq 0\}$, is of the order of $\sqrt{N}$. A direct solution of the system of linear equations by the Gauss factorization algorithm is not the best method when N is large; it is better to use a biconjugate or GMRES method, or simply a conjugate gradient method [69] if the drift term has been treated explicitly in the time discretization; see [81, 106, 107, 113] for descriptions of advanced iterative methods for systems of linear equations; note also that for a more general mesh, the matrices $\mathbf{M}$ and $\mathbf{A}$ are not pentadiagonal, but they are sparse. The iterative methods do not need the matrix $\mathbf{A}$ but only a function which implements $U \to \mathbf{A}U$, i.e., which computes $a(\sum_j u_j w^j, w^i)$, if $U = (u_1, \dots, u_N)^T$.

To compute $\mathbf{A}U$, we use the fact that

$$\mathbf{A}U = \sum_K \mathbf{A}^K U,$$

where $\mathbf{A}^K U$ is the vector whose entries are $a_K(\sum_j u_j w^j), w^i, i = 1, \dots, N$, and where

$$\begin{aligned} a_K(v, w) = &\int_K \frac{1}{2} \sum_{k,l=1}^{2} \Xi_{k,l} S_k S_l \frac{\partial v}{\partial S_k} \frac{\partial w}{\partial S_l} \\ &+ \int_K \sum_{k=1}^{2} \left(-r + \sum_{l=1}^{2} \frac{1}{2} \Xi_{k,l} \right) S_k \frac{\partial v}{\partial S_k} w + r \int_K vw. \end{aligned} \tag{4.38}$$

Hence

$$(\mathbf{A}U)_i = \sum_j u_j \sum_K \mathbf{A}^K_{ij}. \tag{4.39}$$

We shall also use (4.17) and (4.19).

For simplicity only, let us consider only the first term in (4.38), so a_K becomes

$$a_K(v, w) = \frac{1}{2} \int_K \sum_{k,l=1}^{2} \Xi_{k,l} S_k S_l \frac{\partial v}{\partial S_k} \frac{\partial w}{\partial S_l},$$

and

$$\mathbf{A}^K_{i,j} = \frac{1}{2} \int_K \sum_{k,l=1}^{2} \Xi_{k,l} S_k S_l \frac{\partial w^i}{\partial S_k} \frac{\partial w^j}{\partial S_l}.$$

But ∇w^i is constant on K and $S_k = \sum_{\nu=1}^{3} S_{k,\nu} \lambda^K_\nu$, so from (4.19) and (4.20),

$$\begin{aligned} \mathbf{A}^K_{i,j} &= \frac{1}{2} \sum_{k,l=1}^{2} \Xi_{k,l} \frac{\partial w^i}{\partial S_k} \frac{\partial w^j}{\partial S_l} \sum_{\nu_1=1}^{3} \sum_{\nu_2=1}^{3} S_{k,\nu_1} S_{l,\nu_2} \int_K \lambda^K_{\nu_1} \lambda^K_{\nu_2} \\ &= \frac{|K|}{24} \sum_{k,l=1}^{2} \Xi_{k,l} \frac{\partial w^i}{\partial S_k} \frac{\partial w^j}{\partial S_l} \sum_{\nu_1=1}^{3} \sum_{\nu_2=1}^{3} S_{k,\nu_1} S_{l,\nu_2} (1 + \delta_{\nu_1 \nu_2}). \end{aligned} \tag{4.40}$$

The summation (4.39) should *not* be programmed directly like

$$\begin{aligned} &\text{for } i = 1, \dots, N \\ &\quad \text{for } j = 1, \dots, N \\ &\quad\quad \text{for } K \in \mathcal{T}_h \\ &\quad\quad\quad (\mathbf{A}U)_i \mathrel{+}= \mathbf{A}^K_{ij} u_j, \end{aligned} \tag{4.41}$$

because the numerical complexity will then be of order $N^2 N_T$, where N_T is the number of triangles in $\mathcal{T}_h$. One should notice that the sums commute, i.e.,

$$
\begin{aligned}
&\text{for } K \in \mathcal{T}_h\\
&\quad \text{for } j = 1, \dots, N\\
&\quad\quad \text{for } i = 1, \dots, N\\
&\quad\quad\quad (\mathbf{A}U)_i \mathrel{+}= \mathbf{A}^K_{ij} u_j,
\end{aligned}
\tag{4.42}
$$

and then see that $\mathbf{A}^K_{ij}$ is zero when q^i or q^j is not in K, so that effectively one has

$$
\begin{aligned}
&\text{for } K \in \mathcal{T}_h\\
&\quad \text{for } jloc = 1, 2, 3\\
&\quad\quad \text{for } iloc = 1, 2, 3\\
&\quad\quad\quad (\mathbf{A}U)_{i_{iloc}} \mathrel{+}= \mathbf{A}^K_{i_{iloc} i_{jloc}} u_{i_{jloc}}.
\end{aligned}
\tag{4.43}
$$

This technique is called *assembling*. The complexity is $O(N_t)$ now. It has brought up the fact that vertices of triangle K have global numbers (their positions in the array that store them) and local numbers, their positions in the triangle K, i.e., 1, 2, or 3. The notation i_{iloc} refers to the map from local to global.

Therefore, one should compute $\mathbf{A}^K_{i_{iloc} i_{jloc}}$; from (4.40) and (4.17), we have that

$$
\mathbf{A}^K_{i_{iloc}, i_{jloc}} = \frac{E^{iloc} E^{jloc}}{96|K|} \sum_{k,l=1}^{2} \Xi_{k,l} n_k^{i_{iloc}} n_l^{i_{jloc}} \sum_{\nu_1=1}^{3} \sum_{\nu_2=1}^{3} S_{k,\nu_1} S_{l,\nu_2} (1 + \delta_{\nu_1 \nu_2}),
\tag{4.44}
$$

where $n_k^{i_{iloc}}$ is the kth coordinate of $n^{i_{iloc}}$. It helps a little to note that $E^i \vec{n}^i$ is the edge $\overrightarrow{q^{i+} q^{i++}}$ rotated counterclockwise by 90° in the triangle (q^i, q^{i+}, q^{i++}) provided that the numbering is counterclockwise too:

$$
E^i \vec{n}^i = (q_2^{i++} - q_2^{i+}, q_1^{i+} - q_1^{i++})^T.
$$

We split the computation of $U \mapsto (\alpha \mathbf{M} + \mathbf{A})U$ into two parts corresponding, respectively, to the bilinear forms

$$
a_1(v, w) = \int_Q \frac{1}{2} \sum_{k,l=1}^{2} \Xi_{k,l} S_k S_l \frac{\partial v}{\partial S_k} \frac{\partial w}{\partial S_l} + r \int_Q vw
$$

and

$$
a_2(v, w) = \int_Q \sum_{k=1}^{2} \left(-r + \sum_{l=1}^{2} \frac{1}{2} \Xi_{k,l} \right) S_k \frac{\partial v}{\partial S_k} w
$$

to enable an explicit treatment of the nonsymmetric part of $\mathbf{A}$. Doing so, it is possible to use the well-known conjugate gradient algorithm [69, 81] to solve the systems of linear equations at each time step, because the matrix of the system is symmetric and positive definite. It is also possible to use a fully implicit method, but then one must choose other iterative methods like BICGSTAB [113] or GMRES [107, 106].

Thus the main part of the program will be as follows.

ALGORITHM 4.10. Two-dimensional Black–Scholes.

```
void aMul(Vec& au, Grid& g, Vec& u, double alpha, double s11, double s12,
          double s22,double a)
{
  double k11,k12,k22;
  for(int i=0; i<g.nv;i++)
    au[i] = 0;                                              //   init au
  for(int k=0; k<g.nt;k++)                         //   loop on triangles
    {
      k11=0;
      k12=0;
      k22=0;
      for (int iv=0;iv<3;iv++)
        {
          int i = g(k,iv);
          k11+= s11*pow(g.v[i].x,2);
          k12+= 2*s12*g.v[i].x*g.v[i].y;
          k22+= s22*pow(g.v[i].y,2);
          for (int jv=0;jv<3;jv++)
            {
              int j= g(k,jv);
              k11+= s11*g.v[i].x*g.v[j].x;
              k12+= s12*g.v[i].x*g.v[j].y;
              k12+= s12*g.v[i].y*g.v[j].x;
              k22+= s22*g.v[i].y*g.v[j].y;
            }
        }
      for(int iloc = 0; iloc < 3; iloc++)
        {
          int i = g(k,iloc);
          int ip = g(k,(iloc+1)%3);
          int ipp= g(k,(iloc+2)%3);

          for(int jloc = 0; jloc<3; jloc++)
            {
              int j = g(k,jloc);
              int jp = g(k,(jloc+1)%3);
              int jpp= g(k,(jloc+2)%3);
              double aijk = a*(k22*(g.v[jpp].x - g.v[jp].x)
                                  *(g.v[ipp].x - g.v[ip].x)
                                    +k11*(g.v[jpp].y - g.v[jp].y)
                                  *(g.v[ipp].y - g.v[ip].y)
                                    +k12*((g.v[jpp].x - g.v[jp].x)
                                  *(g.v[ipp].y - g.v[ip].y)
                                    + (g.v[jpp].y - g.v[jp].y)
                                  *(g.v[ipp].x - g.v[ip].x)))/g.t[k].area/96.;
              if (!g.v[i].where)
                au[i] += aijk * u[j];
            }
          if (!g.v[i].where)
            au[i] += (u[i]*2.+u[ip]+u[ipp])* g.t[k].area * alpha/ 12.;
        }

    }
}
```

```
void bMul(Vec& bu, Grid& g, Vec& u, double b1, double b2, double s11,
                         double s12, double s22)
{
  double rs1,rs2;
  double s11_,s12_,s22_;
  rs1=b1/12;
  rs2=b2/12;
  s11_=-s11/24;
  s12_=-s12/24;
  s22_=-s22/24;

  for(int i=0; i<g.nv; i++)
    bu[i] = 0;
  for(int k=0; k<g.nt;k++)
    {
      double Kgradu1=0;
      double Kgradu2=0;
      for(int iloc = 0; iloc < 3; iloc++)
        {
          int i = g(k,iloc);
          int ip = g(k,(iloc+1)%3);
          int ipp= g(k,(iloc+2)%3);

          Kgradu1+=u[i]*(g.v[ipp].y-g.v[ip].y)/2;
          Kgradu2-=u[i]*(g.v[ipp].x-g.v[ip].x)/2;
        }

      for(int iloc = 0; iloc < 3; iloc++)
        {
          int i = g(k,iloc);
          bu[i]+=rs1* Kgradu1*g.v[i].x+rs2* Kgradu2*g.v[i].y;
          bu[i]+=s11_* Kgradu1*g.v[i].x+s12_* Kgradu1*g.v[i].y;
          bu[i]+=s12_* Kgradu2*g.v[i].x+s22_* Kgradu2*g.v[i].y;
          for(int jloc = 0; jloc<3; jloc++)
            {
              int j = g(k,jloc);
              bu[i]+=rs1* Kgradu1*g.v[j].x+rs2* Kgradu2*g.v[j].y;
              bu[i]+=s11_* Kgradu1*g.v[j].x+s12_* Kgradu1*g.v[j].y;
              bu[i]+=s12_* Kgradu2*g.v[j].x+s22_* Kgradu2*g.v[j].y;
            }
        }
    }
}
```

With a fixed time step, the number of conjugate gradient iterations scales like the number of nodes. To avoid such an unpleasant behavior, one has to use a preconditioned conjugate gradient method with a good preconditioner. Although it is beyond the scope of this book, we advocate the use of multigrid or algebraic multigrid preconditioners; see [96] and references therein. In Figures 4.11 and 4.12, we plot the price of a put 0.7 years to maturity, with payoff functions given by (2.65) and (2.64).

The full program is given in the appendix (§4.10).

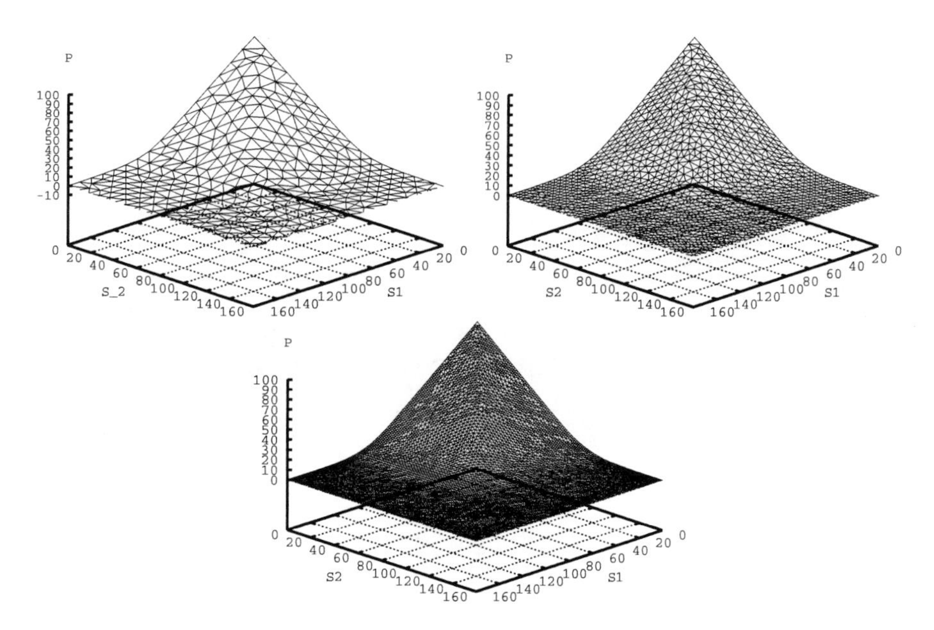

Figure 4.11. *The price of a put option on a basket of two assets* 0.7 *years to maturity. The coefficients are* $r = 0.1$, $\Xi_{11} = \Xi_{22} = 0.04$, $\Xi_{12} = -0.012$, P_0 *given by* (2.65) *with* $K = 100$ *computed with three different meshes with, respectively,* 497, 1969, *and* 7733 *nodes. The time step is* 0.01 *years.*

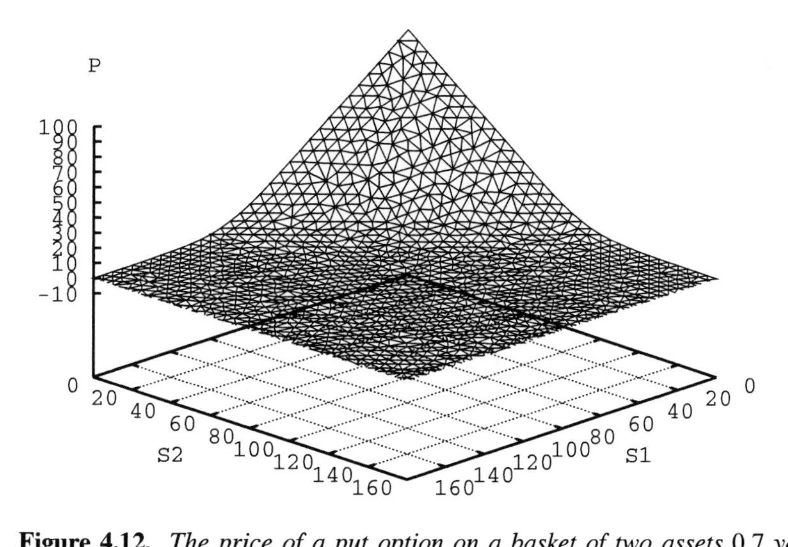

Figure 4.12. *The price of a put option on a basket of two assets* 0.7 *years to maturity. The coefficients are* $r = 0.1$, $\Xi_{11} = \Xi_{22} = 0.04$, $\Xi_{12} = -0.012$, P_0 *given by* (2.64) *with* $K = 100$ *computed with a mesh with* 1969 *nodes. The time step is* 0.01 *years.*

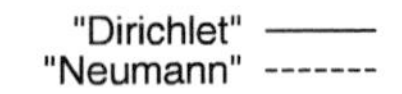

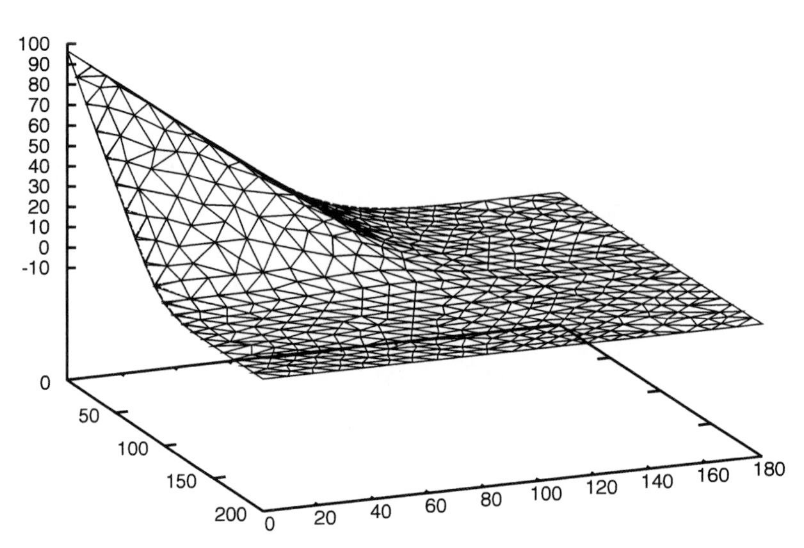

Figure 4.13. *The price of a put option on a basket of two assets computed with different boundary conditions: first the finite element method described in §4.7.1; second the same method but imposing Dirichlet conditions at $S_i = 0$", $i = 1, 2$, whose data are obtained by solving one-dimensional Black–Scholes equations. The two surfaces are indistinguishable.*

Project 4.1. *For a basket option with two assets, compare several boundary conditions. For some of them, one can use a finite difference discretization. Use a Gauss–Seidel method to solve the systems of linear equations. In Figure* 4.13, *we have plotted the price of a put option obtained with two kinds of boundary conditions.*

Project 4.2. *Test the program given in the appendix below* (§4.10), *and run it (no special boundary condition is imposed on the lines $S_1 = 0$, $S_2 = 0$). Modify the method by imposing the traces on the lines $S_1 = 0$ and $S_2 = 0$ (these trace functions are obtained by solving one-dimensional parabolic problems). Does it give better results?*

Project 4.3. *In the program given in the appendix below, replace the conjugate gradient by a biconjugate gradient, BICGSTAB, or GMRES (the source codes for these methods are freely distributed on many web sites), and treat the first order terms implicitly.*

Project 4.4. *Take a basket European put option, and make the change of variables $S_i \to S_i/(1 + S_i)$, $i = 1, 2$, which maps $(\mathbb{R}_+)^2$ onto the unit square. Modify the program given in the appendix below in order to solve the new boundary value problem and compare. A mesh adapted to the case (see Chapter 5) may be generated with freeFEM (http://www.freefem.org).*

4.7.2 A Stochastic Volatility Model

We consider the partial differential equation (2.87) obtained in §2.7. The partial differential equation is rewritten in terms of $\tau = T - t$ and in divergence form as

$$\partial_\tau P + rP - \partial_x\left(\frac{x^2y^2}{2}\partial_x P\right) - \partial_x\left(\frac{\rho\beta x|y|}{2}\partial_y P\right) - \partial_y\left(\frac{\rho\beta x|y|}{2}\partial_x P\right) - \partial_y\left(\frac{\beta^2}{2}\partial_y P\right)$$
$$-\left(rx - xy^2 - \frac{\rho\beta}{2}\frac{x|y|}{y}\right)\partial_x P - \left(\alpha m - \alpha y - \frac{\rho\beta}{2}|y|\right)\partial_y P = 0. \tag{4.45}$$

A semi-implicit in time finite difference discretization is applied:

$$\frac{P^m - P^{m-1}}{\Delta t} + rP^m - \partial_x\left(\frac{x^2y^2}{2}\partial_x P^m\right) - \partial_x\left(\frac{\rho\beta x|y|}{2}\partial_y P^m\right) - \partial_y\left(\frac{\rho\beta x|y|}{2}\partial_x P^m\right)$$
$$-\partial_y\left(\frac{\beta^2}{2}\partial_y P^m\right) - \left(rx - xy^2 - \frac{\rho\beta}{2}\frac{x|y|}{y}\right)\partial_x P^{m-1} - \left(\alpha m - \alpha y - \frac{\rho\beta}{2}|y|\right)\partial_y P^{m-1} = 0.$$

Let V_h be the space of continuous piecewise linear functions on a triangulation of the square $\Omega := (0, Lx) \times (-Ly, Ly)$ which are equal to zero on the Dirichlet boundaries of the problem. We consider the following finite element discretization:

$$\forall v_h \in V_h, \quad \int_\Omega \left(\frac{P_h^m}{\Delta t} + rP_h^m\right) v_h$$
$$+\int_\Omega \left[\frac{x^2y^2}{2}\partial_x P_h^m \partial_x v_h + \frac{\rho\beta x|y|}{2}\partial_y P_h^m \partial_x v_h + \frac{\rho\beta x|y|}{2}\partial_x P_h^m \partial_y v_h + \frac{\beta^2}{2}\partial_y P_h^m\right]\partial_y v_h$$
$$= \int_\Omega \left[\frac{P_h^{m-1}}{\Delta t} + \left(rx - xy^2 - \frac{\rho\beta}{2}\frac{x|y|}{y}\right)\partial_x P_h^{m-1} + \left(\alpha m - \alpha y - \frac{\rho\beta}{2}|y|\right)\partial_y P_h^{m-1}\right] v_h.$$

For $\rho = 0$, the aim is to approximate P in the domain $(0, \bar{S}) \times (-1.5, 1.5)$ for t smaller than 1. We choose $\bar{S} = 800$. For computing the solution, we discretize (2.87) in the larger domain $(0, 800) \times (-\bar{y}, \bar{y})$ with $\bar{y} > 1.5$. We use artificial homogeneous Dirichlet conditions on the boundaries $y = \pm\bar{y}$. These conditions, which are obviously not satisfied by P, induce nevertheless small errors on P in the smaller domain $(0, 800) \times (-1.5, 1.5)$ because the advection terms are strong near the top and bottom boundaries, and directed outward, so the bad effects of the wrong artificial boundary conditions are limited to a boundary layer near $y = \pm\bar{y}$. We take $\bar{y} = 3$. In Figure 4.14, we plot the price of the put option one year to maturity performed with the parameters

$$r = 0.05, \quad \rho = 0, \quad \alpha = 1, \quad \nu = 0.5, \quad m = 0.2, \quad K = 100, \quad \bar{S} = 800, \quad \bar{y} = 3,$$

with a time step of 6 days. There is a Neumann boundary condition on the boundary $x = \bar{S}$. No boundary condition is needed on $x = 0$ because of the degeneracy of the equation.

Remark 4.6. *The choice of $\alpha = 1$ is not quite realistic from a financial viewpoint if the asset is linked to stocks, because the mean reversion rate is generally larger. When the asset*

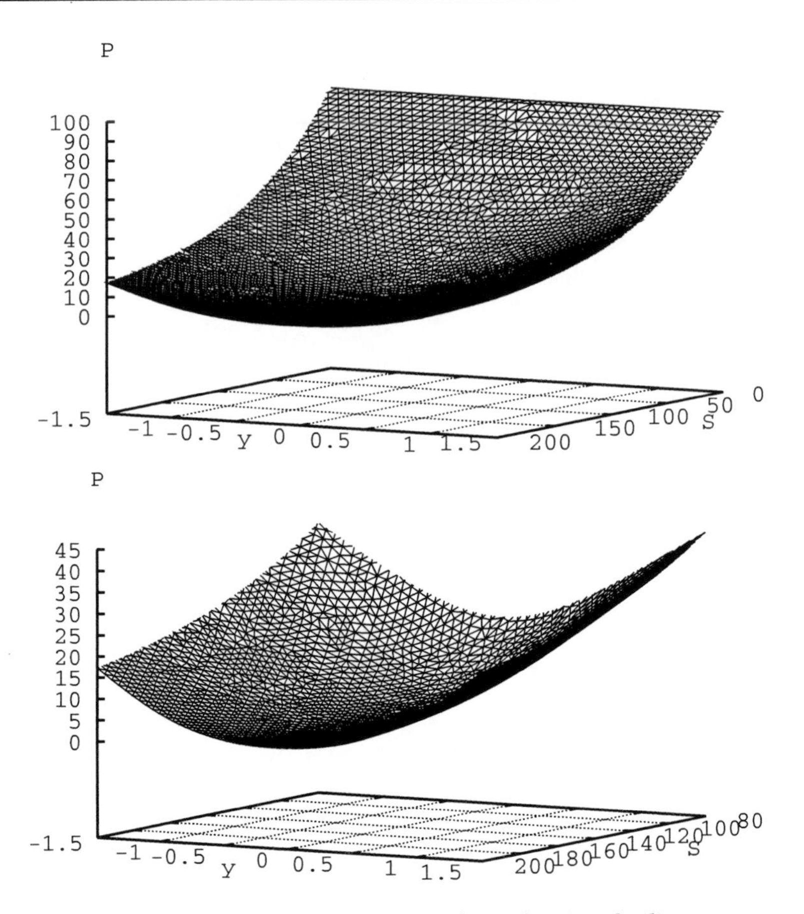

Figure 4.14. *The price of a put option with stochastic volatility one year to maturity: two views.*

corresponds to interest rates, smaller values of α are reasonable. When the mean reversion rate is large, it is possible to carry out an asymptotic expansion of the solution as in [51], *and we believe that the variational setting introduced above permits us to fully justify these expansions.*

We give below the new functions `aMul` and `bMul` used for the systems of linear equations.

ALGORITHM 4.11. Stochastic volatility.

```
void aMul2(Vec& au, Grid& g, Vec& u, double alpha, double s11,
           double s12, double s22,double a)
{
  double k11,k12,k22;
  int k,i,ip,ipp,j,jp,jpp;
  for(i=0; i<g.nv;i++)
```

```
    au[i] = 0;
  for(k=0; k<g.nt;k++)
    {
      double int_Kx2y2=0;
      double int_Kxabsy=0;
      for (int iloc=0;iloc<3;iloc++)
        {
          i= g(k,iloc);
          for (int jloc=0;jloc<3;jloc++)
            {
              j = g(k,jloc);
              for (int lloc=0;lloc<3;lloc++)
                {
                  int l = g(k,lloc);
                  for (int mloc=0;mloc<3;mloc++)
                    {
                      int m = g(k,mloc);
                      int c[3];
                      c[0]=0;
                      c[1]=0;
                      c[2]=0;
                      c[iloc]++;c[jloc]++;c[lloc]++;c[mloc]++;

                      int n1=1;int n2=1;int n3=1;int n4=1;
                      for(int p=1;p<=c[0];p++)
                        n1*=p;
                      for(int p=1;p<=c[1];p++)
                        n2*=p;
                      for(int p=1;p<=c[2];p++)
                        n3*=p;
                      for(int p=1;p<=c[0]+c[1]+c[2]+2;p++)
                        n4*=p;
                      int_Kx2y2+=double(2*n1*n2*n3)/n4*g.v[i].x*g.v[j]
                                  .x*g.v[l].y*g.v[m].y;
                    }
                }
            }
        }
                                                          //      int_Kx2y2*=2;

      for (int iloc=0;iloc<3;iloc++)
        {
          i= g(k,iloc);
          int_Kxabsy+=2*g.v[i].x*abs(g.v[i].y)/12;
          for (int jloc=iloc+1;jloc<3;jloc++)
            {
              j = g(k,jloc);
              int_Kxabsy+=(g.v[i].x*abs(g.v[j].y)+g.v[j].x*abs(g.v[i].y))/12;
            }
        }
      k11=s11*int_Kx2y2;
      k12=s12*int_Kxabsy;
      k22=s22;
      for(int iloc = 0; iloc < 3; iloc++)
        {
          i = g(k,iloc);
          ip = g(k,(iloc+1)%3);
```

```
          ipp= g(k,(iloc+2)%3);

          for(int jloc = 0; jloc<3; jloc++)
            {
              j = g(k,jloc);
              jp = g(k,(jloc+1)%3);
              jpp= g(k,(jloc+2)%3);
              double aijk =  a*(k22*(g.v[jpp].x - g.v[jp].x)
                                 *(g.v[ipp].x - g.v[ip].x)
                                   +k11*(g.v[jpp].y - g.v[jp].y)
                                 *(g.v[ipp].y - g.v[ip].y)
                                   +k12*((g.v[jpp].x - g.v[jp].x)
                                 *(g.v[ipp].y - g.v[ip].y)
                                   + (g.v[jpp].y - g.v[jp].y)*(g.v[ipp].x
                                   - g.v[ip].x))) /g.t[k].area/4.;

              if (!g.v[i].where)
                au[i] += aijk * u[j];
            }
          if (!g.v[i].where)
            au[i] += (u[i]*2.+u[ip]+u[ipp])* g.t[k].area * alpha/ 12.;
        }

    }
}


void bMul2(Vec& bu, Grid& g, Vec& u,
          double alpha, double rhobeta2, double alpham, double r)
{
  double rs1,rs2,rs3;
  rs1=r/12;
  rs2=-alpha/12;
  rs3=rhobeta2/12;
  for(int i=0; i<g.nv; i++)
    bu[i] = 0;
  for(int k=0; k<g.nt;k++)
    {
      double dxu=0;
      double dyu=0;
      for(int iloc = 0; iloc < 3; iloc++)
        {
          int i = g(k,iloc);
          int ip = g(k,(iloc+1)%3);
          int ipp= g(k,(iloc+2)%3);
          dxu-=u[i]*(g.v[ipp].y-g.v[ip].y)/2;
          dyu+=u[i]*(g.v[ipp].x-g.v[ip].x)/2;
        }

      for(int iloc = 0; iloc < 3; iloc++)
        {
          int i = g(k,iloc);
          for (int jloc=0;jloc<3;jloc++)
            {
              int j = g(k,jloc);
              for (int lloc=0;lloc<3;lloc++)
```

```
            {
              int l = g(k,lloc);
              for (int mloc=0;mloc<3;mloc++)
                {
                  int m = g(k,mloc);
                  int c[3];
                  c[0]=0;
                  c[1]=0;
                  c[2]=0;
                  c[iloc]++;c[jloc]++;c[lloc]++;c[mloc]++;

                  int n1=1;int n2=1;int n3=1;int n4=1;
                  for(int p=1;p<=c[0];p++)
                    n1*=p;
                  for(int p=1;p<=c[1];p++)
                    n2*=p;
                  for(int p=1;p<=c[2];p++)
                    n3*=p;
                  for(int p=1;p<=c[0]+c[1]+c[2]+2;p++)
                    n4*=p;
                  if (!g.v[i].where)
                    bu[i]-=dxu*double(2*n1*n2*n3)/n4
                                *g.v[j].x*g.v[l].y*g.v[m].y;
                }
            }
        }
      if (!g.v[i].where)
        {
          double signy= (g.v[i].y>0)?1:-1;
          bu[i]+=(rs1- signy*rs3 )* dxu*g.v[i].x+(rs2-signy*rs3)
                 * dyu*g.v[i].y+alpham*dyu/3;
          for(int jloc = 0; jloc<3; jloc++)
            {
              int j = g(k,jloc);
              signy= (g.v[j].y>0)?1:-1;
              if (!g.v[i].where)
                bu[i]+=(rs1- signy*rs3 )* dxu*g.v[j].x
                       +(rs2-signy*rs3)* dyu*g.v[j].y;
            }
        }
    }                                                          //  i
  }                                                       //  triangles
}
```

4.7.3 Matrix Storage: The Compressed Sparse Row Format

Since the matrices **A**, **M** do not depend on time and since the matrix-vector products have to be performed many times, it is much more efficient to compute the matrices once and for all and store them. Of course, only the nonzero entries must be stored; a popular format for sparse matrices is called the compressed sparse row format (also called Morse format):

- the nonzero entries of the matrix **A** are stored in a large vector

  ```
  double * ent_a,
  ```

  and the nonzero entries of a given row are contiguous in the vector;

- the addresses in the vector `ent_a` of the first nonzero entry of each row are stored in a vector

  ```
  int * first_in_row_a,
  ```

  whose size is the number of rows;

- the column indices corresponding to the nonzero entries are stored in a vector

  ```
  int * col_a,
  ```

  whose size matches that of `ent_a`.

The code for assembling the symmetric part of **A** and storing it in the compressed sparse row format uses the standard template library of C++, particularly the container *map*.

ALGORITHM 4.12. Compressed sparse row storage.

```
void build_a(Grid& g, double alpha, double s11,
             double s12, double s22,double a,  map<pair_int, double,
             std::less<pair_int>  > & entries)
{
  double k11,k12,k22;
  int k,i,ip,ipp,j,jp,jpp;
  if (a>0)
    {
      for(k=0; k<g.nt;k++)
        {
          double int_Kx2y2=0;
          double int_Kxabsy=0;
          for (int iloc=0;iloc<3;iloc++)
            {
              i= g(k,iloc);
              for (int jloc=0;jloc<3;jloc++)
                {
                  j = g(k,jloc);
                  for (int lloc=0;lloc<3;lloc++)
                    {
                      int l = g(k,lloc);
                      for (int mloc=0;mloc<3;mloc++)
                        {
                          int m = g(k,mloc);
                          int c[3];
                          c[0]=0;
                          c[1]=0;
                          c[2]=0;
                          c[iloc]++;c[jloc]++;c[lloc]++;c[mloc]++;
```

```
                    int n1=1;int n2=1;int n3=1;int n4=1;
                    for(int p=1;p<=c[0];p++)
                      n1*=p;
                    for(int p=1;p<=c[1];p++)
                      n2*=p;
                    for(int p=1;p<=c[2];p++)
                      n3*=p;
                    for(int p=1;p<=c[0]+c[1]+c[2]+2;p++)
                      n4*=p;
                    int_Kx2y2+=double(2*n1*n2*n3)/n4*g.v[i].x*g.v[j].x
                                *g.v[l].y*g.v[m].y;
                  }
              }
          }
      }
    for (int iloc=0;iloc<3;iloc++)
      {
        i= g(k,iloc);
        int_Kxabsy+=2*g.v[i].x*abs(g.v[i].y)/12;
        for (int jloc=iloc+1;jloc<3;jloc++)
          {
            j = g(k,jloc);
            int_Kxabsy+=(g.v[i].x*abs(g.v[j].y)+g.v[j].x
                          *abs(g.v[i].y))/12;
          }
      }
    k11=s11*int_Kx2y2;
    k12=s12*int_Kxabsy;
    k22=s22;
    for(int iloc = 0; iloc < 3; iloc++)
      {
        i = g(k,iloc);
        ip = g(k,(iloc+1)%3);
        ipp= g(k,(iloc+2)%3);
        for(int jloc = 0; jloc<3; jloc++)
          {
            j = g(k,jloc);
            jp = g(k,(jloc+1)%3);
            jpp= g(k,(jloc+2)%3);
            double aijk = a*(k22*(g.v[jpp].x - g.v[jp].x)
                          *(g.v[ipp].x - g.v[ip].x)
                            +k11*(g.v[jpp].y - g.v[jp].y)
                          *(g.v[ipp].y - g.v[ip].y)
                            +k12*((g.v[jpp].x - g.v[jp].x)
                          *(g.v[ipp].y - g.v[ip].y)
                            + (g.v[jpp].y - g.v[jp].y)*(g.v[ipp].x
                            - g.v[ip].x))) /g.t[k].area/4.;

            pair_int auxp;
            auxp[0]=i;
            auxp[1]=j;
            pair <map<pair_int,double, less<pair_int > >::iterator,
                  bool> pit;
            pair< pair_int,double> val;
            val.first=auxp;
            val.second=aijk;
```

```
                    pit=entries.insert(val);
                    if (pit.second==false)
                      (*(pit.first)).second += aijk;
                  }
              }

          }
      }
    if(alpha>0)
      for(k=0; k<g.nt;k++)
        for(int iloc = 0; iloc < 3; iloc++)
          {
            i = g(k,iloc);
            pair_int auxp;
            auxp[0]=i;
            auxp[1]=i;
            pair <map<pair_int,double, less<pair_int > >::iterator,
                  bool> pit;
            pair< pair_int,double> val;
            val.first=auxp;
            val.second=g.t[k].area * alpha/ 6.;
            pit=entries.insert(val);
            if (pit.second==false)
              (*(pit.first)).second +=  val.second;
            for (int jloc=1;jloc<3;jloc++)
              {
                ip = g(k,(iloc+jloc)%3);
                auxp[0]=i;
                auxp[1]=ip;
                val.first=auxp;
                val.second=g.t[k].area * alpha/ 12.;
                pit=entries.insert(val);
                if (pit.second==false)
                  (*(pit.first)).second +=  val.second;
              }
          }
}


int main()
{
  Grid g("mesh6_VS.msh");                             //    triangulated square
  map<pair_int, double, std::less<pair_int>  >  entries;
  for(int i=0;i<g.nv;i++)     //     a hack to put axis at Neumann conditions
    if(g.v[i].where == 2||g.v[i].where==4 )
      g.v[i].where = 0;
 ...

  double * ent_stiff_sym;
  int * col_stiff_sym;
  int * first_in_row_stiff_sym;
  int size_stiff_sym;
  int k,j;
  map<pair_int,double, less<pair_int > >::iterator it;
  build_a(g, r+1./dt,s11,s12,s22,1.,entries);
  ent_stiff_sym=new double[entries.size()];
```

```
  col_stiff_sym=new int[entries.size()];
  first_in_row_stiff_sym=new int[g.nv];

  k=-1;
  j=-1;
  for(it=entries.begin();it!=entries.end();it++)
    {
      j++;
      ent_stiff_sym[j]= (*it).second;
      col_stiff_sym[j]=  (*it).first[1];
      if ( (*it).first[0]!=k)
        {
          k++;
          first_in_row_stiff_sym[k]=j;
        }
    }
  size_stiff_sym=entries.size();
  entries.clear();

...
}
```

Assembling the mass matrix and the nonsymmetric part of **A** are done similarly. The matrix-vector product is as follows.

ALGORITHM 4.13. Matrix-vector product.

```
void Mul(Vec& bu, Grid& g, Vec& u, double * ent, int * col,
         int * first_in_raw, int & size_of_ent)
{
  int i;
  for( i=0; i<g.nv;i++)
    bu[i] = 0;
  for( i=0;i<g.nv-1;i++)
    if (!g.v[i].where)
      for(int j=first_in_raw[i];j<first_in_raw[i+1];j++)
        bu[i]+=         ent[j]*u[col[j] ];
  if (!g.v[i].where)
    for(int j=first_in_raw[i];j<size_of_ent;j++)
      bu[i]+= ent[j]*u[col[j] ];
}
```

4.8 Programming in Dimension $d > 2$

There are more and more financial products built on several assets, so the numerical solution of the Black–Scholes equation in higher dimension is a current area of research.

In three dimensions the finite element method is used intensely in engineering, so some of the engineering software can be used. For example, Figure 4.15 shows the solution obtained with `ff3d` [40] by the finite element method with quadrangles of the Black–Scholes equation for a European put with $r = 0$, $\sigma_i = 0.1 * i$, $\sigma_{ij} = -0.1$, $i, j = 1, 2, 3$, and payoff

$$P(x, y, z, T) = |x - 0.5|_+ + |y - 0.5|_+ + |z - 0.5|_+.$$

The computational domain is the intersection of the cube $(0, 1)^3$ with the unit sphere centered at 0.

4.9 High Dimensions: An Introduction to Galerkin Methods with Sparse Tensor Product Spaces

For parabolic problems in space dimensions $d > 3$, the finite element and finite difference methods fail, because they require typically $O(h^{-d})$ degrees of freedom for an accuracy of $O(h)$ in the energy norm. For this reason, a popular cliché says that only Monte-Carlo methods can be applied for high-dimensional problems. Yet quite recent developments have shown that it is possible in some cases to use deterministic Galerkin methods or grid-based methods for parabolic problems in dimensions d for $4 \le d \le 20$: these methods are based either on sparse grids [118, 65, 63] or sparse tensor product approximation spaces [64, 114]. For a recent survey on sparse grids, a good reference is [23]. Here we give a brief survey of a paper by Petersdoff and Schwab [114]. The full results contained there are of great interest but rather technical and beyond the scope of the present book.

For convenience, we restrict ourselves to sparse tensor product finite element spaces constructed with one-dimensional piecewise affine functions. The construction of the approximation space involves wavelets: for simplicity, we focus on a very simple example of wavelets.

4.9.1 Wavelets in $\mathbb{R}$

In the interval $I = [0, 1]$, we define the mesh $\mathcal{T}^\ell$ whose nodes are $x_j^\ell = j2^{-\ell-1}$, $j = 0, \dots, 2^{\ell+1}$. We define V^ℓ as the space of piecewise linear continuous functions on the mesh $\mathcal{T}^\ell$ vanishing at 0 and 1. The dimension of V^ℓ is $N^\ell = 2^{\ell+1} - 1$. We define $M^\ell = N^\ell - N^{\ell-1} = 2^\ell$. For $\ell > 0$, we define the wavelets ψ_j^ℓ, $j = 1, \dots, M^\ell$, at level ℓ, by

$$\begin{aligned}
&\forall 1 < j < M^\ell,\\
&\psi_j^\ell(x_{2j-1}^\ell) = 2c_\ell, \quad \psi_j^\ell(x_{(2j-1\pm1)}^\ell) = -c_\ell, \quad \psi_j^\ell(x_k^\ell) = 0 \quad \text{for} \begin{cases} k < 2j-1, \\ k > 2j, \end{cases}\\
&\psi_1^\ell(x_1^\ell) = 2c_\ell, \quad \psi_1^\ell(x_{(2)}^\ell) = -c_\ell, \quad \psi_1^\ell(x_k^\ell) = 0 \quad \text{for} \begin{cases} k = 0, \\ k = 3, \dots, N^\ell + 1, \end{cases}\\
&\psi_{M^\ell}^\ell(x_{N^\ell}^\ell) = 2c_\ell, \quad \psi_{M^\ell}^\ell((x_{N^\ell-1}^\ell) = -c_\ell, \quad \psi_{M^\ell}^\ell(x_k^\ell) = 0 \quad \text{for} \begin{cases} k < M^\ell, \\ k = N^\ell + 1, \end{cases}
\end{aligned} \tag{4.46}$$

where the normalization constant is chosen in order to achieve $\|\psi_j^\ell\|_{L^2} = 1$ for $1 < j < M^\ell$. The support of ψ_j^ℓ has diameter less than $2^{2-\ell}$.

The first three levels are displayed on Figure 4.16. For $\ell > 0$, we define $W^\ell = span\{\psi_j^\ell,\ 1 \le j \le M^\ell\}$ and $W^0 = V^0$: we have $V^\ell = W^\ell \oplus V^{\ell-1}$, and $V^\ell = \bigoplus_{k=0}^\ell W^k$. Each function v of V^ℓ has the representation $v = \sum_{k=0}^\ell \sum_{j=1}^{M^k} v_j^k \psi_j^k$, and it is easy to check that most of the wavelets belonging to levels k and ℓ with $|k - \ell| > 1$ are orthogonal for the L^2 and H^1 scalar products.

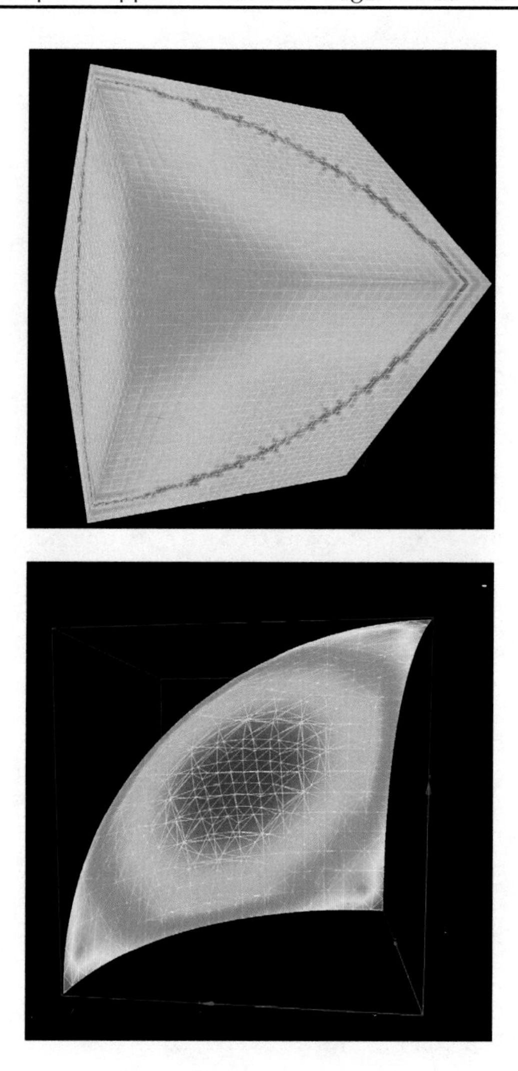

Figure 4.15. *A put option built on three underlying assets. On the top the payoff and the computational domain projected on a cube. On the bottom the solution on the sphere* $B(0, 0.5)$.

The following properties can also be checked:

$$
\begin{aligned}
c\|v\|^2_{L^2(I)} &\le \sum_{k=0}^{\ell}\sum_{j=1}^{M^k} |v^k_j|^2 \le C\|v\|^2_{L^2(I)},\\
c|v|^2_{H^1(I)} &\le \sum_{k=0}^{\ell}\sum_{j=1}^{M^k} 2^{2k}|v^k_j|^2 \le C|v|^2_{H^1(I)},
\end{aligned}
\tag{4.47}
$$

with c and C independent of ℓ.

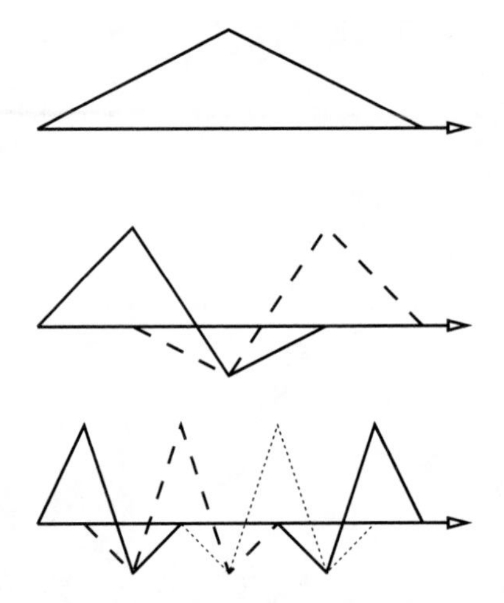

Figure 4.16. *The wavelet basis: the first three levels.*

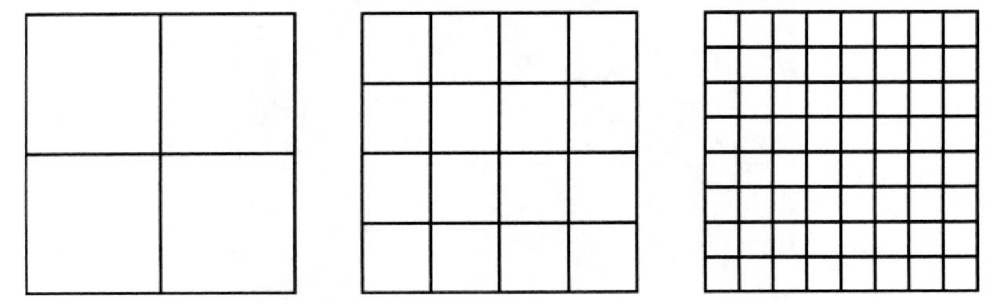

Figure 4.17. *A representation of the full tensor product space for $d = 2$ (for $\ell = 0, 1, 2$).*

4.9.2 Sparse Tensor Product Spaces

For a parabolic equation in $\Omega = (0, 1)^d$, a natural choice for a discrete space is the tensor product space $\tilde{V}_h = V^\ell \otimes \cdots \otimes V^\ell = \sum_{0 \le k_i \le \ell} W^{k_1} \otimes \cdots \otimes W^{k_d}$. The dimension of $\tilde{V}_h$ is $(2^{\ell+1} + 1)^d$, so it grows very rapidly with d. We shall use instead the sparse tensor product space $V_h = \sum_{k_1 + \cdots + k_d \le \ell} W^{k_1} \otimes \cdots \otimes W^{k_d}$, whose dimension is $O(\ell^d 2^\ell)$. The space V_h is considerably smaller than $\tilde{V}_h$ and can be used for practical computations for $d \le 20$.

A schematic representation of the spaces $\tilde{V}_h$ and V_h are displayed in Figures 4.17 and 4.18.

Consider the discretization of an elliptic Dirichlet problem in Ω: the discretization error of the Galerkin method with the space $\tilde{V}_h$ (resp., V_h) is of the same order as the best fit error when approximating the solution of the continuous problem by a function of $\tilde{V}_h$ (resp., V_h). We know that $\inf_{v_h \in \tilde{V}_h} \|v - v_h\|_{H^1(\Omega)} \le Ch|v|_{H^2(\Omega)}$, where $h = 2^{-\ell}$, and $|v|^2_{H^2(\Omega)} = \sum_{k_1 + \cdots + k_d = 2} \|\frac{\partial^2 v}{\partial x_1^{k_1} \ldots \partial x_d^{k_d}}\|^2_{L^2(\Omega)}$. Since V_h is much smaller than $\tilde{V}_h$ a similar estimate is not true. However, the following estimate has been proved (see [64, 114]):

$$\inf_{v_h \in V_h} \|v - v_h\|_{H^1(\Omega)} \le Ch|v|_{\mathcal{H}^2(\Omega)},$$

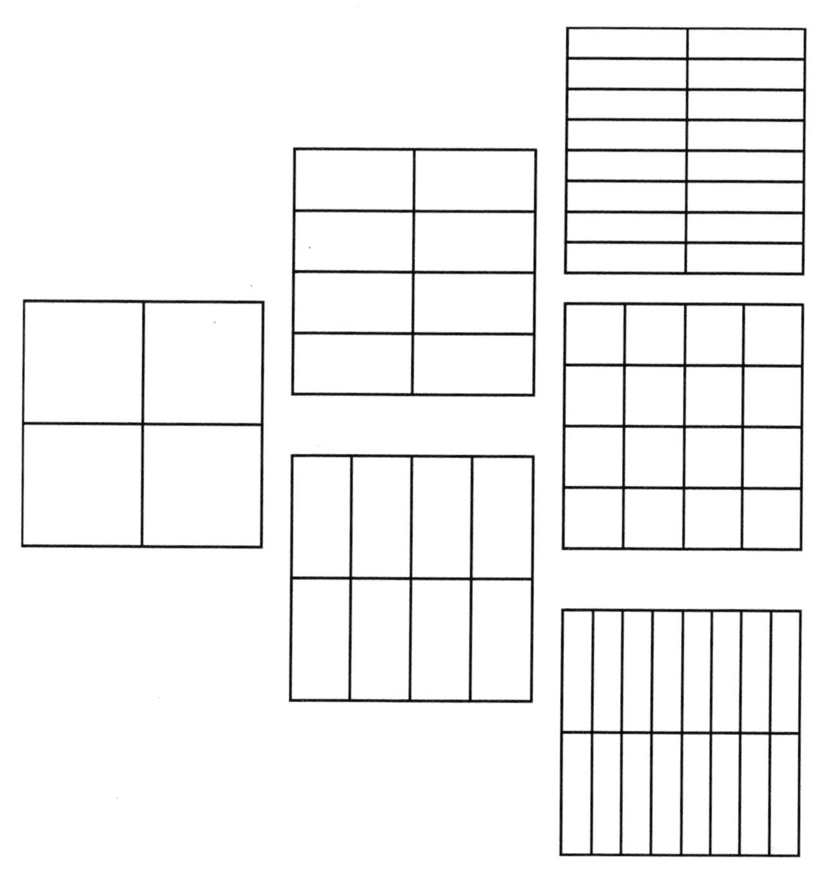

Figure 4.18. *A representation of the sparse tensor product space for $d = 2$ (for $\ell = 0, 1, 2$).*

where

$$\mathcal{H}^2(\Omega) = \left\{ v, \frac{\partial^{k_1+\cdots+k_d} v}{\partial x_1^{k_1} \ldots \partial x_d^{k_d}} \in L^2(\Omega), 0 \le k_i \le 2, 1 \le i \le d \right\}$$

is endowed with its natural norm and seminorm. We see that the Galerkin method with the space V_h converges linearly in h provided the solution to the continuous problem belongs to $\mathcal{H}^2(\Omega)$ (which is a much smaller space than $H^2(\Omega)$).

When dealing with a homogeneous parabolic problem such as (2.61) with smooth coefficients, Petersdoff and Schwab [114] made use of the smoothing property of the parabolic operator and designed a method based on sparse tensor product spaces as above, even if the Cauchy data are not smooth: indeed, for any Cauchy data, the solution to the problem belongs to $\mathcal{H}^2(\Omega)$ for all positive time, so the sparse tensor product space will be large enough to approximate the solution at $t > 0$ in an optimal way. Yet, when the Cauchy data are not smooth enough, the $\mathcal{H}^2$-norm of the solution blows up as $t \to 0$. This is the case for options with nonsmooth payoff functions. To compensate for this, Petersdoff and Schwab [114] proposed using a time stepping with a very nonuniform time grid suitably refined near $t = 0$. This yields a sophisticated method which enables one to price basket options with up to twenty assets. Similar ideas have been used for pricing options under stochastic volatility in [71].

4.10 Appendix: The Full Program for Two-Dimensional Black–Scholes

ALGORITHM 4.14. Two-dimensional Black–Scholes.

```
                                                    //    file :  BS2DfemCG.cpp
#include <iostream>
#include <fstream>
#include <math.h>
#include <cmath>
#include <stdlib.h>
#include <assert.h>
#include "RNM.hpp"
#define NDEBUG                            //    uncomment when debugging is over
using namespace std;

typedef KN<double> Vec;

class Vertex
{
public:
  double x, y;                                                  //    coordinates
  int where;                                              //    on which boundary
};

class Triangle
{
public:
  Vertex* v[3];                              //    the 3 vertices of the triangle
  int where;                                                //    in which region
  double area;
};

class Grid
{ public:
  int nt, nv;                          //    nb of triangles,vertices, and edges
  KN<Vertex> v;                                                //    all vertices
  KN<Triangle> t;                                             //    all triangles
  Grid(const char *path );        //    reads a triangulation in freeFEM format
  int no(Triangle* tt) const { return tt-( Triangle*)t;} //    the place in
                                                //    array t of triangle tt
  int no(Vertex* tt) const { return tt-( Vertex*)v;}      //    the place in
                                               //    array v of Vertex tt 32
  int operator()(int k,int iloc) const { return no(t[k].v[iloc]);}
                                                     //    same as no(vertex)
};

Grid::Grid(const char *path ):v(),t()
{                                 //    reads a triangulation in freeFEM format
  int i0,i1,i2;
```

```
  ifstream file(path);
  if(!file) cout<<"can t find triangulation file"<<endl;
  file >> nv >> nt;
  v.init(nv);
  t.init(nt);
  for(int i=0; i<nv; i++ )
    file >> v[i].x >> v[i].y >> v[i].where;
  for(int i=0; i<nt; i++ )
    {
      file >> i0 >> i1 >> i2 >> t[i].where;
      t[i].v[0] = &v[i0-1];
      t[i].v[1] = &v[i1-1];
      t[i].v[2] = &v[i2-1];
      t[i].area = ((t[i].v[1]->x - t[i].v[0]->x) * (t[i].v[2]->y
                   - t[i].v[0]->y) -
                   (t[i].v[2]->x - t[i].v[0]->x) * (t[i].v[1]->y
                   - t[i].v[0]->y))/2;
    }
}

void bMul(Vec& bu, Grid& g, Vec& u, double alpha, double b1, double b2,
          double s11, double s12, double s22)
{
  double rs1,rs2;
  double s11_,s12_,s22_;
  rs1=b1/12;
  rs2=b2/12;
  s11_=-s11/24;
  s12_=-s12/24;
  s22_=-s22/24;

  for(int i=0; i<g.nv; i++)
    bu[i] = 0;
  for(int k=0; k<g.nt;k++)
    {
      double Kgradu1=0;
      double Kgradu2=0;
      for(int iloc = 0; iloc < 3; iloc++)
        {
          int i = g(k,iloc);
          int ip = g(k,(iloc+1)%3);
          int ipp= g(k,(iloc+2)%3);

          Kgradu1+=u[i]*(g.v[ipp].y-g.v[ip].y)/2;
          Kgradu2-=u[i]*(g.v[ipp].x-g.v[ip].x)/2;
        }

      for(int iloc = 0; iloc < 3; iloc++)
        {
          int i = g(k,iloc);
          bu[i]+=rs1* Kgradu1*g.v[i].x+rs2* Kgradu2*g.v[i].y;
          bu[i]+=s11_* Kgradu1*g.v[i].x+s12_* Kgradu1*g.v[i].y;
          bu[i]+=s12_* Kgradu2*g.v[i].x+s22_* Kgradu2*g.v[i].y;
          for(int jloc = 0; jloc<3; jloc++)
            {
              int j = g(k,jloc);
              bu[i]+=rs1* Kgradu1*g.v[j].x+rs2* Kgradu2*g.v[j].y;
```

```
                bu[i]+=s11_* Kgradu1*g.v[j].x+s12_* Kgradu1*g.v[j].y;
                bu[i]+=s12_* Kgradu2*g.v[j].x+s22_* Kgradu2*g.v[j].y;
              }
          }
      }
}

void aMul(Vec& au, Grid& g, Vec& u, double alpha, double s11, double s12,
          double s22,double a)
{
  double k11,k12,k22;
  for(int i=0; i<g.nv;i++)
    au[i] = 0;                                                  //   init au
  for(int k=0; k<g.nt;k++)                                 //   loop on triangles
    {
      k11=0;
      k12=0;
      k22=0;
      for (int iv=0;iv<3;iv++)
        {
          int i = g(k,iv);
          k11+= s11*pow(g.v[i].x,2);
          k12+= 2*s12*g.v[i].x*g.v[i].y;
          k22+= s22*pow(g.v[i].y,2);
          for (int jv=0;jv<3;jv++)
            {
              int j= g(k,jv);
              k11+= s11*g.v[i].x*g.v[j].x;
              k12+= s12*g.v[i].x*g.v[j].y;
              k12+= s12*g.v[i].y*g.v[j].x;
              k22+= s22*g.v[i].y*g.v[j].y;
            }
        }
      for(int iloc = 0; iloc < 3; iloc++)
        {
          int i = g(k,iloc);
          int ip = g(k,(iloc+1)%3);
          int ipp = g(k,(iloc+2)%3);

          for(int jloc = 0; jloc<3; jloc++)
            {
              int j = g(k,jloc);
              int jp = g(k,(jloc+1)%3);
              int jpp = g(k,(jloc+2)%3);
              double aijk = a*(k22*(g.v[jpp].x - g.v[jp].x)
                                  *(g.v[ipp].x - g.v[ip].x)
                                     +k11*(g.v[jpp].y - g.v[jp].y)
                                  *(g.v[ipp].y - g.v[ip].y)
                                     +k12*((g.v[jpp].x - g.v[jp].x)
                                  *(g.v[ipp].y - g.v[ip].y)
                                     + (g.v[jpp].y - g.v[jp].y)
                                  *(g.v[ipp].x - g.v[ip].x))) /g.t[k].area/96.;
              if (!g.v[i].where)
                au[i] += aijk * u[j];
            }
          if (!g.v[i].where)
            au[i] += (u[i]*2.+u[ip]+u[ipp])* g.t[k].area * alpha/ 12.;
```

```
        }

    }
}

void solvecg(Grid& g, Vec& f, Vec& u, int nIter, double precise,
             double alpha, double s11,double s12,double s22)
{
  int nv = g.nv;
  Vec au(nv), ag(nv), grad(nv), hh(nv), diag(nv);
  double normOldGrad = 1e60;

  for(int m=0; m<nIter ; m++)
    {
      aMul(au, g, u, alpha,s11,s12,s22,1);

      double normGrad = 0;
      for(int i=0;i<nv; i++)
        if(!g.v[i].where)
           {
           grad[i] = (au[i] - f[i]);
           normGrad += pow(grad[i],2);
         }

      double  gh =0, gamma = normGrad / normOldGrad;
      normOldGrad = normGrad;
      for(int i=0;i<nv; i++)
        if(!g.v[i].where)
           {
           hh[i] = gamma * hh[i] - grad[i];
           gh += grad[i] * hh[i];
         }

      aMul(ag,g,hh,alpha,s11,s12,s22,1);
      double rho = 0;
      for(int i=0;i<nv; i++)
        if(!g.v[i].where) rho += hh[i] * ag[i];
      rho = - gh / rho ;

      for(int i=0;i<nv; i++)
        if(!g.v[i].where) u[i] += rho * hh[i];

      if(m==0)  precise = normGrad * pow(precise,2);
      if(normGrad < precise)
        {          cout << "        nb iter=" <<m<<"
normGrad = " <<normGrad<< endl;
        return;
        }
    }
}

void myexit() { cout<<"program ended at myexit()"<<endl;}

int main()
{
  atexit(myexit);                                   //    for debugging
  Grid g("mesh2.msh");                          //    triangulated square
```

```
  for(int i=0;i<g.nv;i++)             //    a hack to have Neumann conditions
    if(g.v[i].where != 0) g.v[i].where = 0;

  const double T =0.7,                                    //     financial data
    r=0.05,
    K1=100,
    K2=100,
    s1=0.2,
    s2=0.2,
    s11=s1*s1,
    s22=s2*s2,
    s12=-s1*s2*0.3;



  const int itermax=70;
  double dt = T/itermax;
  double t=0;
  Vec u0(g.nv), u1(g.nv);
  Vec f(g.nv),x(g.nv),f1(g.nv),f2(g.nv);

  for(int i=0; i<g.nv;i++)
    {                                                //    set payoff at maturity
      double a = (g.v[i].x > g.v[i].y )? K1-g.v[i].x : K2-g.v[i].y ;
      u0[i] = a>0 ?  a : 0;
      u1[i]=0;
    }

  for(int timeIter=0; timeIter < itermax; timeIter++)         //     time loop
    {
      t+=dt;
      aMul(f2,g,u0,1./dt,0.,0.,0.,0);                        //    the mass matrix
      bMul(f1,g,u0,0,r,r,s11,s12,s22);      //    the nonsymmetric part of the
                                                          //     stiffness matrix
      for(int i=0;i<g.nv;i++)
        f[i]=f1[i]+f2[i];                     //   add the two contributions

      cout<<"timeiter = "<<timeIter+1<<"          temps = "<<t<<'\t';
      solvecg(g,f, u1,200, 1e-5, r+1./dt,s11,s12,s22);       //    solve linear
                                                            //    system by cg

      for(int i=0; i<g.nv;i++)                                     //    update
        u0[i] =u1[i];
    }
  ofstream plot("plot2");
  for(int it=0;it<g.nt;it++)
    plot <<g.v[g(it,0)].x <<" "<<g.v[g(it,0)].y << " " << u0[g(it,0)] << endl
         <<g.v[g(it,1)].x <<" "<<g.v[g(it,1)].y << " " <<u0[g(it,1)]  << endl
         <<g.v[g(it,2)].x <<" "<<g.v[g(it,2)].y << " " << u0[g(it,2)] << endl
         <<g.v[g(it,0)].x <<" "<<g.v[g(it,0)].y << " " << u0[g(it,0)] << endl
         <<endl<<endl;
  return 0;
}
```

Chapter 5

Adaptive Mesh Refinement

This chapter is devoted to automatic mesh refinements with criteria based on a posteriori estimates of the finite element discretization errors of the Black–Scholes equation.

The main idea consists of finding local error indicators which can be computed explicitly from the solution of the discrete problem, and such that their Hilbertian sum is equivalent to the global error. These indicators are said to be optimal if the constants of the norm-equivalence inequalities are independent of the error. Moreover, since they are local, they provide a good representation of the error distribution.

The result (Theorem 5.6) leads to a numerical method which puts the discretization nodes where they are needed; for a given accuracy, the method is fast because it has fewer unknowns than with a uniform mesh. For example, this may be important for calibration problems where the Black–Scholes equation is solved a large number of times.

This chapter uses many of the technicalities of the finite element method and may be difficult for nonspecialists.

We have chosen to follow the same strategy as in the enlightening paper by Bernardi, Bergam, and Mghazli [14]. Therefore we need a finite element method with a mesh in the variable S that can vary in time, and this chapter provides such a tool. We consider two families of error indicators, both of residual type. The first family is global with respect to the price variable and local with respect to time: it gives relevant information in order to refine the mesh in time. The second family is local with respect to both price and time variables, and provides an efficient tool for mesh adaption in the price variable at each time step.

Other approaches for mesh adaption for parabolic problems and finite element methods are available in, e.g., [45, 46, 47].

This chapter is rather technical but essentially self-contained; to this end, there are some repetitions concerning Sobolev spaces and variational methods and other crucial notions for constructing the error indicators. The technical proofs are all given separately in the appendix at the end of this chapter. They can of course be skipped.

5.1 The Black–Scholes Equation and Some Discretizations

5.1.1 The Black–Scholes Equation and Its Variational Formulation

We consider the Black–Scholes equation for a European put with a local volatility σ:

$$\begin{aligned} \frac{\partial u}{\partial t} - \frac{\sigma^2 S^2}{2}\frac{\partial^2 u}{\partial S^2} - rS\frac{\partial u}{\partial S} + ru &= 0 && \text{in } \mathbb{R}_+ \times (0, T], \\ u_{|t=0} &= u_0 && \text{in } \mathbb{R}_+. \end{aligned} \tag{5.1}$$

Here t is the time to maturity and S is the price of the underlying asset. The volatility σ is a function of S and t and the interest rate r is a function of t. For a vanilla put, the payoff function is

$$u_0(S) = (S - k)_-, \tag{5.2}$$

where K is the strike. What follows can be generalized to any payoff function vanishing for S large enough.

To simplify the discussion, we assume that the volatility $\sigma(S, t)$ and the interest rate $r(t)$ are smooth functions. We need to make the following assumptions: there exist constants $0 < \sigma_{\min} \le \sigma_{\max}$, $0 < C_\sigma$, and $R \ge 0$ such that

$$\sigma_{\min} \le \sigma(S, t) \le \sigma_{\max} \quad \text{in } \mathbb{R}_+ \times [0, T], \tag{5.3}$$

$$\left| S\frac{\partial \sigma}{\partial S} \right| \le C_\sigma \quad \text{in } \mathbb{R}_+ \times [0, T], \tag{5.4}$$

$$0 \le r(t) \le R \quad \text{in } [0, T]. \tag{5.5}$$

For the purpose of discretization, we truncate the domain in the variable S: we introduce a large constant $\bar{S}$, and instead of (5.1), we consider

$$\begin{aligned} \frac{\partial u}{\partial t} - \frac{\sigma^2 S^2}{2}\frac{\partial^2 u}{\partial S^2} - rS\frac{\partial u}{\partial S} + ru &= 0 && \text{in } \Omega \times (0, T], \\ u_{|S=\bar{S}} &= 0 && \text{in } (0, T], \\ u_{|t=0} &= u_0 && \text{in } \Omega, \end{aligned} \tag{5.6}$$

where $\Omega = (0, \bar{S})$.

In what follows, we use the space $L^2(\Omega)$ of square integrable functions on Ω. We denote by $(\cdot, \cdot)$ the inner product in $L^2(\Omega)$ and by $\|\cdot\|$ the associated norm. We introduce the weighted Sobolev space V:

$$V = \left\{ v : \; v \in L^2(\Omega), \, S\frac{\partial v}{\partial S} \in L^2(\Omega) \right\}. \tag{5.7}$$

Endowed with the inner product and norm

$$(v, w)_V = \int_\Omega \left(v(S)w(S) + S^2 \frac{\partial v}{\partial S}(S)\frac{\partial w}{\partial S}(S) \right) dS, \quad \|v\|_V = (v, v)_V^{\frac{1}{2}}, \tag{5.8}$$

V is a Hilbert space. This space has the following properties:

1. V is separable.
2. Denoting by $\mathcal{D}(\mathbb{R}_+)$ the space of infinitely differentiable functions with compact support in $\mathbb{R}_+$, and by $\mathcal{D}(\bar{\Omega})$ the space containing the restrictions of the functions of $\mathcal{D}(\mathbb{R}_+)$ to Ω, $\mathcal{D}(\bar{\Omega})$ is densely embedded in V.
3. V is densely embedded in $L^2(\Omega)$.
4. The seminorm

$$|v|_V = \left(\int_\Omega S^2 \left(\frac{\partial v}{\partial S}(S) \right)^2 dS \right)^{\frac{1}{2}} \tag{5.9}$$

   is in fact a norm in V, equivalent to $\|\cdot\|_V$; more precisely, we have the following Hardy inequality: for all $v \in V$,

$$\|v\|_{L^2(\Omega)} \le 2|v|_V. \tag{5.10}$$

Denoting by $\mathcal{D}(\Omega)$ the space of infinitely differentiable functions with compact support in Ω, we define V_0 as the closure of $\mathcal{D}(\Omega)$ in V. It is easy to prove that V_0 is the subspace of V containing the functions vanishing at $\bar{S}$. For simplicity, we also denote by $(\cdot, \cdot)$ the duality pairing between V_0', the dual space of V_0, and V_0, and we define $\|\cdot\|_{V_0'}$ by

$$\|w\|_{V_0'} = \sup_{v \in V_0} \frac{(w, v)}{|v|_V}. \tag{5.11}$$

We define $C^0([0,T]; L^2(\Omega))$ as the space of continuous functions with values in $L^2(\Omega)$, and $L^2(0,T; V_0)$ as the space of square integrable functions with values in V_0.

We call $a_t(v, w)$ the bilinear form:

$$a_t(v, w) = \left(\frac{\sigma^2}{2} S \frac{\partial u}{\partial S}, S \frac{\partial v}{\partial S} \right) + \left(\left(-r + \sigma^2 + S\sigma \frac{\partial \sigma}{\partial S} \right) S \frac{\partial u}{\partial S}, v \right) + r\,(u, v). \tag{5.12}$$

It is clear from the assumptions above on r and σ that for all $t \in [0, T]$, a_t is a continuous bilinear form on $V \times V$. Let μ be the best positive constant such that, for all $v, w \in V_0$,

$$|a_t(v, w)| \le \mu |v|_V |w|_V. \tag{5.13}$$

The boundary value problem (5.6) has the following equivalent variational formulation (see [90]):

Find $u \in C^0([0,T]; L^2(\Omega)) \cap L^2(0,T; V_0)$ satisfying

$$u_{|t=0} = u_0 \quad \text{in } \Omega, \tag{5.14}$$

$$\text{for a.e. } t \in (0,T), \quad \forall v \in V_0, \quad \left(\frac{\partial u}{\partial t}(t), v \right) + a_t(u(t), v) = 0. \tag{5.15}$$

From the assumptions above on r and σ, we have the following Gårding inequality.

Lemma 5.1 (Gårding's inequality). *There exists a nonnegative constant λ such that*

$$\forall t \in [0,T], \ \forall v \in V_0, \qquad a_t(v, v) \ge \frac{1}{4} \sigma_{\min}^2 |v|_V^2 - \lambda \|v\|^2. \tag{5.16}$$

Proof. Take, for example,

$$\lambda = \max\left(0, \frac{1}{\sigma_{\min}^2}(R + \sigma_{\max}^2 + C_\sigma \sigma_{\max})^2 - \inf_{t\in(0,T)} r(t)\right). \qquad \square$$

Using Lemma 5.1 and abstract results due to Lions and Magenes [90], it is possible to prove that the problem (5.14), (5.15) admits a unique solution. Moreover, introducing the norm

$$[[v]](t) = \left(e^{-2\lambda t}\|v(t)\|^2 + \frac{1}{2}\sigma_{\min}^2 \int_0^t e^{-2\lambda\tau}|v(\tau)|_V^2 d\tau\right)^{\frac{1}{2}}, \tag{5.17}$$

we have, by taking v equal to $u(t)e^{-2\lambda t}$ in (5.15) and integrating in time,

$$[[u]](t) \le \|u_0\|. \tag{5.18}$$

From this, we deduce that

$$\left\| e^{-\lambda t}\frac{\partial u}{\partial t}\right\|_{L^2(0,T;V_0')} \le \sqrt{2}\frac{\mu}{\sigma_{\min}}\|u_0\|. \tag{5.19}$$

5.1.2 The Time Semidiscrete Problem

We introduce a partition of the interval $[0, T]$ into subintervals $[t_{n-1}, t_n]$, $1 \le n \le N$, such that $0 = t_0 < t_1 < \cdots < t_N = T$. We denote by Δt_n the length $t_n - t_{n-1}$, and by Δt the maximum of the Δt_n, $1 \le n \le N$. We also define the regularity parameter $\rho_{\Delta t}$:

$$\rho_{\Delta t} = \max_{2\le n\le N}\frac{\Delta t_n}{\Delta t_{n-1}}. \tag{5.20}$$

For a continuous function f on $[0, T]$, we introduce the notation $f^n = f(t_n)$. The semidiscrete problem arising from an implicit Euler scheme is the following:

Find $(u^n)_{0\le n\le N} \in L^2(\Omega) \times V_0^N$ satisfying

$$u^0 = u_0, \tag{5.21}$$

$$\forall n, 1 \le n \le N, \quad \forall v \in V_0, \quad \left(u^n - u^{n-1}, v\right) + \Delta t_n a_{t_n}(u^n, v) = 0. \tag{5.22}$$

For Δt smaller than $1/(2\lambda)$, the existence and uniqueness of $(u^n)_{0\le n\le N}$ is a consequence of the Lax–Milgram lemma. We call $u_{\Delta t}$ the function which is affine on each interval $[t_{n-1}, t_n]$, and such that $u_{\Delta t}(t_n) = u^n$.

From the standard identity $(a - b, a) = \frac{1}{2}|a|^2 + \frac{1}{2}|a - b|^2 - \frac{1}{2}|b|^2$, a few calculations show that

$$(1 - 2\lambda\Delta t_n)\|u^n\|^2 + \frac{1}{2}\Delta t_n \sigma_{\min}^2 |u^n|_V^2 \le \|u^{n-1}\|^2. \tag{5.23}$$

Multiplying equation (5.23) by $\prod_{i=1}^{n-1}(1 - 2\lambda\Delta t_i)$ and summing the equations on n, we obtain

$$\left(\prod_{i=1}^{n}(1 - 2\lambda\Delta t_i)\right)\|u^n\|^2 + \frac{1}{2}\sigma_{\min}^2\sum_{m=1}^{n}\Delta t_m\left(\prod_{i=1}^{m-1}(1 - 2\lambda\Delta t_i)\right)|u^m|_V^2 \le \|u^0\|^2. \tag{5.24}$$

Introducing the discrete norm for the sequence $(v^m)_{1\le m\le n}$,

$$[[(v^m)]]_n = \left(\left(\prod_{i=1}^{n}(1-2\lambda\Delta t_i)\right)\|v^n\|^2 + \frac{1}{2}\sigma_{\min}^2\sum_{m=1}^{n}\Delta t_m\left(\prod_{i=1}^{m-1}(1-2\lambda\Delta t_i)\right)|v^m|_V^2\right)^{\frac{1}{2}}, \tag{5.25}$$

we have the discrete analogue of (5.18):

$$[[(u^m)]]_n \le \|u^0\|. \tag{5.26}$$

In what follows, we will need an equivalence relation between $[[(u^m)]]_n$ and $[[u_{\Delta t}]](t_n)$.

Lemma 5.2. *There exists a positive real number $\alpha \le \frac{1}{2}$ such that the following equivalence property holds for $\Delta t \le \frac{\alpha}{\lambda}$ and for any family $(v^n)_{0\le n\le N}$ in V_0^{N+1}:*

$$\frac{1}{8}[[(v^m)]]_n^2 \le [[v_{\Delta t}]]^2(t_n) \le \max(2, 1+\rho_{\Delta t})[[(v^m)]]_n^2 + \frac{1}{2}\sigma_{\min}^2\Delta t_1|v^0|_V^2. \tag{5.27}$$

From (5.26) and (5.27), we deduce that for all n, $1 \le n \le N$,

$$[[u_{\Delta t}]](t_n) \le c(u_0), \tag{5.28}$$

where

$$c(u_0) = \left(\max(2, 1+\rho_{\delta t})\|u_0\|^2 + \frac{1}{2}\sigma_{\min}^2\Delta t_1|u_0|_V^2\right)^{\frac{1}{2}}. \tag{5.29}$$

5.1.3 The Fully Discrete Problem

We now describe the full discretization of (5.6). For each n, $0 \le n \le N$, let $(\mathcal{T}_{nh})$ be a family of grids of Ω. As usual, $h^{(n)}$ denotes the maximal size of the intervals in $\mathcal{T}_{nh}$. For a given element $\omega \in \mathcal{T}_{nh}$, let h_ω be the diameter of ω and let $S_{\min}(\omega)$, $S_{\max}(\omega)$ be the endpoints of ω. We assume that there exists a constant ρ_h such that, for two adjacent elements ω and ω' of $(\mathcal{T}_{nh})$,

$$h_\omega \le \rho_h h_{\omega'}. \tag{5.30}$$

For each h, we define the discrete spaces by

$$V_{nh} = \left\{v_h \in V,\ \forall\omega \in \mathcal{T}_{nh},\ v_{h|\omega} \in \mathcal{P}_1\right\}, \qquad V_{nh}^0 = V_{nh} \cap V_0. \tag{5.31}$$

The grids $\mathcal{T}_{nh}$ for different values of n are not independent: indeed, each triangulation $\mathcal{T}_{nh}$ is derived from $\mathcal{T}_{n-1,h}$ by cutting some elements of $\mathcal{T}_{n-1,h}$ into smaller intervals or, on the contrary, by gluing together elements of $\mathcal{T}_{n-1,h}$. This enables us to use simple Lagrange interpolation operators to map a discrete function of $V_{n-1,h}$ to a function of V_{nh} and to compute exactly (w_h^{n-1}, v_h^n) if $w_h^{n-1} \in V_{n-1,h}$ and $v_h^n \in V_{nh}$.

Assuming that $u_0 \in V_{0h}$, the fully discrete problem reads as follows:
Find $(u_h^n)_{0\le n\le N}$, $u_h^n \in V_{nh}^0$, satisfying

$$u_h^0 = u_0, \tag{5.32}$$

$$\forall n, 1 \le n \le N, \quad \forall v_h \in V_{nh}^0, \quad \left(u_h^n - u_h^{n-1}, v_h\right) + \Delta t_n a_{t_n}(u_h^n, v_h) = 0. \tag{5.33}$$

As above, for Δt smaller than $1/(2\lambda)$, the existence and uniqueness of $(u_h^n)_{0\le n\le N}$ is a consequence of the Lax–Milgram lemma, and we have the *stability* estimate

$$[[(u_h^m)]]_n \le \|u^0\|. \tag{5.34}$$

We call $u_{h,\Delta t}$ the function which is affine on each interval $[t_{n-1}, t_n]$, and such that $u_{h,\Delta t}(t_n) = u_h^n$.

5.2 Error Indicators for the Black–Scholes Equation

5.2.1 An Upper Bound for the Error

We now intend to bound the error $[[u - u_{h,\Delta t}]](t_n)$, $1 \le n \le N$, as a function of error indicators which can be computed from $u_{h,\Delta t}$. We are going to use the triangular inequality

$$[[u - u_{h,\Delta t}]](t_n) \le [[u - u_{\Delta t}]](t_n) + [[u_{\Delta t} - u_{h,\Delta t}]](t_n),$$

and we begin by evaluating $[[u - u_{\Delta t}]](t_n)$.

With this aim, we make a further assumption on the coefficients: we assume that σ, $S\frac{\partial\sigma}{\partial S}$ are Lipschitz continuous with respect to t uniformly with respect to S and that r is Lipschitz continuous on $[0, T]$. Thanks to the previous set of assumptions on the coefficients, we can introduce three constants L_1, L_2, and L_3 such that, for all t and t' in $[0, T]$,

$$\begin{aligned}
&\left\|\sigma^2(\cdot, t) - \sigma^2(\cdot, t')\right\|_{L^\infty(0,\bar S)} \le L_1|t' - t|,\\
&\left\|-r(t) + r(t') + \frac{\sigma^2(\cdot, t) - \sigma^2(\cdot, t')}{2} + S\left(\sigma(\cdot, t)\frac{\partial\sigma}{\partial S}(\cdot, t) - \sigma(\cdot, t')\frac{\partial\sigma}{\partial S}(\cdot, t')\right)\right\|_{L^\infty(0,\bar S)}\\
&\le L_2|t' - t|,\\
&|r(t) - r(t')| \le L_3|t' - t|.
\end{aligned} \tag{5.35}$$

Proposition 5.3. *Assume that the function u_0 belongs to V_{1h}. Then there exists a constant $\alpha \le \frac{1}{2}$ such that if $\Delta t \le \frac{\alpha}{\lambda}$, the following a posteriori error estimate holds between the solutions of problems* (5.15) *and* (5.22)*:*

$$\begin{aligned}
&[[u - u_{\Delta t}]](t_n)\\
&\le c\left(\frac{L}{\sigma_{\min}^2}c(u_0)\Delta t + \frac{\mu}{\sigma_{\min}^2}(1 + \rho_{\Delta t})[[u_{\Delta t} - u_{h,\Delta t}]](t_n) + \frac{\mu}{\sigma_{\min}^2}\left(\sum_{m=1}^n \eta_m^2\right)^{\frac{1}{2}}\right),
\end{aligned} \tag{5.36}$$

where

$$\eta_m^2 = \Delta t_m e^{-2\lambda t_{m-1}}\frac{\sigma_{\min}^2}{2}|u_h^m - u_h^{m-1}|_V^2, \tag{5.37}$$

and c is a positive constant, $L = 4L_1 + 2L_2 + L_3$, where L_1, L_2, L_3 are given by (5.35), *and $c(u_0)$ is given by* (5.29).

Corollary 5.4. *If the assumptions of Proposition* 5.3 *are satisfied, there exists a positive constant $\alpha \leq \frac{1}{2}$ such that if $\Delta t \leq \frac{\alpha}{\lambda}$, the following a posteriori error estimate holds between the solutions of problems* (5.15) *and* (5.22)*:*

$$\left\| \frac{\partial}{\partial t}(u - u_{\Delta t}) \right\|_{L^2(0,t_n,V_0')} \leq c \left(\frac{\mu + \sigma_{\min}^2}{\sigma_{\min}} \right) \begin{pmatrix} \frac{L}{\sigma_{\min}^2} c(u_0)\Delta t \\ + \frac{\mu}{\sigma_{\min}^2}(1 + \rho_{\Delta t})[[u_{\Delta t} - u_{h,\Delta t}]](t_n) + \frac{\mu}{\sigma_{\min}^2} \left(\sum_{m=1}^{n} \eta_m^2 \right)^{\frac{1}{2}} \end{pmatrix}. \tag{5.38}$$

Proposition 5.5. *Assume that $u_0 \in V_{1h}$. Then the following a posteriori error estimate holds between the solution $(u^n)_{0 \leq n \leq N}$ of problem* (5.21), (5.22) *and the solution $(u_h^n)_{0 \leq n \leq N}$ of problem* (5.32), (5.33)*: there exists a constant c such that, for all t_n, $1 \leq n \leq N$,*

$$[[(u_{\Delta t} - u_{h,\Delta t})]]^2(t_n) \leq \frac{c}{\sigma_{\min}^2} \max(2, 1 + \rho_{\Delta t}) \sum_{m=1}^{n} \Delta t_m \prod_{i=1}^{m-1} (1 - 2\lambda \Delta t_i) \sum_{\omega \in \mathcal{T}_{mh}} \eta_{m,\omega}^2, \tag{5.39}$$

where

$$\eta_{m,\omega} = \frac{h_\omega}{S_{\max}(\omega)} \left\| \frac{u_h^m - u_h^{m-1}}{\Delta t_m} - rS \frac{\partial u_h^m}{\partial S} + r u_h^m \right\|_{L^2(\omega)}. \tag{5.40}$$

Remark 5.1. *One could also take for $\eta_{m,\omega}$ the larger indicator*

$$\eta_{m,\omega} = \begin{pmatrix} \frac{h_\omega}{S_{\max}(\omega)} \left\| \frac{u_h^m - u_h^{m-1}}{\Delta t_m} - rS \frac{\partial u_h^m}{\partial S} + r u_h^m \right\|_{L^2(\omega)} \\ + \frac{1}{4} h_\omega^{\frac{1}{2}} \sum_{i=1}^{2} \sigma^2(t_m, \xi_i) \xi_i \left| \left[\frac{\partial u_h^m}{\partial S} \right] (\xi_i) \right| \end{pmatrix}, \tag{5.41}$$

where ξ_i, $i = 1, 2$, are the two endpoints of ω and where $\left[\frac{\partial u_h^m}{\partial S}\right](\xi_i)$ is the jump of $\frac{\partial u_h^m}{\partial S}$ at ξ_i. This larger indicator is not necessary for parabolic problems in two dimensions (including time), but in more than two dimensions, (5.40) *does not yield an upper bound for the error, and* (5.41) *is compulsory. This will be explained in Remark* 5.2 *in the appendix below.*

Combining the results of Propositions 5.3 and 5.5 leads to the following full a posteriori error estimate.

Theorem 5.6. *Assume that $u_0 \in V_{1h}$ and that $\lambda \Delta t \leq \alpha$ as in Lemma* 5.2. *Then the following a posteriori error estimate holds between the solution u of problem* (5.14), (5.15) *and the solution $u_{h,\Delta t}$ of problem* (5.32), (5.33)*: there exists a constant c such that, for all t_n, $1 \le n \le N$,*

$$[[u - u_{h,\Delta t}]](t_n) \le c \left(\begin{array}{l} \dfrac{L}{\sigma_{\min}^2} c(u_0)\Delta t \\ + \dfrac{\mu}{\sigma_{\min}^2} \left(\displaystyle\sum_{m=1}^{n} \eta_m^2 + \dfrac{\Delta t_m}{\sigma_{\min}^2} g(\rho_{\Delta t}) \prod_{i=1}^{m-1} (1 - 2\lambda \Delta t_i) \sum_{\omega \in \mathcal{T}_{mh}} \eta_{m,\omega}^2 \right)^{\frac{1}{2}} \end{array} \right), \tag{5.42}$$

where $L = 4L_1 + 2L_2 + L_3$, L_1, L_2, L_3 are given by (5.35)*, $c(u_0)$ is given by* (5.29)*, η_m is given by* (5.37)*, and $\eta_{m,\omega}$ is given by* (5.40)*, and*

$$g(\rho_{\Delta t}) = (1 + \rho_{\Delta t})^2 \max(2, 1 + \rho_{\Delta t}).$$

5.2.2 An Upper Bound for the Error Indicators

The program is now to prove separate bounds for each indicator η_n and $\eta_{n,\omega}$. We begin with η_n. For that, we introduce the notation $[[v^n]]$ for $(v^n)_{1 \le n \le N}$, $v^n \in V_0$:

$$[[v^n]]^2 = \frac{\sigma_{\min}^2}{2} \Delta t_n \prod_{i=1}^{n-1} (1 - 2\lambda \Delta t_i) |v^n|_V^2. \tag{5.43}$$

Proposition 5.7. *Assume that u^0 belongs to V_0, and that $\lambda \Delta t \le \alpha$ as in Lemma 5.2. The following estimate holds for the indicator η_n, $2 \le n \le N$:*

$$\eta_n \le c \left(\begin{array}{l} [[u^n - u_h^n]] + \sqrt{\rho_{\Delta t}} [[u^{n-1} - u_h^{n-1}]] \\ + \dfrac{e^{-\lambda t_{n-1}}}{\sigma_{\min}} \left(\left\| \dfrac{\partial}{\partial t}(u - u_{\Delta t}) \right\|_{L^2(t_{n-1}, t_n; V_0')} + \|u - u_{\Delta t}\|_{L^2(t_{n-1}, t_n; V_0)} \right) \\ + \left(\dfrac{L}{\sigma_{\min}^2} (\max(1, \rho_{\Delta t}))^{\frac{1}{2}} + \dfrac{\lambda \mu}{\sigma_{\min}^2} \right) \Delta t_n ||u^0|| \end{array} \right), \tag{5.44}$$

and

$$\eta_1 \le c \left(\begin{array}{l} [[u^1 - u_h^1]] + \dfrac{1}{\sigma_{\min}} \left(\left\| \dfrac{\partial}{\partial t}(u - u_{\Delta t}) \right\|_{L^2(0, t_1; V_0')} + \|u - u_{\Delta t}\|_{L^2(0, t_1; V_0)} \right) \\ + \dfrac{L + \lambda \mu}{\sigma_{\min}^2} \Delta t_1 \|u^0\| + \dfrac{L}{\sigma_{\min}} (\Delta t_1)^{\frac{3}{2}} |u^0|_V \end{array} \right), \tag{5.45}$$

where c is a positive constant.

The most important property of estimate (5.44) is that, up to the last term, which depends on the data, all the terms on the right-hand side of (5.44) are local in time. More precisely, they involve the solution in the interval $[t_{n-1}, t_n]$.

We need to define some more notation before stating the upper bound result for $\eta_{n,\omega}$. For $\omega \in \mathcal{T}_{n,h}$, let K_ω be the union of ω and the element that shares a node with ω, and let $V_0(K_\omega)$ be the closure of $\mathcal{D}(K_\omega)$ in $V(K_\omega) = \{v \in L^2(K_\omega); S\frac{\partial v}{\partial S} \in L^2(K_\omega)\}$ endowed with the norm $\|v\|_{V(K_\omega)} = (\int_{K_\omega} v^2(S)+S^2(\frac{\partial v}{\partial S}(S))^2)^{\frac{1}{2}}$. We also define $\|v\|_{V_0(K_\omega)} = (\int_{K_\omega} S^2(\frac{\partial v}{\partial S}(S))^2)^{\frac{1}{2}}$ for $v \in V_0(K_\omega)$. We denote by $V_0'(K_\omega)$ the dual space of $V_0(K_\omega)$ endowed with dual norm.

Proposition 5.8. *The following estimate holds for the indicator $\eta_{n,\omega}$ defined in* (5.40) *for all $\omega \in \mathcal{T}_{n,h}$, $1 \le n \le N$:*

$$\eta_{n,\omega} \le C\left(\left\|\frac{u^{n-1} - u_h^{n-1} - u^n + u_h^n}{\Delta t_n}\right\|_{V_0'(K_\omega)} + \mu\left\|S\frac{\partial(u^n - u_h^n)}{\partial S}\right\|_{L^2(K_\omega)}\right). \tag{5.46}$$

5.3 Conclusion

In §5.2.1 we have bounded the norm of the error produced by the finite element method by a Hilbert sum involving the error indicators η_m and $\eta_{m,\omega}$, which are, respectively, local in t and local in t and S. Conversely, in §5.2.2, we have seen that the error indicators can be bounded by local norms of the error. This shows that the error indicators are both reliable and efficient, or in other words that the error produced by the method is well approached by these indicators. Furthermore, since the indicators are local, they tell us where the mesh should be refined.

It is now possible to build a computer program which adapts the mesh so as to reduce the error to a given number ϵ. From the result of an initial computation $u_{h,\Delta t}$ we can adapt separately the meshes in the variables t and S so that the Hilbert sum in (5.42) decreases. The process is repeated until the desired accuracy is obtained.

5.4 A Taste of the Software

The software for the finite element method with adaptive mesh refinement based on the error indicators presented above is surely more complex and longer than the simple program presented in §4.4. For that reason, we will not reproduce it entirely here. Instead, we focus on two points:

- the program for a backward Euler scheme with the mesh in the S variable varying in time;
- the computation of the indicators η_m.

An `Euler_scheme` class is defined for the backward Euler scheme as follows.

ALGORITHM 5.1. Euler_scheme.

```
class Euler_Scheme
{
private:
  vector<vector<double> >S_nodes,S_steps;      //  the meshes in S (one for
                                                   //    each time step)
  vector<double> grid_t;                             //    the mesh in t
  vector<int> change_grid;                    //    at each time step, tells
                                   //    whether the S-mesh varies or not
  double rate (double);                           //    the function for the
                                                     //    interest rate
  double vol(double,double);               //    the local volatility function
  double eps;                                     //    a small parameter
protected:
public:
  Euler_Scheme(const  vector<vector<double> > &g_grid_S,
               const  vector<vector<double> > &g_S_steps,
               vector<double> & g_grid_t, vector<int> & g_change_grid)
    :S_nodes(g_grid_S),S_steps(g_S_steps), grid_t(g_grid_t),
     change_grid(g_change_grid){eps=1e-9;};             //    the constructor
  void Time_Step(int i,  vector<KN<double> >  &P);          //    a time step
  void build_rhs(KN<double> &u, const KN<double> &u_p, const  vector<double>
           & steps_p, const  vector<double> & nodes_p,const  vector<double>
           & steps, const  vector<double> & nodes);
             //    computes the RHS of the linear system at each time step
  void build_rhs(KN<double> &u, const KN<double> &u_p,
             //    computes the RHS of the linear system at each time step
                              //    assuming that the S-mesh does not vary
  double build_time_error_indicator(const KN<double> &u, const KN<double> &u_p,
                                    const  vector<double> & steps_p,
                                    const  vector<double> & nodes_p,
                                    const  vector<double> & steps,
                                    const  vector<double> & nodes);
              //   computes the time error indicator at a given time step
  double build_time_error_indicator(const KN<double> &u, const KN<double> &u_p,
              //   computes the time error indicator at a given time step
                                          //    if the S-mesh does not vary
  double build_time_error_indicator(int it, const  vector< KN<double> >& P);
                    //   the loop for computing the time error indicators
  void build_S_indicator(const KN<double> &u, const KN<double> &u_p,
                         const  vector<double> & steps_p,
                         const  vector<double> & nodes_p,
                         const  vector<double> & steps,
                         const  vector<double> & nodes,
                         const double rt , const double dt,
                         KN<double> &indic );
               //   computes the S-error indicators at a given time step
  void build_S_indicator(int it,  const vector< KN<double> >& P,
                         vector< KN<double> >& indic);
               //   computes the S-error indicators at a given time step
                                          //    if the S-mesh does not vary
}
;
```

A time step of the method is implemented in the function `Time_Step:` it consists of building the matrix $\mathbf{B} = \mathbf{M} + \Delta t_m \mathbf{A}^m$, computing its LU factorization, constructing the right-hand side of the system of linear equations, and solving this system.

ALGORITHM 5.2. Time step.

```
void Euler_Scheme::Time_Step(int it,  vector< KN<double> >& P)
{
  int i,n;
  double dt,t,S,h_p,h_n,r;
  double a,b,c,d,e;

                              //  constructs the matrix of the linear system
  n=S_steps[it].size();
  MatriceProfile<double> A(n,2);                       //   memory allocation

  t=grid_t[it];
  dt=t-grid_t[it-1];
  r=rate(t);
  e=0.5*dt;

  h_n=S_steps[it][0];
  A(0,0)=e*r*h_n+ h_n/3;
  A(0,1)=h_n/6;
  for(i=1;i< n-1;i++)
    {
      h_p=h_n;
      S=S_nodes[it][i];
      h_n=S_steps[it][i];
      a=pow(S*vol(t,S),2);
      b=a/h_p;
      c=a/h_n;
      d=r*S;
      A(i,i)=e*(b+c+r*(h_p+h_n))+ (h_p+h_n)/3 ;
      A(i,i-1)=e*(-b+d)+h_p/6;
      A(i,i+1)=e*(-c-d)+h_n/6;
    }
  h_p=h_n;
  S=S_nodes[it][i];
  h_n=S_steps[it][i];
  a=pow(S*vol(t,S),2);
  b=a/h_p;
  c=a/h_n;
  d=r*S;
  A(i,i)=e*(b+c+r*(h_p+h_n))+ (h_p+h_n)/3 ;
  A(i,i-1)=e*(-b+d)+h_p/6;
                         //   the matrix of the linear system is constructed

                                              //   builds the right-hand side
  if (change_grid[it])
    build_rhs(P[it],P[it-1],S_steps[it-1],S_nodes[it-1],
              S_steps[it],S_nodes[it]);
  else
    build_rhs(P[it],P[it-1],S_steps[it],S_nodes[it]);          //   simpler
```

```
  A.LU();                                        //   LU factorization of A
  A.Solve(P[it],P[it]);                        //   solves the linear system
}
```

The difficult part is the construction of the right-hand side of the system of linear equations when the mesh in S varies: assume that $\mathcal{T}_{n,h} \neq \mathcal{T}_{n-1,h}$; then we have to compute $\int_\Omega u_h^{n-1} w_i$ for the shape functions of $V_{n,h}^0$. To do it exactly, one has to intersect the two meshes $\mathcal{T}_{n,h}$ and $\mathcal{T}_{n-1,h}$. We reproduce here a function which is not optimized (so that it is not too intricate).

ALGORITHM 5.3. Right-hand side.

```
void    Euler_Scheme::build_rhs(KN<double> &u, const KN<double> &u_p,
                                const  vector<double> & steps_p,
                                const  vector<double> & nodes_p,
                                const  vector<double> & steps,
                                const  vector<double> & nodes)
{
  double h;
  double u_pl,u_pr;
  double phi_l,phi_r;
  double psi_l,psi_r;
  double x_l,x_r;
  int i=0;
  int j=0;
  u=0;
  u_pl=u_p(0);
  x_l=0;
  while (i<steps.size())
    {
      if(nodes_p[j+1]<=nodes[i+1]+eps)
        {
          phi_l=0.0;
          psi_l=1.0;
          while (nodes_p[j+1]<=nodes[i+1]+eps)
            {
              if (j<steps_p.size()-1)
                u_pr=u_p(j+1);
              else
                u_pr=0;
              x_r=nodes_p[j+1];
              h=x_r-x_l;
              phi_r=(nodes_p[j+1]-nodes[i])/steps[i];
              psi_r=1.-phi_r;
              if(i<steps.size()-1)
                u(i+1)+=h*(2* (phi_l*u_pl+phi_r*u_pr)
                        + (phi_l*u_pr+phi_r*u_pl) )/6;
              u(i)+=h*(2* (psi_l*u_pl+psi_r*u_pr)
                      + (psi_l*u_pr+psi_r*u_pl) )/6;
              x_l=x_r;
              u_pl=u_pr;
              phi_l=phi_r;
              psi_l=psi_r;
```

```
              j++;
              if (j==steps_p.size())
                break;
            }
          if(nodes[i+1]>nodes_p[j]+eps)
            {
              x_r=nodes[i+1];
              h=x_r-x_l;
              if(j<steps_p.size()-1)
                u_pr=u_p(j)+(nodes[i+1]-nodes_p[j])/steps_p[j]
                     *(u_p(j+1)-u_p(j));
              else
                u_pr=(1.-(nodes[i+1]-nodes_p[j])/steps_p[j])*u_p(j);
              phi_r=1.0;
              psi_r=0.0;
              if(i<steps.size()-1)
                u(i+1)+=h*(2* (phi_l*u_pl+phi_r*u_pr)
                        + (phi_l*u_pr+phi_r*u_pl) )/6;
              u(i)+=h*(2* (psi_l*u_pl+psi_r*u_pr)
                     + (psi_l*u_pr+psi_r*u_pl) )/6;
              u_pl=u_pr;
              x_l=x_r;
            }
          i++;
        }
      else
        {
          while(nodes[i+1]<=nodes_p[j+1]+eps)
            {
              x_r=nodes[i+1];
              h=x_r-x_l;
              if (j<steps_p.size()-1)
                u_pr=u_p(j)+(nodes[i+1]-nodes_p[j])/steps_p[j]
                     *(u_p(j+1)-u_p(j));
              else
                u_pr=u_p(j)*(1-(nodes[i+1]-nodes_p[j])/steps_p[j]);
              if(i<steps.size()-1)
                u(i+1)+=h*(2* u_pr +u_pl) /6;
              u(i)+=h*(2* u_pl +u_pr) /6;
              x_l=x_r;
              u_pl=u_pr;
              i++;
              if (i==steps.size())
                break;
            }
          if(nodes[i]>nodes_p[j+1]-eps)
            j++;
        }
    }
}
```

We do not reproduce the function for building the right-hand side when the mesh in the variable S does not vary: it follows along the same lines as that presented in §4.4.

The function for evaluating η_m given by (5.37) is `Euler_Scheme::build_time_error_indicator`.

ALGORITHM 5.4. Error indicator for the time mesh.

```
double Euler_Scheme::build_time_error_indicator(const KN<double> &u,
const KN<double> &u_p,
const  vector<double> & steps_p,
const  vector<double> & nodes_p,
const  vector<double> & steps,
const  vector<double> & nodes)
{
  double h;
  int i=0;
  int j=0;
  double u_pr,u_r,e_r;
  double x_l,x_r;
  double u_pl=u_p(0);
  double e_l=u(0)-u_pl;
  double indic=0;
  x_l=0;
  while (i<steps.size())
    {
      if(nodes_p[j+1]<=nodes[i+1]+eps)
        {
          while (nodes_p[j+1]<=nodes[i+1]+eps)
            {
              if (j<steps_p.size()-1)
                u_pr=u_p(j+1);
              else
                u_pr=0;

              x_r=nodes_p[j+1];
              h=x_r-x_l;
              if(i<steps.size()-1)
                u_r=u(i)+(x_r-nodes[i])/steps[i]*(u(i+1)-u(i));
              else
                u_r=u(i)-(x_r-nodes[i])/steps[i]*u(i);
              e_r=u_r-u_pr;
              indic+= (x_r*x_r*e_r-x_l*x_l*e_l)
                      *(e_r-e_l)/h+h*((2*e_r+e_l)*e_r+(2*e_l+e_r)*e_l)/6;
              e_l=e_r;
              u_pl=u_pr;
              x_l=x_r;
              j++;
              if (j==steps_p.size())
                break;
            }
          if(nodes[i+1]>nodes_p[j]+eps)
            {
              x_r=nodes[i+1];
              h=x_r-x_l;
              if (j<steps_p.size()-1)
              u_pr=u_p(j)+(x_r-nodes_p[j])/steps_p[j]*(u_p(j+1)-u_p(j));
              else
                u_pr=u_p(j)*(1.-(x_r-nodes_p[j])/steps_p[j]);
              if (i<steps.size()-1)
```

```
              e_r=u(i+1)-u_pr;
            else
              e_r=-u_pr;
            indic+= (x_r*x_r*e_r-x_l*x_l*e_l)
                    *(e_r-e_l)/h+h*((2*e_r+e_l)*e_r+(2*e_l+e_r)*e_l)/6;
            e_l=e_r;
            u_pl=u_pr;
            x_l=x_r;
          }
        i++;
      }
    else
      {
        while(nodes[i+1]<=nodes_p[j+1]+eps)
          {
            x_r=nodes[i+1];
            h=x_r-x_l;
            if (j<steps_p.size()-1)
              u_pr=u_p(j)+(x_r-nodes_p[j])/steps_p[j]*(u_p(j+1)-u_p(j));
            else
              u_pr=u_p(j)-(x_r-nodes_p[j])/steps_p[j]*u_p(j);
            if (i<steps.size()-1)
              e_r=u(i+1)-u_pr;
            else
              e_r=-u_pr;
            indic+= (x_r*x_r*e_r-x_l*x_l*e_l)
                    *(e_r-e_l)/h+h*((2*e_r+e_l)*e_r+(2*e_l+e_r)*e_l)/6;
            e_l=e_r;
            u_pl=u_pr;
            x_l=x_r;
            i++;
            if (i==steps.size())
              break;
          }
        if(nodes[i]>nodes_p[j+1]-eps)
          j++;
      }
  }
  return indic;
}
```

5.5 Results

5.5.1 Test with a Uniform Volatility

We apply the refinement strategy to the computation of a vanilla European put with payoff 100 and maturity 1 year: the volatility and the interest rates are constant: $\sigma = 0.2$ and $r = 0.04$. Therefore, we can also compute the price of the option by the Black–Scholes formula, and the error due to the discretization.

We compute the option price in the rectangle $[0, 200] \times [0, 1]$.

First Refinement Strategy. We start with a uniform mesh with 20 nodes in t and 80 in S. Along with the discrete solution, we compute the indicators η_m and $\eta_{m,\omega}$, and according to (5.42), we can compare η_m^2 with $\frac{\Delta t_m}{\sigma_{\min}^2} \sum_{\omega \in \mathcal{T}_{mh}} \eta_{m,\omega}^2$, the former term being a global S-discretization error indicator at time t_m. If the time-discretization error indicators tend to dominate the global S-discretization error indicators, we choose to refine the mesh in the t variable, and we divide the time steps for which the error indicators are large into smaller time steps, keeping the S mesh unchanged. If, on the contrary, the global S-discretization error indicators tend to dominate the time-discretization error indicators, then, for each t_m, we generate a finer mesh $\mathcal{T}_m$ by dividing the elements ω such that $\eta_{n,\omega}$ is large into smaller elements. In such a way we obtain a new mesh, and we can compute again a discrete solution and error indicators.

For the refinement in t, we compute $\bar{\zeta} = \max_m \eta_m$ and $\underline{\zeta} = \min_m \eta_m$ and we divide the time interval $[t_{m-1}, t_m]$ by two if $\eta_m > (\bar{\zeta} + \underline{\zeta})/2$. For the refinement in S, we take a similar strategy. This produces a very progressive mesh refinement: we made up to 19 refinements. We plot in Figure 5.1 the meshes obtained after 0, 5, 10, 19 refinements. In Figure 5.2, we plot $\frac{\eta_m}{\sigma_{\min}}$ and $\frac{\Delta t_m}{\sigma_{\min}^2} \sum_{\omega \in \mathcal{T}_{mh}} \eta_{m,\omega}^2$ as functions of time. In Figure 5.3, we have plotted the error between the prices computed by the Black–Scholes formula and by the finite element method, for the four meshes. In Figure 5.4, we have plotted the error indicators $\eta_{m,\omega}$.

Finally, in Table 5.1 we have listed both $\sigma \|u - u_{h,\Delta t}\|_{L^2((0,T);V)}$ and $\left(\sum_m (\eta_m^2 + \frac{\Delta t_m}{\sigma^2} \sum_\omega \eta_{m,\omega}^2)\right)^{1/2}$ for the different meshes. We see that the estimate of the error by means of the error indicators is really reliable, for the estimated error never exceeds $2.1\times$ the actual error. Therefore one can use safely the error indicators for a stopping criterion in the adaption procedure.

The code for the refinement strategy is as follows.

ALGORITHM 5.5. Refinement strategy.

```
int main()
{
                                    //  parameters for the mesh refinement function
  int max_aspect_ratio_t=2;
  int max_aspect_ratio=2;
  double reduction_factor=8.;
                                       //  data structure for the sequance of grids
  vector< vector<double> >grid_t(21);
  vector<vector<vector<double> > >grid_S(21);
  vector< vector<vector<double> > >S_steps(21);
  vector < vector<int> >change_grid(21);
                                                          //  bounds of the domain
  double S_max= 200.0, T=1.;
  double K=100;
                                                  //  construction of a first grid

  int Nt=21;
  int NS=81;
  grid_S[0].resize(Nt);
  S_steps[0].resize(Nt);
  change_grid[0].resize(Nt,0);
  for(int i=0;i<Nt;i++)
    {
      grid_t[0].push_back((T*i)/(Nt-1));
      for(int j=0;j<NS;j++)
```

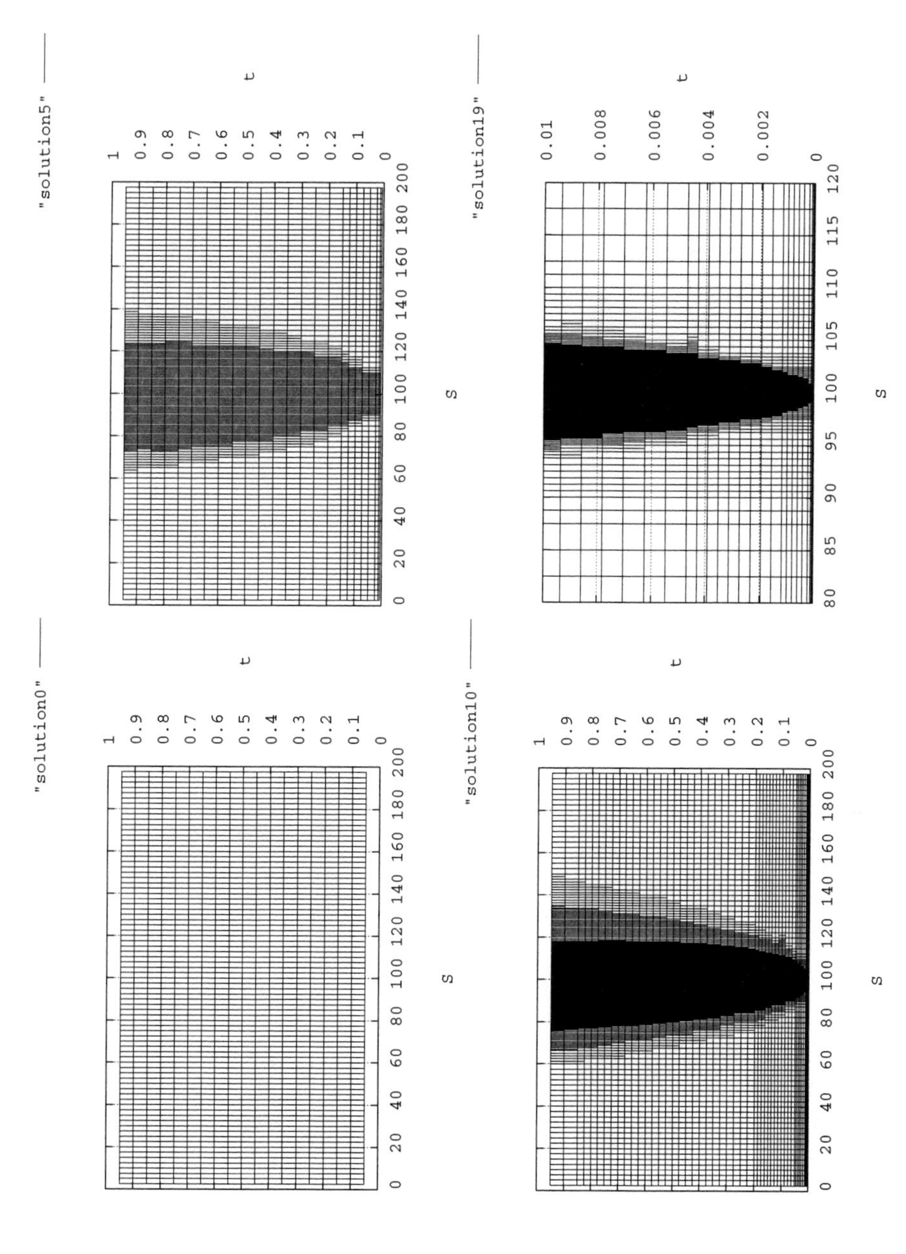

Figure 5.1. *Four successive mesh refinements: the bottom right figure is a zoom of the more refined mesh near the singularity.*

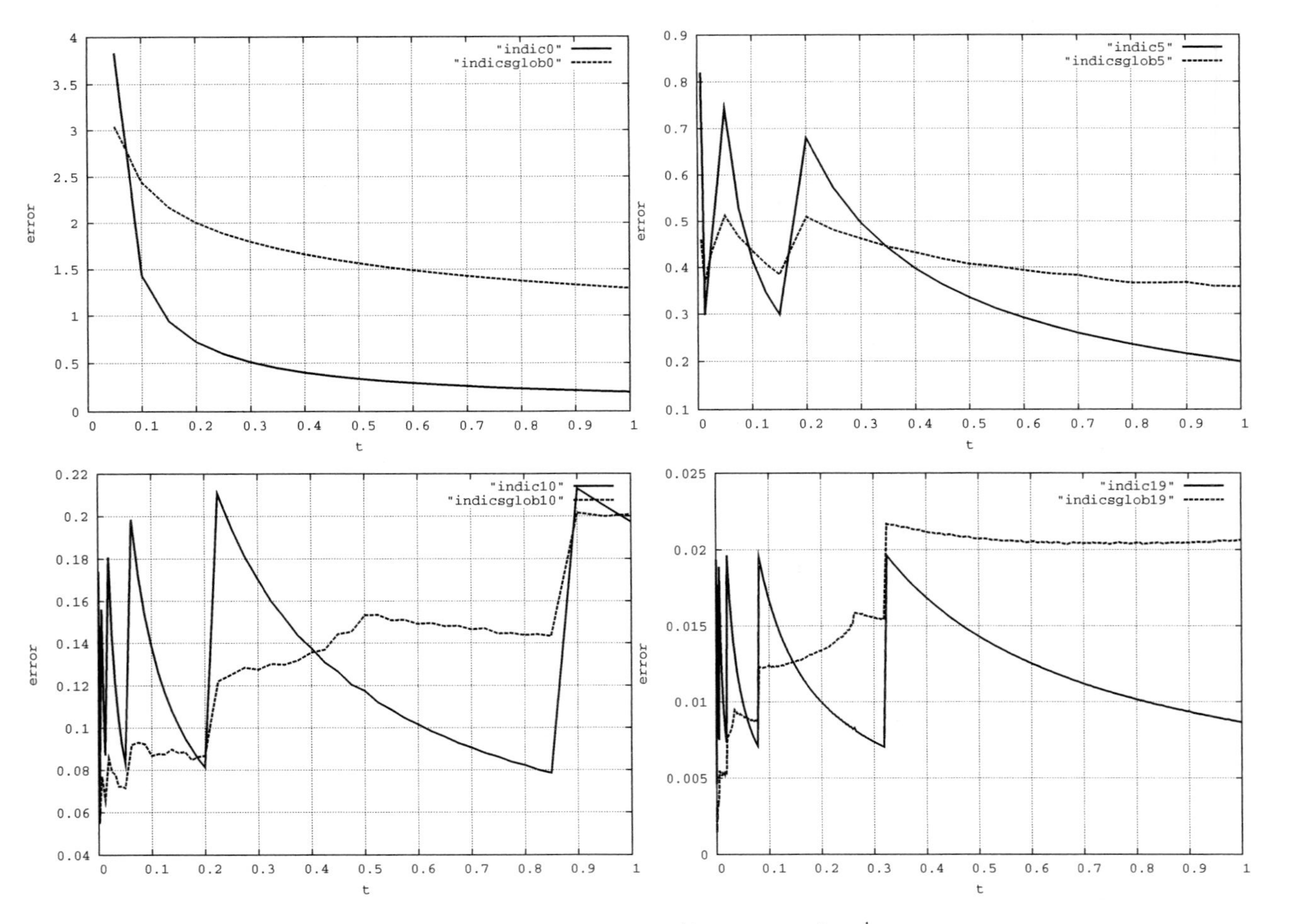

Figure 5.2. *Error indicators* η_m *and* $\left(\frac{\Delta t_m}{\sigma_{\min}^2} \sum_{\omega \in \mathcal{T}_{mh}} \eta_{m,\omega}^2\right)^{\frac{1}{2}}$ *versus time.*

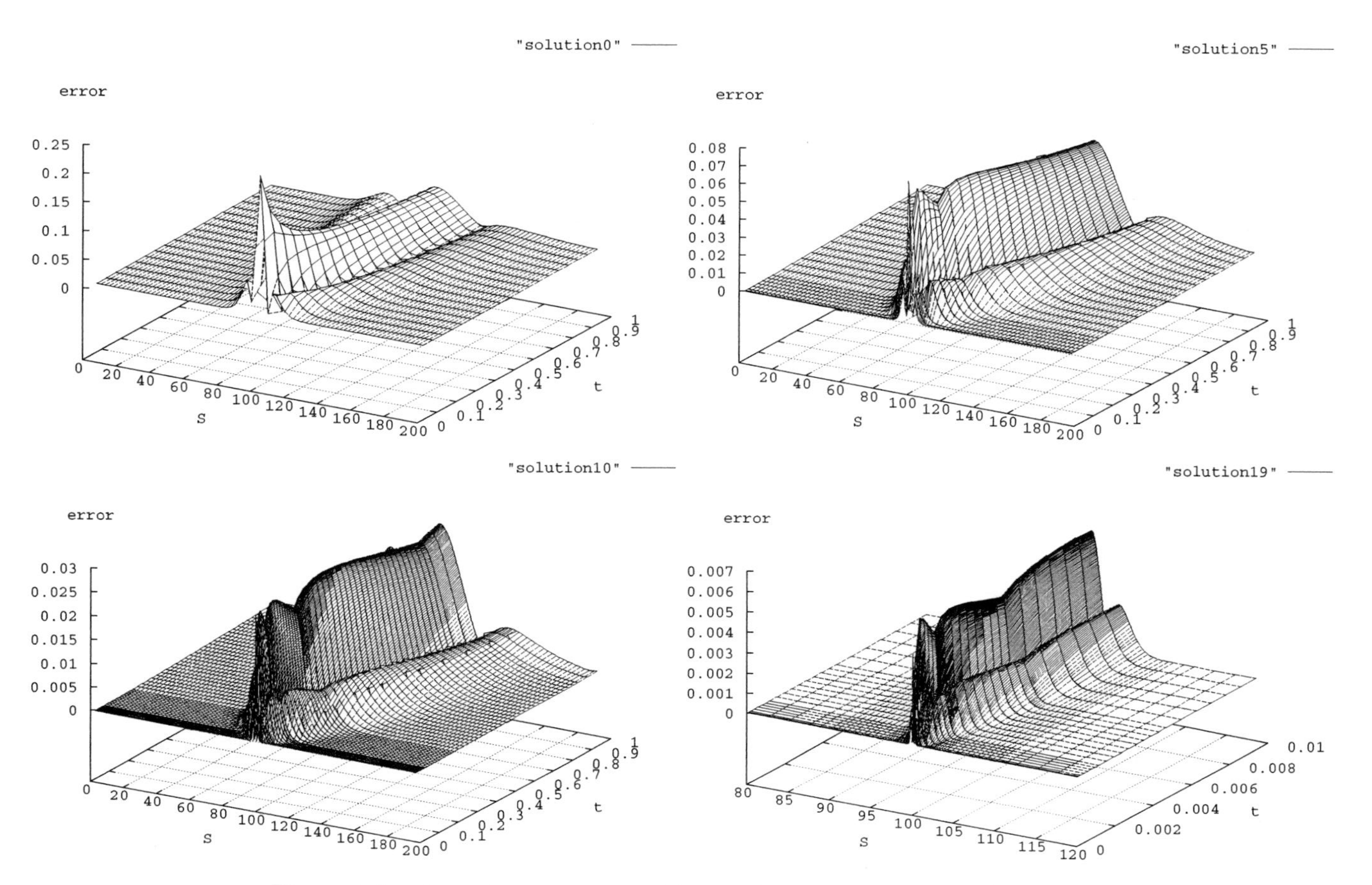

Figure 5.3. *The pointwise errors with the adaptive strategy: the bottom right figure is a zoom.*

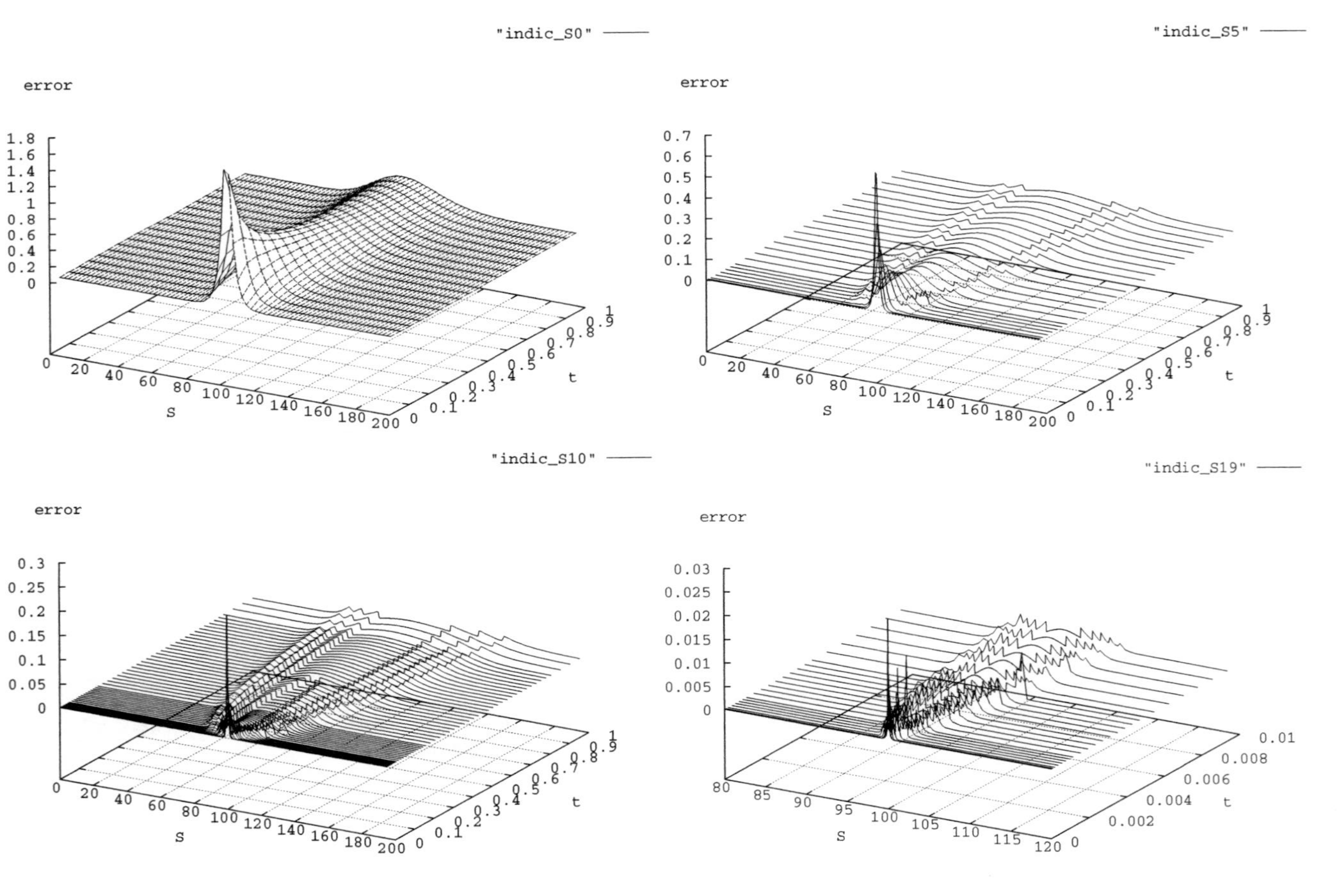

Figure 5.4. *Error indicators $\eta_{m,\omega}$: the bottom right figure is a zoom.*

```
      grid_S[0][i].push_back((S_max*j)/(NS-1));
    for(int j=0;j<NS-1;j++)
      S_steps[0][i].push_back(grid_S[0][i][j+1]-grid_S[0][i][j]);
  }
int ref_it=0;                              //    counts the refinement steps
double  estim_norm_error=10000;                        //    estimated error
while(ref_it<20 &&  estim_norm_error>1)
  {
    Euler_Scheme scheme(grid_S[ref_it], S_steps[ref_it],grid_t[ref_it],
                        change_grid[ref_it]);
                                                   //    construct a scheme
    vector<KN<double> >  P(Nt);                 //    the discrete solution
    vector<KN<double> >  indic_S(Nt);                //    indicators in S
    for (int i=0;i<P.size();i++)                   //   init P and indic_S
      {
        P[i].init(S_steps[ref_it][i].size());
        indic_S[i].init(S_steps[ref_it][i].size());
        for(int j=0; j<S_steps[ref_it][i].size();j++)
          {
            P[i](j)=0;
            indic_S[i](j)=0;
          }
      }
    int  iS=0;
    while (grid_S[ref_it][0][iS]<K)
      {
        P[0](iS)=K-grid_S[ref_it][0][iS];                //   the Cauchy datum
        iS++;
      }
    vector<double> indic_t;                          //   indicators in time
    vector<double> indic_S_global;    //   sum of the S indicators at each
                                                               //    time step
    double norm_indic_t=0, norm_indic_S_global=0;
    for(int i=1;i<Nt;i++)
      {
        scheme.Time_Step(i,P);                //   a time step of the scheme
        double aux1=sqrt(scheme.build_time_error_indicator(i,P))
                    * vol(0.5,100);
        indic_t.push_back(aux1);          //   time error indicator :  eta_i
        norm_indic_t+=aux1*aux1;
        scheme.build_S_indicator_1(i,P,indic_S);                //    S error
                                      //    indicator at time step i:  eta_{i,ω}
        double aux=scheme.build_global_S_indicator(i,indic_S[i],
                                                   grid_S[ref_it][i]);
                        //   sum of the S error indicators at time step i
        indic_S_global.push_back(aux);
        norm_indic_S_global+=aux*aux;
      }
    estim_norm_error=sqrt( norm_indic_t)+ sqrt( norm_indic_S_global);

     //   next refinement :  in t or S ? compares the error indicators in
                                                             //    S and t
    int which_refine=ref_it%2;
    if ( norm_indic_t/norm_indic_S_global>2||
         norm_indic_t/norm_indic_S_global<0.5)
      which_refine=norm_indic_t<norm_indic_S_global;
         //    refinement :  the parameter which_refine tells whether to
                                                   //    refine in t or S
```

```
        adaption_tS(grid_t[ref_it],indic_t,grid_t[ref_it+1],1,
max_aspect_ratio_t, grid_S[ref_it], change_grid[ref_it], S_steps[ref_it],
                    indic_S, grid_S[ref_it+1],S_steps[ref_it+1],
change_grid[ref_it+1], reduction_factor, max_aspect_ratio,  indic_S_global,
                    which_refine);
        Nt=grid_t[ref_it+1].size();
        ref_it++;
      }
}
```

Table 5.1. $\sigma \|u - u_{h,\Delta t}\|_{L^2((0,T);V)}$ *and* $\left(\sum_m (\eta_m^2 + \frac{\Delta t_m}{\sigma^2} \sum_\omega \eta_{m,\omega}^2)\right)^{\frac{1}{2}}$ *for the different meshes.*

error	5.67	5.66	5.67	4.66	3.73	3.25	3.26	2.53	2.53	1.95	1.45	1.06
estim. err.	12.27	8.56	6.62	5.38	4.58	4.19	3.39	2.95	2.56	2.21	1.85	1.59
error	1.06	0.77	0.77	0.57	0.57	0.41	0.41	0.30				
estim. err.	1.48	1.29	1.03	0.90	0.77	0.67	0.52	0.44				

More Aggressive Refinement Strategies. It is of course possible to use more aggressive refinement strategies; i.e., one can refine a time step or an element more than twice. The advantage is that fewer refinement steps are needed, but the mesh so produced may be too fine in some regions. In our tests, starting with a mesh of 20×80, and allowing to split the refined elements into up to eight subelements (depending on the indicator), it is possible to diminish the maximum norm of the error to less than 10^{-2} in around five refinement steps.

5.5.2 Result with a Nonuniform Local Volatility

Here we take the volatility to be $\sigma(S,t) = 0.05 + 0.25\ 1_{\{\frac{\|S-100\|^2}{400} + \frac{\|t-0.5\|^2}{0.01} \leq 1\}}$. The result is plotted in Figure 5.5. We see that the mesh is automatically refined in the zone where the volatility jumps.

Project 5.1. *Run the program described above. Adapt it to a barrier option with realistic data. If time allows make the change* $S \to S/(1+S)$ *which maps* $\mathbb{R}_+$ *onto* $(0,1)$, *compute a European put in this formulation, and compare.*

5.6 Mesh Adaption for a Put on a Basket of Two Assets

To illustrate the power of mesh adaption, we take the example of a European put on a basket of two assets: we solve (2.61), with P_0 given by (2.65). We take $\sigma_{11} = \sigma_{22} = 0.2$, $\frac{2\rho}{1+\rho^2} = -0.6$, and $r = 0.05$. We use an Euler scheme in time and triangular continuous and piecewise affine finite elements in the price variables. In Figure 5.6, we plot the contours of the solution nine months to maturity, computed by using a quasi-uniform mesh. It is

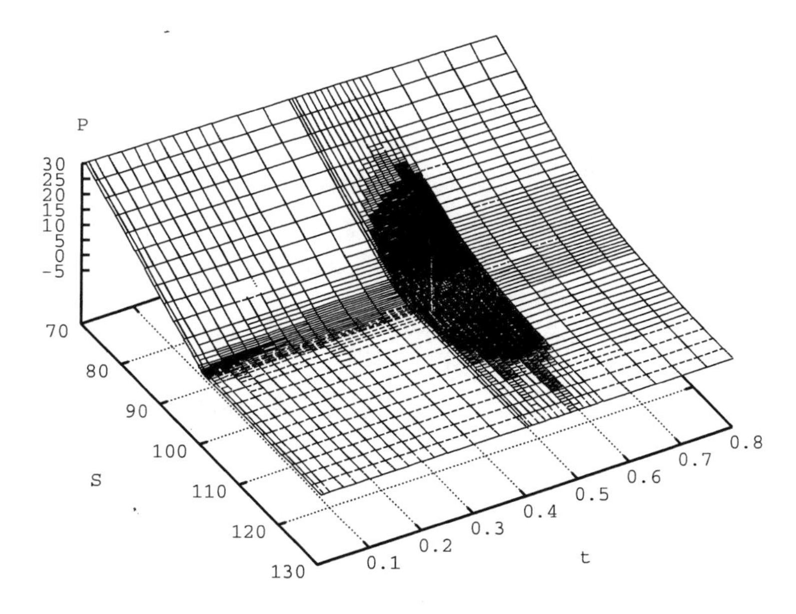

Figure 5.5. *The solution computed with a piecewise constant local volatility.*

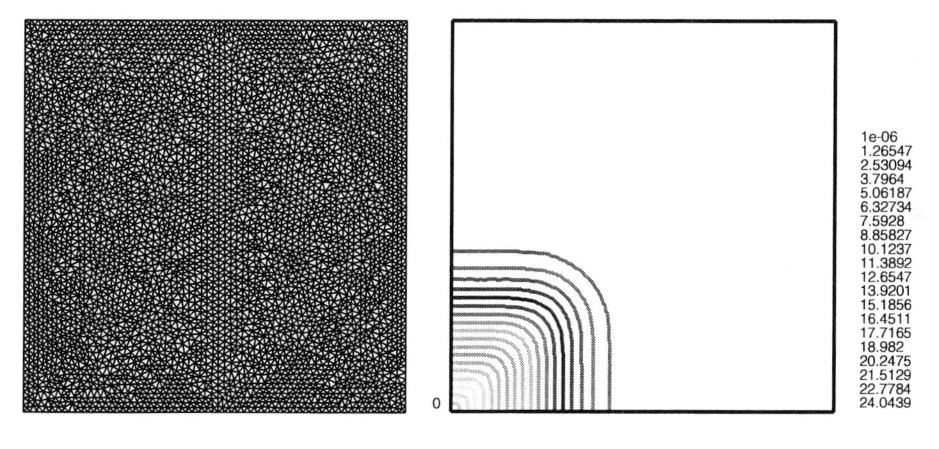

Figure 5.6. *Left: a quasi-uniform mesh in the variables S_1, S_2. Right: contours of $P(S_1, S_2, 0.25)$ computed on this mesh.*

possible to use an adaption strategy. One way do it is to generalize the approach presented above to the two-dimensional case. The approach presented here is different: by and large, starting from a possibly coarse grid, the idea is to adaptively construct a regular mesh (the angles of the triangles are bounded from below by a fixed constant) in the metric generated by an approximated Hessian of the computed solution. Although the theory for this strategy is not as clean as the one presented above, this method gives generally very good results. We have used the freeware `freefem++`, by Pironneau and Hecht [100], which is available at http://www.freefem.org. This software permits one to use two-dimensional finite elements and mesh adaption by means of a user-friendly dedicated language.

Adapted meshes at different times and the contours of the solution nine months to maturity are plotted in Figure 5.7. We see that the meshes need many fewer nodes than the quasi-uniform mesh for a nice accuracy. The mesh is refined only in the zones where the solution has large second order derivatives. Note that no adaption in time has been performed here, although we have made clear that this is crucial for pricing accurately near maturity.

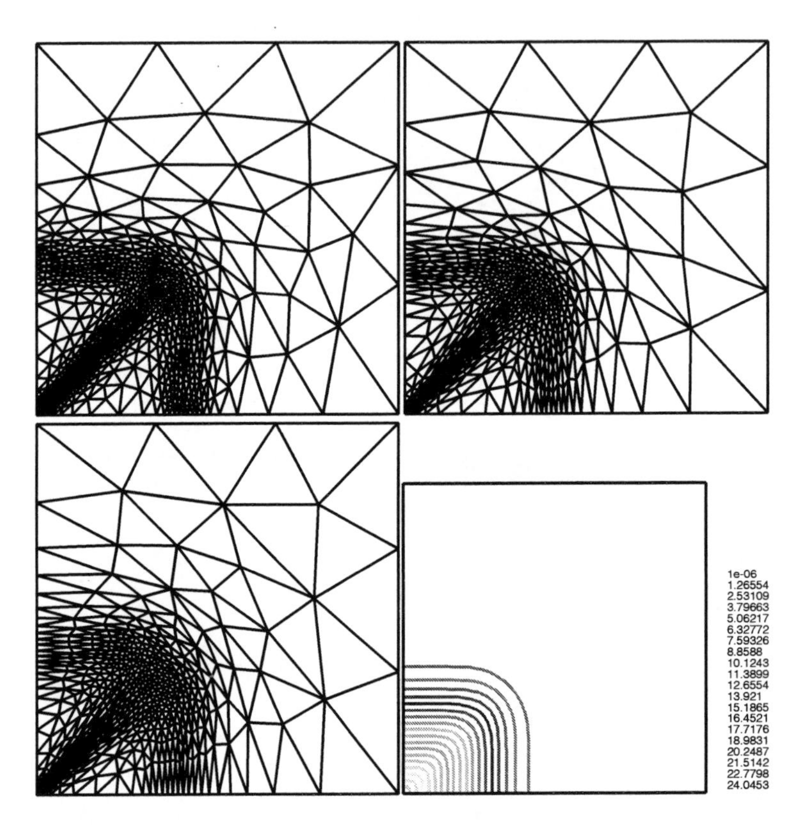

Figure 5.7. *The adapted meshes three, six, and nine months to maturity, and the contours of P nine months to maturity.*

5.7 Appendix: Proofs

***Proof of Lemma* 5.2.** From the definition of $v_{\Delta t}$, an easy computation yields

$$\begin{aligned}&\frac{e^{2\lambda t_{m-1}}}{\Delta t_m}\int_{t_{m-1}}^{t_m} e^{-2\lambda\tau}|v_{\Delta t}|_V^2(\tau)d\tau\\ &=\int_0^1 e^{-2\lambda\Delta t_m\tau}\left(|v^m|_V^2\tau^2+|v^{m-1}|_V^2(1-\tau)^2+2(v^{m-1},v^m)_V\tau(1-\tau)\right)d\tau.\end{aligned}\tag{5.47}$$

If $\Delta t_m = 0$, the right-hand side in (5.47) is equal to $\frac{1}{3}(|v^m|_V^2 + |v^{m-1}|_V^2 + (v^m, v^{m-1})_V) \geq \frac{1}{4}|v^m|_V^2$, using the inequality $ab \geq -\frac{a^2}{4} - b^2$. Furthermore, $e^{-2\lambda\Delta t_m \tau}$ is continuous with respect to Δt_m, so there exists a positive constant $\alpha_1 < \frac{1}{2}$ such that if $\Delta t \leq \frac{\alpha_1}{\lambda}$, then $\frac{e^{2\lambda t_{m-1}}}{\Delta t_m}\int_{t_{m-1}}^{t_m} e^{-2\lambda\tau}|v_{\Delta t}|_V^2(\tau)d\tau \geq \frac{1}{8}|v^m|_V^2$, or in an equivalent manner,

$$\int_{t_{m-1}}^{t_m} e^{-2\lambda\tau}|v_{\Delta t}|_V^2(\tau)d\tau \geq \frac{\Delta t_m}{8} e^{-2\lambda t_{m-1}}|v^m|_V^2.$$

This and the estimate $\prod_{i=1}^{m-1}(1 - 2\lambda\Delta t_i) \leq e^{-2\lambda t_{m-1}}$, which holds as soon as $2\lambda\Delta t < 1$, yield

$$\frac{1}{2}\sigma_{\min}^2 \sum_{m=1}^{n} \Delta t_m \left(\prod_{i=1}^{m-1}(1 - 2\lambda\Delta t_i)\right)|v^m|_V^2 \leq 8\left(\frac{1}{2}\sigma_{\min}^2 \int_0^{t_n} e^{-2\lambda\tau}|v_{\Delta t}|_V^2(\tau)d\tau\right). \quad (5.48)$$

We also have

$$\left(\prod_{i=1}^{n}(1 - 2\lambda\Delta t_i)\right)\|v^n\|^2 \leq e^{-2\lambda t_n}|v^n|_V^2.$$

This and (5.48) yield the upper estimate for $[[(v^m)]]_n$.

For the other estimate, we see that (5.47) also yields

$$\begin{aligned}
&\frac{e^{2\lambda t_{m-1}}}{\Delta t_m}\int_{t_{m-1}}^{t_m} e^{-2\lambda\tau}|v_{\Delta t}|_V^2(\tau)d\tau \\
&\leq |v^m|_V^2 \int_0^1 e^{-2\lambda\Delta t_m \tau}\tau d\tau + |v^{m-1}|_V^2 \int_0^1 e^{-2\lambda\Delta t_m\tau}(1-\tau)d\tau \leq \frac{1}{2}(|v^{m-1}|_V^2 + |v^m|_V^2).
\end{aligned}$$

Therefore,

$$\int_0^{t_n} e^{-2\lambda\tau}|v_{\Delta t}|_V^2(\tau)d\tau \leq \frac{1}{2}\sum_{m=1}^{n}\Delta t_m e^{-2\lambda t_{m-1}}(|v^{m-1}|_V^2 + |v^m|_V^2).$$

We know that there exists a constant $\alpha_2 < \frac{1}{2}$ such that

$$\Delta t \leq \frac{\alpha_2}{\lambda} \quad\Rightarrow\quad e^{-2\lambda t_{m-1}} \leq 2\prod_{i=1}^{m-1}(1 - 2\lambda\Delta t_i),$$

and

$$\begin{aligned}
\int_0^{t_n} e^{-2\lambda\tau}|v_{\Delta t}|_V^2(\tau)d\tau &\leq \sum_{m=1}^{n}\Delta t_m \prod_{i=1}^{m-1}(1 - 2\lambda\Delta t_i)(|v^{m-1}|_V^2 + |v^m|_V^2) \\
&\leq (1 + \rho_{\Delta t})\sum_{m=1}^{n}\Delta t_m \prod_{i=1}^{m-1}(1 - 2\lambda\Delta t_i)|v^m|_V^2 + \Delta t_1|v^0|_V^2,
\end{aligned}$$

where $\rho_{\Delta t}$ is defined by (5.20). We have also that if $\Delta t \leq \frac{\alpha_2}{\lambda}$, then

$$e^{-2\lambda t_n} \leq 2\left(\prod_{i=1}^{n}(1 - 2\lambda\Delta t_i)\right), \quad (5.49)$$

so

$$e^{-2\lambda t_n}\|v^n\|^2 \leq 2\left(\prod_{i=1}^{n}(1-2\lambda\Delta t_i)\right)\|v^n\|^2.$$

Finally, (5.27) follows by taking $\alpha = \min(\alpha_1, \alpha_2)$. □

***Proof of Proposition* 5.3.** We begin by plugging $u_{\Delta t}$ into (5.15); we obtain that for all $t \in (t_{n-1}, t_n]$, and $v \in V_0$,

$$\begin{aligned}
&\left(\frac{\partial}{\partial t}u_{\Delta t}(t), v\right) + a_t(u_{\Delta t}(t), v)\\
&= \left(\frac{u^n - u^{n-1}}{\Delta t_n}, v\right) + \big(a_t(u_{\Delta t}(t), v) - a_{t_n}(u_{\Delta t}(t), v)\big) + a_{t_n}(u_{\Delta t}(t) - u^n, v) + a_{t_n}(u^n, v)\\
&= \big(a_t(u_{\Delta t}(t), v) - a_{t_n}(u_{\Delta t}(t), v)\big) + a_{t_n}(u_{\Delta t}(t) - u^n, v).
\end{aligned}$$

Subtracting from (5.15) leads to

$$\begin{aligned}
&\left(\frac{\partial}{\partial t}(u - u_{\Delta t})(t), v\right) + a_t((u - u_{\Delta t})(t), v)\\
&\qquad = -\big(a_t(u_{\Delta t}(t), v) - a_{t_n}(u_{\Delta t}(t), v)\big) - a_{t_n}(u_{\Delta t}(t) - u^n, v).
\end{aligned}$$

We now take $v(t) = (u - u_{\Delta t})(t)e^{-2\lambda t}$, integrate on (t_{n-1}, t_n), and sum up with respect to n. Since $(u - u_{\Delta t})(0) = 0$, we obtain

$$\begin{aligned}
&[[u - u_{\Delta t}]]^2\\
&\leq -2\sum_{m=1}^{n}\int_{t_{m-1}}^{t_m}(a_\tau(u_{\Delta t}(\tau), v) - a_{t_m}(u_{\Delta t}(\tau), v))d\tau - 2\sum_{m=1}^{n}\int_{t_{m-1}}^{t_m} a_{t_m}(u_{\Delta t}(\tau) - u^m, v)d\tau.
\end{aligned}$$

We evaluate separately each term on the right-hand side of this inequality:

$$\begin{aligned}
&\left|\int_{t_{m-1}}^{t_m}(a_\tau(u_{\Delta t}(\tau), v) - a_{t_m}(u_{\Delta t}(\tau), v))d\tau\right|\\
&\leq \Delta t_m\frac{2L_1}{\sigma_{\min}^2}\int_{t_{m-1}}^{t_m}\frac{\sigma_{\min}^2}{2}\left|\left(S\frac{\partial u_{\Delta t}}{\partial S}, S\frac{\partial v}{\partial S}\right)\right| + \frac{2L_2}{\sigma_{\min}^2}\int_{t_{m-1}}^{t_m}\sigma_{\min}^2\left|\left(S\frac{\partial u_{\Delta t}}{\partial S}, v\right)\right|\\
&+\frac{2L_3}{\sigma_{\min}^2}\int_{t_{m-1}}^{t_m}\sigma_{\min}^2|(u_{\Delta t}, v)|\\
&\leq \Delta t_m\frac{2L_1 + L_2 + L_3/2}{\sigma_{\min}^2}\int_{t_{m-1}}^{t_m}\frac{\sigma_{\min}^2}{2}|u_{\Delta t}|_V|u - u_{\Delta t}|_V e^{-2\lambda\tau}d\tau\\
&\leq \Delta t_m\frac{2L_1 + L_2 + L_3/2}{\sigma_{\min}^2}\left(\int_{t_{m-1}}^{t_m}\frac{\sigma_{\min}^2}{2}|u_{\Delta t}|_V^2 e^{-2\lambda\tau}d\tau\right)^{\frac{1}{2}}\left(\int_{t_{m-1}}^{t_m}\frac{\sigma_{\min}^2}{2}|u - u_{\Delta t}|_V^2 e^{-2\lambda\tau}d\tau\right)^{\frac{1}{2}},
\end{aligned}$$

where we have used (5.10). Calling $L = 4L_1 + 2L_2 + L_3$, and using (5.28), we obtain

$$
\begin{aligned}
2&\left|\sum_{m=1}^{n}\int_{t_{m-1}}^{t_m}(a_\tau(u_{\Delta t}(\tau), v) - a_{t_m}(u_{\Delta t}(\tau), v))d\tau\right| \\
&\leq \frac{L}{\sigma_{\min}^2}c(u_0)\Delta t\left(\sum_{m=1}^{n}\int_{t_{m-1}}^{t_m}\frac{\sigma_{\min}^2}{2}|u - u_{\Delta t}|_V^2 e^{-2\lambda\tau}d\tau\right)^{\frac{1}{2}} \\
&\leq \tfrac{L}{\sigma_{\min}^2}c(u_0)\,\Delta t\,[[u - u_{\Delta t}]](t_n).
\end{aligned}
\tag{5.50}
$$

Dealing now with the second term, we have from a Cauchy–Schwarz inequality

$$
\begin{aligned}
&\left|\int_{t_{m-1}}^{t_m} a_{t_m}(u_{\Delta t}(\tau) - u^m, v)d\tau\right| \\
&\quad\leq \mu\left(\int_{t_{m-1}}^{t_m}|u_{\Delta t}(\tau) - u^m|_V^2 e^{-2\lambda\tau}d\tau\right)^{\frac{1}{2}}\left(\int_{t_{m-1}}^{t_m}|v|_V^2 e^{2\lambda\tau}d\tau\right)^{\frac{1}{2}} \\
&\quad\leq \frac{\sqrt{2}\mu}{\sigma_{\min}}\left(\int_{t_{m-1}}^{t_m}|u_{\Delta t}(\tau) - u^m|_V^2 e^{-2\lambda\tau}d\tau\right)^{\frac{1}{2}}\left(\int_{t_{m-1}}^{t_m}\frac{\sigma_{\min}^2}{2}|u - u_{\Delta t}|_V^2 e^{-2\lambda\tau}d\tau\right)^{\frac{1}{2}}.
\end{aligned}
$$

But $u_{\Delta t}(\tau) - u^m = \frac{t^m - \tau}{\Delta t_m}(u^{m-1} - u^m)$, so

$$
\left(\int_{t_{m-1}}^{t_m}|u_{\Delta t}(\tau) - u^m|_V^2 e^{-2\lambda\tau}d\tau\right)^{\frac{1}{2}} \leq \left(\frac{\Delta t_m}{3}\right)^{\frac{1}{2}} e^{-\lambda t_{m-1}}|u^{m-1} - u^m|_V.
$$

Adding and subtracting u_h^{m-1} and u_h^m yields

$$
\begin{aligned}
&\int_{t_{m-1}}^{t_m}|u_{\Delta t}(\tau) - u^m|_V^2 e^{-2\lambda\tau}d\tau \\
&\quad\leq \Delta t_m e^{-2\lambda t_{m-1}}\left(|u_h^{m-1} - u_h^m|_V^2 + |u^{m-1} - u_h^{m-1}|_V^2 + |u^m - u_h^m|_V^2\right).
\end{aligned}
$$

Using (5.49) then (5.27) yields that the sum over m of the last two terms can be bounded by

$$
2(1 + \rho_{\Delta t})\sum_{m=1}^{n}\Delta t_m\prod_{i=1}^{m-1}(1 - 2\lambda\Delta t_i)|u^m - u_h^m|_V^2 \leq \frac{32}{\sigma_{\min}^2}(1 + \rho_{\Delta t})[[u_{\Delta t} - u_{h,\Delta t}]]^2(t_n).
$$

Therefore

$$
\begin{aligned}
&\left|2\sum_{m=1}^{n}\int_{t_{m-1}}^{t_m} a_{t_m}(u_{\Delta t}(\tau) - u^m, v)d\tau\right| \\
&\quad\leq \frac{4\mu}{\sigma_{\min}^2}\left(\begin{array}{l} 16(1 + \rho_{\Delta t})[[u_{\Delta t} - u_{h,\Delta t}]]^2(t_n) \\ +\displaystyle\sum_{m=1}^{n}\Delta t_m e^{-2\lambda t_{m-1}}\frac{\sigma_{\min}^2}{2}|u_h^m - u_h^{m-1}|_V^2 \end{array}\right)^{\frac{1}{2}}[[u - u_{\Delta t}]](t_n),
\end{aligned}
$$

which gives the desired estimate. □

***Proof of Corollary* 5.4.** It is clear that

$$e^{-\lambda t}\left(\frac{\partial}{\partial t}(u-u_{\Delta t})(t),v\right)$$
$$=e^{-\lambda t}\left(-a_t((u-u_{\Delta t})(t),v)-\left(a_t(u_{\Delta t}(t),v)-a_{t_n}(u_{\Delta t}(t),v)\right)-a_{t_n}(u_{\Delta t}(t)-u^n,v)\right),$$

and reusing the steps of the preceding proof, we obtain that

$$|e^{-\lambda t}a_t((u-u_{\Delta t})(t),v)|\le \mu e^{-\lambda t}|(u-u_{\Delta t})(t)|_V|v|_V,$$

which implies that

$$\int_0^{t_n} e^{-\lambda t}|a_t((u-u_{\Delta t})(t),v)|dt\le\frac{\sqrt{2}\mu}{\sigma_{\min}}[[u-u_{\Delta t}]](t_n)\|v\|_{L^2(0,t_n;V_0)},$$

and we can use (5.36). Also,

$$e^{-\lambda t}\left|a_t(u_{\Delta t}(t),v)-a_{t_n}(u_{\Delta t}(t),v)\right|\le\frac{L\Delta t_n}{2\sqrt{2}\sigma_{\min}}\left(\frac{\sigma_{\min}}{\sqrt{2}}|u_{\Delta t}|_V e^{-\lambda t}\right)|v|_V$$

and

$$\sum_{m=1}^n\int_{t_{m-1}}^{t_m} e^{-\lambda t}\left|a_t(u_{\Delta t}(t),v)-a_{t_m}(u_{\Delta t}(t),v)\right|\le\frac{L\Delta t}{2\sqrt{2}\sigma_{\min}}c(u_0)\|v\|_{L^2(0,t_n;V_0)}.$$

Finally,

$$|e^{-\lambda t}a_{t_n}(u_{\Delta t}(t)-u^n,v)|\le\mu e^{-\lambda t}|u_{\Delta t}(t)-u^n|_V|v|_V$$

and

$$\|e^{-\lambda t}(u_{\Delta t}(t)-u^n)\|^2_{L^2(t_{n-1},t_n;V_0)}$$
$$\le\Delta_{t_n}e^{-2\lambda t_{n-1}}(|u^{n-1}-u_h^{n-1}|_V^2+|u^n-u_h^n|_V^2+|u_h^n-u_h^{n-1}|_V^2),$$

which implies that

$$\sum_{m=1}^n\int_{t_{m-1}}^{t_m}|e^{-\lambda t}a_{t_m}(u_{\Delta t}(t)-u^m,v)|$$
$$\le\frac{\sqrt{2}\mu}{\sigma_{\min}}\left(\begin{array}{l}16(1+\rho_{\Delta t})[[u_{\Delta t}-u_{h,\Delta t}]]^2(t_n)\\ +\displaystyle\sum_{m=1}^n\Delta t_m e^{-2\lambda t_{m-1}}\frac{\sigma^2_{\min}}{2}|u_h^m-u_h^{m-1}|_V^2\end{array}\right)^{\frac{1}{2}}\|v\|_{L^2(0,t_n;V_0)}.$$

Combining all these estimates, we have proved the desired result. □

***Proof of Proposition* 5.5.** For any $v\in V_0$ and for any $v_h\in V_{nh}$, we have

$$\begin{aligned}&(u^n-u_h^n,v)+\Delta t_n a_{t_n}(u^n-u_h^n,v)\\&=(u^{n-1}-u_h^{n-1},v_h)+(u^n-u_h^n,v-v_h)+\Delta t_n a_{t_n}(u^n-u_h^n,v-v_h).\end{aligned}\tag{5.51}$$

By integrating by parts, we see that

$$\begin{aligned}
&\left(u^n - u_h^n, v - v_h\right) + \Delta t_n a_{t_n}(u^n - u_h^n, v - v_h) \\
&= \left(u^{n-1} - u_h^{n-1}, v - v_h\right) - \left(u_h^n - u_h^{n-1}, v - v_h\right) \\
&+ \Delta t_n \sum_{\omega \in \mathcal{T}_{n,h}} \left(\int_\omega \left(rS\frac{\partial u_h^n}{\partial S} - ru_h^n \right)(v - v_h)dS - \frac{1}{4}\sum_{i=1}^{2} \sigma^2(\xi_i, t_n)\xi_i^2 \left[\frac{\partial u_h^n}{\partial S}\right](\xi_i)(v - v_h)(\xi_i) \right),
\end{aligned} \tag{5.52}$$

where ξ_i, $i = 1, 2$, are the endpoints of ω. Calling $(S_i)_{i=0,\dots,N_{nh}}$ the mesh points of $\mathcal{T}_{nh}$, $0 = S_0 < S_1 < \cdots < S_{N_{nh}} = \bar{S}$, we choose $v_h \in V_h$ such that $v_h(S_i) = v(S_i)$ for $S_i \neq 0$, and $v_h(0)$ such that $\int_0^{S_1}(v - v_h) = 0$. We can prove that

$$\|v - v_h\|_{L^2(\omega)} \le C\frac{h_\omega}{S_{\max}(\omega)} \left\| S\frac{\partial v}{\partial S} \right\|_{L^2(\omega)} \tag{5.53}$$

and

$$\|S(v - v_h)\|_{L^\infty(\omega)} \le Ch_\omega^{\frac{1}{2}} \left\| S\frac{\partial v}{\partial S} \right\|_{L^2(\omega)}. \tag{5.54}$$

With this choice,

$$-\frac{1}{4}\sum_{i=1}^{2} \sigma^2(\xi_i, t_n)\xi_i^2 \left[\frac{\partial u_h^n}{\partial S}\right](\xi_i)(v - v_h)(\xi_i) = 0.$$

Remark 5.2. *Note that such a choice for v_h, based essentially on Lagrange interpolation, is not possible for a parabolic problem in more than two dimensions (counting the time dimension), because the functions of V are generally not continuous. This explains Remark 5.1.*

Therefore

$$\begin{aligned}
&\left| \begin{array}{l} -\displaystyle\int_\omega \frac{u_h^n - u_h^{n-1}}{\Delta t_n}(v - v_h)dS + \int_\omega \left(rS\frac{\partial u_h^n}{\partial S} - ru_h^n \right)(v - v_h)dS \\ -\displaystyle\frac{1}{4}\sum_{i=1}^{2} \sigma^2(\xi_i, t_n)\xi_i^2 \left[\frac{\partial u_h^n}{\partial S}\right](\xi_i)(v - v_h)(\xi_i) \end{array} \right| \\
&\le C\left(\frac{h_\omega}{S_{\max}(\omega)} \left\| \frac{u_h^n - u_h^{n-1}}{\Delta t_n} - rS\frac{\partial u_h^n}{\partial S} + ru_h^n \right\|_{L^2(\omega)} \right) \left\| S\frac{\partial v}{\partial S} \right\|_{L^2(\omega)} \\
&= C\eta_{n,\omega}\|S\tfrac{\partial v}{\partial S}\|_{L^2(\omega)}.
\end{aligned}$$

Finally, taking $v = (u^n - u_h^n)$ in (5.51) yields

$$\begin{aligned}
&(1 - \lambda\Delta t_n)\|u^n - u_h^n\|^2 + \frac{1}{4}\Delta t_n \sigma_{\min}^2 |u^n - u_h^n|_V^2 \\
&\le \left(\begin{array}{l} \frac{1}{2}\|u^{n-1} - u_h^{n-1}\|^2 + \frac{1}{2}\|u^n - u_h^n\|^2 + \frac{1}{8}\Delta t_n \sigma_{\min}^2 |u^n - u_h^n|_V^2 \\ +2\frac{C^2}{\sigma_{\min}^2}\Delta t_n \displaystyle\sum_{\omega \in \mathcal{T}_{nh}} \eta_{n,\omega}^2 + \frac{1}{8}\Delta t_n \sigma_{\min}^2 |u^n - u_h^n|_V^2 \end{array} \right),
\end{aligned}$$

and therefore

$$(1-2\lambda\Delta t_n)\|u^n-u_h^n\|^2+\frac{1}{4}\Delta t_n\sigma_{\min}^2|u^n-u_h^n|_V^2\le\|u^{n-1}-u_h^{n-1}\|^2+\frac{4C^2}{\sigma_{\min}^2}\Delta t_n\sum_{\omega\in\mathcal{T}_{nh}}\eta_{n,\omega}^2.$$

Multiplying the previous equation by $\prod_{i=1}^{n-1}(1-2\lambda\Delta t_i)$ and summing up over n, we obtain that for a constant c,

$$[[(u^m-u_h^m)]]_n^2\le\frac{c}{\sigma_{\min}^2}\sum_{m=1}^{n}\Delta t_m\prod_{i=1}^{m-1}(1-2\lambda\Delta t_i)\sum_{\omega\in\mathcal{T}_{mh}}\eta_{m,\omega}^2.$$

Therefore, using (5.27),

$$[[(u_{\Delta t}-u_{h,\Delta t})]]^2(t_n)\le\frac{c}{\sigma_{\min}^2}\max(2,1+\rho_{\Delta t})\sum_{m=1}^{n}\Delta t_m\prod_{i=1}^{m-1}(1-2\lambda\Delta t_i)\sum_{\omega\in\mathcal{T}_{mh}}\eta_{m,\omega}^2.\qquad\square$$

***Proof of Proposition* 5.7.** We apply the triangular inequality to η_n:

$$\eta_n\le\sqrt{\Delta t_n}e^{-\lambda t_{n-1}}\frac{\sigma_{\min}}{\sqrt{2}}(|u^n-u^{n-1}|_V+|u^n-u_h^n|_V+|u^{n-1}-u_h^{n-1}|_V).$$

We know from (5.49) that

$$\sqrt{\Delta t_n}e^{-\lambda t_{n-1}}\frac{\sigma_{\min}}{\sqrt{2}}|u^n-u_h^n|_V\le\sqrt{2}[[u^n-u_h^n]],$$

and using (5.20),

$$\sqrt{\Delta t_n}e^{-\lambda t_{n-1}}\frac{\sigma_{\min}}{\sqrt{2}}|u^{n-1}-u_h^{n-1}|_V\le\sqrt{2\rho_{\Delta t}}[[u^{n-1}-u_h^{n-1}]].$$

It remains to estimate $\Delta t_n e^{-2\lambda t_{n-1}}\frac{\sigma_{\min}^2}{2}|u^n-u^{n-1}|_V^2$. For this, we see that

$$\begin{aligned}\Delta t_n e^{-2\lambda t_{n-1}}&\frac{\sigma_{\min}^2}{2}|u^n-u^{n-1}|_V^2\\&\le 2\Delta t_n e^{-2\lambda t_{n-1}}(a_{t_n}(u^n-u^{n-1},u^n-u^{n-1})+\lambda\|u^n-u^{n-1}\|^2)\\&=-4e^{-2\lambda t_{n-1}}\int_{t_{n-1}}^{t_n}a_{t_n}(u_{\Delta t}(\tau)-u^n,u^n-u^{n-1})d\tau\\&\quad+2\lambda\Delta t_n e^{-2\lambda t_{n-1}}\|u^n-u^{n-1}\|^2\\&=I+II+III+IV,\end{aligned}\tag{5.55}$$

where

$$
\begin{aligned}
I &= 4e^{-2\lambda t_{n-1}} \int_{t_{n-1}}^{t_n} \frac{\partial}{\partial t}(u - u_{\Delta t})(\tau)(u^n - u^{n-1})d\tau, \\
II &= 4e^{-2\lambda t_{n-1}} \int_{t_{n-1}}^{t_n} a_\tau(u - u_{\Delta t}(\tau), u^n - u^{n-1})d\tau), \\
III &= 4e^{-2\lambda t_{n-1}} \int_{t_{n-1}}^{t_n} a_\tau(u_{\Delta t}(\tau), u^n - u^{n-1}) - a_{t_n}(u_{\Delta t}(\tau), u^n - u^{n-1})d\tau, \\
IV &= 2\lambda \Delta t_n e^{-2\lambda t_{n-1}} \|u^n - u^{n-1}\|^2.
\end{aligned} \tag{5.56}
$$

We are going to deal separately with the four terms in the right-hand side of (5.55). We have

$$
\begin{aligned}
|I| &= 4e^{-2\lambda t_{n-1}} \left| \int_{t_{n-1}}^{t_n} \frac{\partial}{\partial t}(u - u_{\Delta t})(\tau)(u^n - u^{n-1})d\tau \right| \\
&\le \frac{4\sqrt{2\Delta t_n}}{\sigma_{\min}} e^{-2\lambda t_{n-1}} \left\| \frac{\partial}{\partial t}(u - u_{\Delta t}) \right\|_{L^2(t_{n-1},t_n;V_0')} \frac{\sigma_{\min}}{\sqrt{2}} |u^n - u^{n-1}|_V \\
&= \frac{4\sqrt{2}}{\sigma_{\min}} e^{-\lambda t_{n-1}} \left\| \frac{\partial}{\partial t}(u - u_{\Delta t}) \right\|_{L^2(t_{n-1},t_n;V_0')} \eta_n .
\end{aligned}
$$

Similarly,

$$
\begin{aligned}
|II| &= 4e^{-2\lambda t_{n-1}} \int_{t_{n-1}}^{t_n} a_\tau(u - u_{\Delta t}(\tau), u^n - u^{n-1})d\tau \\
&\le \frac{4\sqrt{2\mu\Delta t_n}}{\sigma_{\min}} e^{-2\lambda t_{n-1}} \|u - u_{\Delta t}\|_{L^2(t_{n-1},t_n;V_0)} \frac{\sigma_{\min}}{\sqrt{2}} |u^n - u^{n-1}|_V \\
&= \frac{4\sqrt{2\mu}}{\sigma_{\min}} e^{-\lambda t_{n-1}} \|u - u_{\Delta t}\|_{L^2(t_{n-1},t_n;V_0)} \eta_n .
\end{aligned}
$$

We write the fourth term as

$$
\begin{aligned}
IV &= 2\Delta t_n \lambda e^{-2\lambda t_{n-1}} \|u^n - u^{n-1}\|^2 \\
&= 2\lambda e^{-2\lambda t_{n-1}} \begin{pmatrix} \displaystyle\int_{t_{n-1}}^{t_n} \left(u^n - u^{n-1} - \Delta t_n \frac{\partial u}{\partial t}\right)(u^n - u^{n-1})d\tau \\ \displaystyle + \int_{t_{n-1}}^{t_n} \Delta t_n \frac{\partial u}{\partial t}(u^n - u^{n-1})d\tau \end{pmatrix} .
\end{aligned}
$$

But, using (5.19),

$$
\begin{aligned}
&2\lambda e^{-2\lambda t_{n-1}} \left| \int_{t_{n-1}}^{t_n} \Delta t_n \frac{\partial u}{\partial t}(u^n - u^{n-1}) d\tau \right| \\
&\quad \le \frac{2\sqrt{2}\lambda}{\sigma_{\min}} \Delta t_n e^{\lambda \Delta t_n} \left\| e^{-\lambda t} \frac{\partial u}{\partial t} \right\|_{L^2(t_{n-1},t_n;V_0')} \eta_n \\
&\quad \le \frac{4\mu\lambda}{\sigma_{\min}^2} \Delta t_n e^{\lambda \Delta t_n} \|u^0\| \eta_n \\
&\quad \le \frac{4\mu\lambda}{\sigma_{\min}^2} \Delta t_n e^{\alpha} \|u^0\| \eta_n
\end{aligned}
$$

and

$$
\begin{aligned}
&2\lambda e^{-2\lambda t_{n-1}} \left| \int_{t_{n-1}}^{t_n} \left(u^n - u^{n-1} - \Delta t_n \frac{\partial u}{\partial t} \right) (u^n - u^{n-1}) d\tau \right| \\
&\quad = 2\lambda e^{-2\lambda t_{n-1}} \Delta t_n \left| \int_{t_{n-1}}^{t_n} \left(\frac{\partial u_{\Delta t}}{\partial t} - \frac{\partial u}{\partial t} \right) (\tau)(u^n - u^{n-1}) d\tau \right| \\
&\quad \le \frac{2\sqrt{2}\lambda}{\sigma_{\min}} e^{-\lambda t_{n-1}} \Delta t_n \left\| \frac{\partial}{\partial t}(u - u_{\Delta t}) \right\|_{L^2(t_{n-1},t_n;V_0')} \eta_n \\
&\quad \le \frac{2\sqrt{2}\alpha}{\sigma_{\min}} e^{-\lambda t_{n-1}} \left\| \frac{\partial}{\partial t}(u - u_{\Delta t}) \right\|_{L^2(t_{n-1},t_n;V_0')} \eta_n .
\end{aligned}
$$

Let us deal with III. Exactly as for (5.50),

- if $n > 1$,

$$
\begin{aligned}
|III| &= 4e^{-2\lambda t_{n-1}} \left| \int_{t_{n-1}}^{t_n} a_\tau(u_{\Delta t}(\tau), u^n - u^{n-1}) - a_{t_n}(u_{\Delta t}(\tau), u^n - u^{n-1}) d\tau \right| \\
&\le \frac{\sqrt{2}L}{\sigma_{\min}} e^{-\lambda t_{n-1}} \Delta t_n \|u_{\Delta t}\|_{L^2(t_{n-1},t_n;V_0)} \eta_n \\
&\le \frac{2L}{\sigma_{\min}} \left(\prod_{i=1}^{n-1} (1 - 2\lambda \Delta t_i) \right)^{\frac{1}{2}} \Delta t_n \|u_{\Delta t}\|_{L^2(t_{n-1},t_n;V_0)} \eta_n \\
&\le \frac{\sqrt{2}L}{\sigma_{\min}} (\max(1, \rho_{\Delta t}))^{\frac{1}{2}} \Delta t_n
\begin{pmatrix}
\Delta t_n \prod_{i=1}^{n-1} (1 - 2\lambda \Delta t_i) |u^n|_V^2 \\
+ \Delta t_{n-1} \prod_{i=1}^{n-2} (1 - 2\lambda \Delta t_i) |u^{n-1}|_V^2
\end{pmatrix}^{\frac{1}{2}} \eta_n \\
&\le \frac{2L}{\sigma_{\min}^2} (\max(1, \rho_{\Delta t}))^{\frac{1}{2}} \Delta t_n [[(u^m)]]_n \eta_n \\
&\le \frac{2L}{\sigma_{\min}^2} (\max(1, \rho_{\Delta t}))^{\frac{1}{2}} \Delta t_n ||u^0|| \eta_n ;
\end{aligned}
$$

- if $n = 1$,

$$
\begin{aligned}
|III| &= 4\left|\int_0^{t_1} a_\tau(u_{\Delta t}(\tau), u^1 - u^0) - a_{t_1}(u_{\Delta t}(\tau), u^1 - u^0)d\tau\right| \\
&\le \frac{\sqrt{2}L}{\sigma_{\min}}\Delta t_1 \|u_{\Delta t}\|_{L^2(0,t_1;V_0)}\eta_1 \\
&\le \frac{L}{\sigma_{\min}}(\Delta t_1)^{\frac{3}{2}}(|u^1|_V^2 + |u^0|_V^2)^{\frac{1}{2}}\eta_1 \\
&\le \frac{2\sqrt{2}L}{\sigma_{\min}^2}\Delta t_1\left([[(u^m)]]_1 + \frac{\sigma_{\min}}{\sqrt{2}}\sqrt{\Delta t_1}|u^0|_V\right)\eta_1 \\
&\le \frac{2\sqrt{2}L}{\sigma_{\min}^2}\Delta t_1\left(\|u^0\| + \frac{\sigma_{\min}}{\sqrt{2}}\sqrt{\Delta t_1}|u^0|_V\right)\eta_1.
\end{aligned}
$$

The desired bounds follow by inserting all these estimates into (5.55). □

***Proof of Proposition* 5.8.** We use (5.52) with $v_h = 0$ and $v = (\frac{u_h^n - u_h^{n-1}}{\Delta t_n} - rS\frac{\partial u_h^n}{\partial S} + ru_h^n)\psi_\omega$ on ω and $v = 0$ on $(0, \bar{S})\backslash\omega$, where ψ_ω is the bubble function on ω, equal to the product of the two barycentric coordinates associated with the endpoints of ω. This leads to

$$
\begin{aligned}
\|v\|_{L^2(\omega)}^2 &= \left\|\left(\frac{u_h^n - u_h^{n-1}}{\Delta t_n} - rS\frac{\partial u_h^n}{\partial S} + ru_h^n\right)\psi_\Omega^{\frac{1}{2}}\right\|_{L^2(\omega)}^2 \\
&= \frac{1}{\Delta t_n}\left(u^{n-1} - u_h^{n-1}, v\right) - \frac{1}{\Delta t_n}\left(u^n - u_h^n, v\right) + a_{t_n}(u_h^n - u^n, v).
\end{aligned}
\tag{5.57}
$$

We note that v is the product of $\psi_\omega^{\frac{1}{2}}$ by a linear function, and we use the inverse inequalities

$$
\forall w \in \mathcal{P}_1(\omega), \quad \|w\|_{L^2(\omega)} \le c_1\|w\psi_\omega^{\frac{1}{2}}\|_{L^2(\omega)} \quad \text{and} \quad \left\|S\frac{\partial w}{\partial S}\right\|_{L^2(\omega)} \le c_2\frac{S_{\max}(\omega)}{h_\omega}\|w\|_{L^2(\omega)}.
$$

Thus,

$$
\begin{aligned}
&\left\|\left(\frac{u_h^n - u_h^{n-1}}{\Delta t_n} - rS\frac{\partial u_h^n}{\partial S} + ru_h^n\right)\right\|_{L^2(\omega)}^2 \\
&\le c_1^2\left\|\left(\frac{u_h^n - u_h^{n-1}}{\Delta t_n} - rS\frac{\partial u_h^n}{\partial S} + ru_h^n\right)\psi_\Omega^{\frac{1}{2}}\right\|_{L^2(\omega)}^2 \\
&\le C\frac{S_{\max}(\omega)}{h_\omega}\begin{pmatrix} \left\|\dfrac{u^{n-1} - u_h^{n-1} - u^n + u_h^n}{\Delta t_n}\right\|_{V_0'(\omega)} \\ +\mu\left\|S\dfrac{\partial(u^n - u_h^n)}{\partial S}\right\|_{L^2(\omega)} \end{pmatrix}\left\|\left(\frac{u_h^n - u_h^{n-1}}{\Delta t_n} - rS\frac{\partial u_h^n}{\partial S} + ru_h^n\right)\right\|_{L^2(\omega)},
\end{aligned}
$$

and therefore

$$\frac{h_\omega}{S_{\max}(\omega)}\left\|\left(\frac{u_h^n-u_h^{n-1}}{\Delta t_n}-rS\frac{\partial u_h^n}{\partial S}+ru_h^n\right)\right\|_{L^2(\omega)}$$
$$\leq C\left(\left\|\frac{u^{n-1}-u_h^{n-1}-u^n+u_h^n}{\Delta t_n}\right\|_{V_0'(\omega)}+\mu\left\|S\frac{\partial(u^n-u_h^n)}{\partial S}\right\|_{L^2(\omega)}\right). \tag{5.58}$$

Remark 5.3. *If $\eta_{n,\omega}$ was computed by (5.41) as in Remark 5.1, then Proposition 5.8 would hold: indeed, for any endpoint $\xi_i \neq 0$ of ω, we call ω' the other element sharing the node ξ_i with ω, and we use (5.52) with $v_h = 0$ and*

$$v = R_{\xi_i,\mathcal{O}} \quad in\ \mathcal{O} = \omega, \omega' \quad and \quad v = 0 \quad in\ (0, \bar{S})\backslash(\omega \cup \omega'),$$

where $R_{\xi_i,\mathcal{O}}$ is the affine function in $\mathcal{O}$ taking the value $1/\xi_i$ at ξ_i and 0 at the other endpoint of $\mathcal{O}$. We see that $\sigma^2(\xi_i,t_n)\xi_i|[\frac{\partial u_h^n}{\partial S}](\xi_i)|$ is bounded by

$$4\left|\int_{\omega\cup\omega'}\left(-\frac{u_h^n-u_h^{n-1}}{\Delta t_n}+rS\frac{\partial u_h^n}{\partial S}-ru_h^n\right)v\,dS\right|$$

plus the same terms as in (5.57). Using (5.30), (5.58) for ω and ω' and the estimate

$$\|R_{\xi_i,\mathcal{O}}\|_{L^2(\mathcal{O})} \leq c\frac{h_{\mathcal{O}}^{\frac{1}{2}}}{\xi_i},$$

we obtain that

$$h_\omega^{\frac{1}{2}}\sum_{i=1}^{2}\sigma^2(\xi_i,t_n)\xi_i\left|\left[\frac{\partial u_h^n}{\partial S}\right](\xi_i)\right| \leq C\left(\begin{array}{l}\left\|\dfrac{u^{n-1}-u_h^{n-1}-u^n+u_h^n}{\Delta t_n}\right\|_{V_0'(K_\omega)}\\ +\mu\left\|S\dfrac{\partial(u^n-u_h^n)}{\partial S}\right\|_{L^2(K_\omega)}\end{array}\right). \quad \square \tag{5.59}$$

Chapter 6

American Options

6.1 Introduction

Unlike European options, American options can be exercised anytime before maturity. Note that since the American option gives more rights to its owner than the European option, its price should be larger.

As before, the price of the underlying asset satisfies

$$dS_\tau = S_\tau \left(r(\tau)d\tau + \sigma(S_\tau, \tau)dW_\tau \right),$$

where the volatility may depend on time and price and where W_τ is a standard Brownian motion. There exists a probability $\mathbb{P}^*$ (the risk neutral probability) under which the discounted price of the asset is a martingale. It can be proven that under the risk neutral probability, the price of the American option of payoff $P_\circ$ and maturity T is

$$P(S_t, t) = \sup_{\tau \in \mathcal{T}_{t,T}} \mathbb{E}^* \left(e^{-\int_t^\tau r(s)ds} P_\circ(S_\tau) \Big| F_t \right), \tag{6.1}$$

where $\mathcal{T}_{t,T}$ denotes the set of stopping times in $[t, T]$ (see [78]).

It is possible to prove that $P(S, t)$ is also the solution to the variational inequality, which is the weak form of the following set of inequalities:

$$\begin{aligned}
\frac{\partial P}{\partial t} + \frac{\sigma^2(S,t)S^2}{2}\frac{\partial^2 P}{\partial S^2} + rS\frac{\partial P}{\partial S} - rP \le 0 \qquad & \text{in } \mathbb{R}_+ \times [0, T), \\
P \ge P_\circ \qquad & \text{in } \mathbb{R}_+ \times [0, T), \\
\left(\frac{\partial P}{\partial t} + \frac{\sigma^2(S,t)S^2}{2}\frac{\partial^2 P}{\partial S^2} + rS\frac{\partial P}{\partial S} - rP \right)(P - P_\circ) = 0 \qquad & \text{in } \mathbb{R}_+ \times [0, T),
\end{aligned} \tag{6.2}$$

with data

$$P|_{t=T} = P_\circ. \tag{6.3}$$

The theory for pricing an American option is not easy: the proof of the abovementioned result can be found in Bensoussan and Lions [13] and Jaillet, Lamberton, and Lapeyre [78].

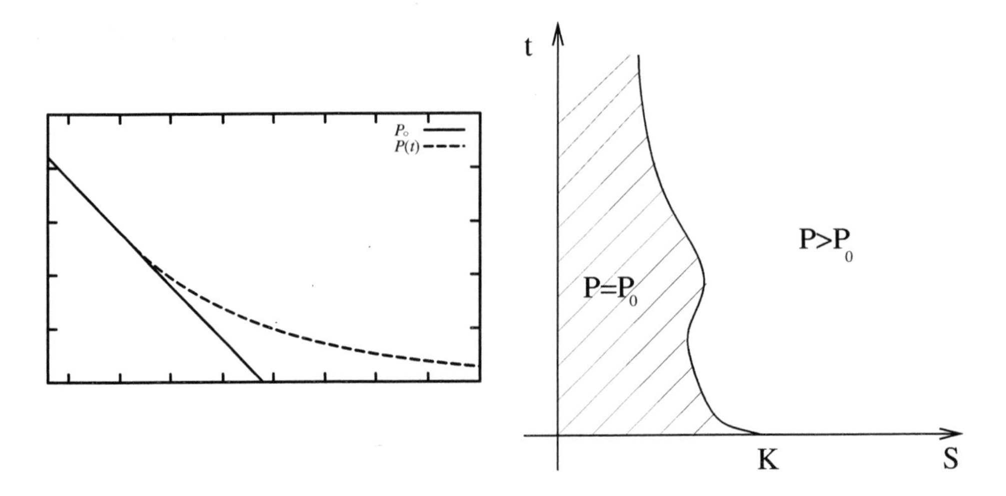

Figure 6.1. *Left: the function $S \mapsto P(S,t)$ at $t = T$ and $t < T$. Right: the region of exercise.*

It can be proved that the price $C(S,t)$ of a European vanilla call is always larger than the payoff $(S-K)_+$. Therefore, the American and European vanilla calls have the same prices. On the contrary, for the American vanilla put, there is a region where $P(S,t)$ coincides with the payoff $P_\circ(S) = (K-S)_+$; see Figure 6.1. If S_t falls in this region, then the put option should be exercised, because it is worth selling the underlying asset in order to buy some risk-free asset. The region where $P(S,t) = P_\circ(S)$ is called the region of exercise.

In what follows, in conformity with the rest of the book, we focus on partial differential equations and their numerical resolutions: the first two sections are devoted to the mathematical analysis of the variational equation and to the free boundary (the boundary of the region of exercise). Then we discuss discretization by finite elements and prove convergence of the method. Later, we consider the solution procedure to the discrete variational inequality, and give a C++ program. Finally, we present two different computations, with a constant and a local volatility: in these tests, we also use a mesh refinement strategy close to the one presented in Chapter 5.

6.2 The Variational Inequality

Calling t the time to maturity, (6.2), (6.3) becomes

$$\begin{aligned} &\frac{\partial P}{\partial t} - \frac{\sigma^2(S,t)S^2}{2}\frac{\partial^2 P}{\partial S^2} - rS\frac{\partial P}{\partial S} + rP \geq 0, \qquad P \geq P_\circ, \\ &\left(\frac{\partial P}{\partial t} - \frac{\sigma^2(S,t)S^2}{2}\frac{\partial^2 P}{\partial S^2} - rS\frac{\partial P}{\partial S} + rP\right)(P - P_\circ) = 0, \end{aligned} \tag{6.4}$$

with Cauchy data

$$P|_{t=0} = P_\circ. \tag{6.5}$$

We focus on the case of a vanilla put, i.e., the payoff function is $P_\circ(S) = (K - S)_+$, but, to a large extent, what follows holds for more general functions.

To write the variational formulation of (6.4), (6.5), we need to use the same Sobolev space V as for the European option, i.e.,

$$V = \left\{ v \in L^2(\mathbb{R}_+) : S\frac{dv}{dS} \in L^2(\mathbb{R}_+) \right\}, \tag{6.6}$$

and we call $\mathcal{K}$ the subset of V:

$$\mathcal{K} = \{v \in V, v \geq P_\circ \text{ in } \mathbb{R}_+\}. \tag{6.7}$$

Since the functions of V are continuous, the inequality in (6.7) has a pointwise meaning. The set $\mathcal{K}$ is a closed and convex subset of V, because convergence in V implies pointwise convergence.

Using the notation defined in Chapter 2, we can formally multiply the first inequality in (6.4) by a smooth nonnegative test function of S and perform some integration by parts. We obtain that

$$0 \leq \frac{d}{dt}\int_{\mathbb{R}_+} P(S,t)\phi(S)dS + a_t(P(t),\phi), \tag{6.8}$$

where the bilinear form a_t is defined in (2.19):

$$\begin{aligned} a_t(v,w) = &\int_{\mathbb{R}_+} \frac{S^2\sigma^2(S,t)}{2}\frac{\partial v}{\partial S}\frac{\partial w}{\partial S}\,dS \\ &+ \int_{\mathbb{R}_+}\left(-r(t) + \sigma^2(S,t) + S\sigma(S,t)\frac{\partial \sigma}{\partial S}(S,t)\right) S\frac{\partial v}{\partial S} w\,dS \\ &+ r(t)\int_{\mathbb{R}_+} vw\,dS. \end{aligned} \tag{6.9}$$

We make the same assumptions as for the European option, namely, (2.20) and (2.21). These imply that the bilinear form a_t is continuous on V uniformly in t; see (2.22) and Gårding's inequality (2.25). Observing that $P_\circ \in V$, the set $\mathcal{K}$ is exactly $\mathcal{K} = P_\circ + \mathcal{K}_\circ$, where $\mathcal{K}_\circ$ is the cone of nonnegative functions in V. Therefore, from (6.8), we see that a variational formulation to (6.4) is as follows:

Find $P \in C^0([0,T]; L^2(\mathbb{R}_+)) \cap L^2(0,T;\mathcal{K})$, such that $\frac{\partial P}{\partial t} \in L^2(0,T;V')$, satisfying

$$P_{|t=0} = P_\circ \quad \text{in } \mathbb{R}_+; \tag{6.10}$$

for a.e. $t \in (0,T)$,

$$\forall v \in \mathcal{K}_\circ, \quad \left(\frac{\partial P}{\partial t}(t), v\right) + a_t(P(t), v) \geq 0, \tag{6.11}$$

or in an equivalent manner,

$$\forall v \in \mathcal{K}, \quad \left(\frac{\partial P}{\partial t}(t), v - P_\circ\right) + a_t(P(t), v - P_\circ) \geq 0; \tag{6.12}$$

and finally

$$\left(\frac{\partial P}{\partial t}(t), P(t) - P_\circ\right) + a_t(P(t), P(t) - P_\circ) = 0. \tag{6.13}$$

Note that (6.12) and (6.13) imply that

$$\forall v \in \mathcal{K}, \quad \left(\frac{\partial P}{\partial t}(t), v - P(t)\right) + a_t(P(t), v - P(t)) \geq 0. \tag{6.14}$$

Conversely, choosing $v = P(t) + w - P_\circ$, with $w \in \mathcal{K}$ in (6.14), implies (6.12). Then using (6.12) and choosing $v = P_\circ$ in (6.14) yields (6.13). Therefore, we have found that the weak formulation is equivalent to the following variational inequality:

Find $P \in C^0([0,T]; L^2(\mathbb{R}_+)) \cap L^2(0,T;\mathcal{K})$, such that $\frac{\partial P}{\partial t} \in L^2(0,T;V')$, satisfying (6.10) and (6.14).

We do not write here the proof of existence and uniqueness for the variational inequality evolution problem (6.10), (6.14). It is given in [2], after the book by Kinderlehrer and Stampacchia [82], which is an excellent reference on the mathematical analysis of variational inequalities. One may also look at the papers by Friedman [56, 57].

The main idea is to observe that

- the price of the American option is always larger than that of the European option, which is positive for $t > 0$. Therefore, $P(t)$ cannot coincide with $P_\circ$ for values of $S > K$;
- it holds that
$$\frac{\partial P_\circ}{\partial t} - \frac{\sigma^2(S,t)S^2}{2}\frac{\partial^2 P_\circ}{\partial S^2} - rS\frac{\partial P_\circ}{\partial S} + rP_\circ = rK1_{S<K} - \frac{\sigma^2(S,t)S^2}{2}\delta_{S=K},$$
so in the region where P and $P_\circ$ coincide, we have
$$\frac{\partial P}{\partial t} - \frac{\sigma^2(S,t)S^2}{2}\frac{\partial^2 P}{\partial S^2} - rS\frac{\partial P}{\partial S} + rP = rK,$$

and to approach the variational inequality by the penalized nonlinear problem:

Find $P_\epsilon \in C^0([0,T]; L^2(\mathbb{R}_+)) \cap L^2(0,T;V)$, such that $\frac{\partial P_\epsilon}{\partial t} \in L^2(0,T;V')$, satisfying (6.10) and

$$\forall v \in V, \quad \left(\frac{\partial P_\epsilon}{\partial t}(t), v\right) + a_t(P_\epsilon(t), v) - rK(1_{\{S<K\}}\mathcal{V}_\epsilon(P_\epsilon(t) - P_\circ), v) = 0, \tag{6.15}$$

with $\mathcal{V}_\epsilon : y \mapsto \mathcal{V}_\epsilon(y) = \mathcal{V}(\frac{y}{\epsilon})$, where $\mathcal{V}$ is a smooth nonincreasing convex function (see Figure 6.2) such that

$$\begin{aligned} &\mathcal{V}(0) = 1, && \\ &\mathcal{V}(y) = 0, && y \geq 1, \\ 0 \geq\ &\mathcal{V}'(y) \geq -2, && 0 \leq y \leq 1. \end{aligned} \tag{6.16}$$

Then the proof of existence for the variational inequality consists of showing that the penalized problem has a unique solution P_ϵ, proving estimates on P_ϵ, in particular that a.e. in t, $P_\epsilon \in \mathcal{K}$, and finally obtaining that when $\epsilon \to 0$, P_ϵ converges to a limit P which is a solution to (6.14). The main arguments are related to the fact that $\mathcal{V}_\epsilon$ is a monotone function and to the weak maximum principle (see [89]). The results are summarized in the following theorem.

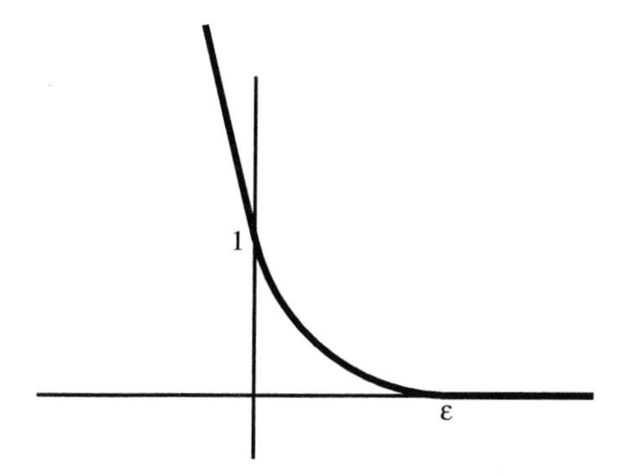

Figure 6.2. *The function $\mathcal{V}_\epsilon$.*

Theorem 6.1. *With σ satisfying assumptions* (2.20) *and* (2.21)*, the problem* (6.10), (6.14) *has a unique solution P which belongs to $C^0([0,+\infty)\times[0,T])$ with $P(0,t)=K$, for all $t\in[0,T]$, and is such that $S\frac{\partial P}{\partial S}, \frac{\partial P}{\partial S}\in L^2(0,T;V)$, $S\frac{\partial P}{\partial S}\in C^0([0,T];L^2(\mathbb{R}_+))$, and $\frac{\partial P}{\partial t}\in L^2(0,T;L^2(\mathbb{R}_+))$.*

The function P is also greater than or equal to P_e, the price of the vanilla European put.

The quantities $\|P\|_{L^2(0,T;V)}$, $\|P\|_{L^\infty(0,T;L^2(\mathbb{R}_+))}$, $\|S\frac{\partial P}{\partial S}\|_{L^2(0,T;V)}$, $\|\frac{\partial P}{\partial S}\|_{L^2(0,T;V)}$, $\|S\frac{\partial P}{\partial S}\|_{L^\infty(0,T;L^2(\mathbb{R}_+))}$, and $\|\frac{\partial P}{\partial t}\|_{L^2(0,T;L^2(\mathbb{R}_+))}$ are bounded independently from σ in the class defined in assumptions (2.20) *and* (2.21).

We have that

$$-1\le\frac{\partial P}{\partial S}\le 0\quad \forall t\in(0,T],\ \text{for a.a. } S>0. \tag{6.17}$$

Furthermore, if P_ϵ is the solution to (6.10), (6.15), we have

$$P_\epsilon(S,t)-\epsilon\le P(S,t)\le P_\epsilon(S,t)\quad \forall S,t, \tag{6.18}$$

and calling μ the function $\mu=\frac{\partial P}{\partial t}+A_tP$, where A_t is the linear operator $V\to V'$, for all $v\in V$, $A_tv=-\frac{\sigma^2(S,t)S^2}{2}\frac{\partial^2 v}{\partial S^2}-rS\frac{\partial v}{\partial S}+rv$, we have

$$0\le\mu\le rK1_{\{P=P_\circ\}}\quad a.e. \tag{6.19}$$

6.3 The Exercise Boundary

Lemma 6.2. *Let P be the solution to* (6.10), (6.14)*. There exists a function $\gamma:(0,T]\to[0,K)$, such that for all $t\in(0,T)$, $\{S;\,P(S,t)=P_\circ(S)\}=[0,\gamma(t)]$. In the open set $\Omega_+=\{(S,t),0<t<T,S>\gamma(t)\}$, we have $\frac{\partial P}{\partial t}+A_tP=0$.*

Proof. The function $S\frac{\partial P}{\partial S}$ belongs to $C^0([0,T];L^2(\mathbb{R}_+))$, and, for any t, we have that $\frac{\partial P}{\partial S}(\cdot,t)\ge-1$ a.e. in S. The set $\{S;\,P(S,t)=P_\circ(S)\}$ is not empty since it contains 0. If it was not connected, then there would exist an interval where $\frac{\partial P}{\partial S}(S,t)<-1$ a.e. This would contradict the bound on $\frac{\partial P}{\partial S}$. Therefore, the set $\{S;\,P(S,t)=P_\circ(S)\}$ is an interval containing 0. □

Theorem 6.3. *For σ satisfying* (2.20) *and* (2.21), *there exists $\gamma_0 > 0$ depending only on $\bar{\sigma}$ such that*

$$\gamma(t) \geq \gamma_0 \quad \forall t \in [0, T]. \tag{6.20}$$

***Proof* (sketched).** Let $\bar{P}$ be the solution of (6.4) with $\sigma = \bar{\sigma}$. The region $\{\bar{P} = P_\circ\}$ is given by $\{(S, t), S \leq \bar{\gamma}(t)\}$, where $\bar{\gamma}$ is a function from $[0, T]$ to $[0, K]$.

It can be proved by using the weak maximum principle that $\frac{\partial^2 \bar{P}}{\partial S^2} \geq 0$ (this relies on the fact that $\bar{\sigma}$ is a constant, so the equation for the penalized problem can be differentiated twice).

We can also prove that $\bar{P}$ is a nondecreasing function of time (by studying the corresponding penalized problem and passing to the limit); therefore $\bar{\gamma}$ is a nonincreasing function of time.

Also, as proven in [82, p. 288], $\bar{\gamma}$ is continuous, so the boundary of $\{\bar{P} = P_\circ\}$ is a negligible set, and $\bar{\mu} = \frac{\partial \bar{P}}{\partial t} + \bar{A}_t \bar{P} = rK 1_{\{\bar{P}=P_\circ\}} = rK 1_{\{(S,t),0<t<T,S<\bar{\gamma}(t)\}}$.

We also have that $P \leq \bar{P}$: indeed, calling $E = \bar{P} - P$,

$$\begin{aligned}
&\frac{\partial E}{\partial t} - \frac{\sigma^2(S,t)S^2}{2}\frac{\partial^2 E}{\partial S^2} - rS\frac{\partial E}{\partial S} + rE - rK\left(1_{\bar{P} \leq P_\circ} - 1_{P \leq P_\circ}\right) \\
&= \left(\frac{S^2}{2}(\bar{\sigma}^2 - \sigma^2(S,t))\frac{\partial^2 \bar{P}}{\partial S^2}\right) + \left(rK(1_{P \leq P_\circ}) - \mu\right).
\end{aligned}$$

The two terms on the right-hand side are nonnegative. The first term is nonnegative thanks to the convexity of $\bar{P}$ with respect to S. The second term is nonnegative thanks to (6.19). Furthermore, $rK(1_{\{\bar{P} \leq P_\circ\}} - 1_{\{P \leq P_\circ\}})E_-$ is nonnegative a.e. thanks to the nonincreasing character of the real function $u \mapsto 1_{u \leq P_\circ(x)}$ (here E_- is the negative part of E). Therefore, a weak maximum principle can be applied and we see that $E_- = 0$ (everywhere since P and $\bar{P}$ are continuous).

Since $P \leq \bar{P}$, we know that $\bar{\gamma} \leq \gamma$. Therefore, if there exists $t_0 < T$ such that $\gamma(t_0) = 0$, then $\bar{\gamma}(t_0) = 0$ and $\bar{\gamma}(t) = 0$ for $t \geq t_0$. It follows that $\bar{P}$ solves the Black–Scholes equation for $t > t_0$, and $\bar{P}(t_0) \leq K$: the maximum principle indicates that $\bar{P}(t) \leq Ke^{-r(t-t_0)}$ for $t > t_0$. This is in contradiction with the fact that $\bar{P} \geq P_\circ$.

The assertion $\gamma(T) = 0$ is also impossible, because we can always look for $\bar{P}$ in a larger time interval.

Since $\bar{\gamma}$ is continuous on $[0, T]$, there exists $\gamma_0 > 0$ such that inequality (6.20) is satisfied. ☐

We now state a regularity result on γ, with the minimal assumptions (2.20) and (2.21).

Theorem 6.4. *For σ satisfying* (2.20) *and* (2.21), *the function $t \mapsto \gamma(t)$ is upper semicontinuous. Moreover, it is right continuous in* $[0, T)$. *Furthermore, for each $t \in (0, T]$, γ has a left limit at t.*

Proof. See [2]. The proof is based on the construction of auxiliary variational inequalities with constant volatilities and uses the weak maximum principle. ☐

As a consequence of Theorem 6.4, we have the following result.

Theorem 6.5. *For σ satisfying* (2.20) *and* (2.21), *the function* $\mu = \frac{\partial P}{\partial t} + A_t P$ *is*

$$\mu = rK1_{\{P=P_\circ\}} = rK1_{\{S\le\gamma(t)\}}. \tag{6.21}$$

In other words, a.e., one of the two conditions $P = P_\circ$ and $\mu = 0$ is not satisfied: we see that there is strict complementarity in (6.4).

Proof. For any time t, both $\lim_{\tau<t}\gamma(\tau)$ and $\lim_{\tau>t}\gamma(\tau)$ exist. Therefore, the function γ is the uniform limit of a sequence of piecewise constant functions γ_k (i.e., γ_k is constant on a finite number of intervals). Thus, calling $\mathcal{J}$ (resp., $\mathcal{J}_k$) the set of points where γ (resp., γ_k) jumps, we have $\mathcal{J} \subset \cup_{k\in\mathbb{N}}\mathcal{J}_k$ because of the uniform convergence and $\mathcal{J}_k$ is finite. Thus the set $\mathcal{J}$ is countable.

Consider now the boundary Γ of the set $\{P = P_\circ\} = \{S \le \gamma(t), t \in [0,T]\}$: we have $\Gamma = (\Gamma \cap \{(S,t), t \in [0,T]\backslash\mathcal{J}\}) \cup (\Gamma \cap \{(S,t), t \in \mathcal{J}\})$. The second set is negligible, since $\mathcal{J}$ is countable. For the first set, we have $\Gamma \cap \{(S,t), t \in [0,T]\backslash\mathcal{J}\} = \{(\gamma(t),t), t \in [0,T]\backslash\mathcal{J}\}$, so it is also negligible. Thus, Γ is negligible.

Therefore, the set $\{P = P_\circ\}$ has the same measure as its interior, on which $\mu = rK$. This proves (6.21). □

For additional information on γ, we need more assumptions on the volatility.

Assumption 6.1. *We assume that for a constant $C_\sigma > 0$,*

$$\left|S^2\frac{\partial^2\sigma}{\partial S^2}\right| + \left|\frac{\partial\sigma}{\partial t}\right| + \left|S\frac{\partial^2\sigma}{\partial S\partial t}\right| \le C_\sigma \tag{6.22}$$

a.e. in $(0,T)\times\mathbb{R}_+$.

Lemma 6.6. *With σ satisfying assumptions* (2.20), (2.21), *and* (6.22), *the solution P to* (6.10), (6.14) *satisfies* $\frac{\partial^2 P}{\partial S^2} \ge 0$ *a.e.*

From Lemma 6.6, we can prove the following result.

Theorem 6.7. *Under the assumptions of Lemma* 6.6, *the function γ is continuous.*

Proof. The proof uses the strong maximum principle and is given in [2]. □

When Theorem 6.7 applies, the graph Γ of γ is a free boundary and the reaction term μ is $rK1_{\{P=P_\circ\}}$: P is a solution to the partial differential equation on one side of Γ and coincides with $P_\circ$ on the other side, and a.e. one of the two conditions $P = P_\circ$ and $\mu = 0$ is not satisfied.

Since the option should be exercised when $P = P_\circ$, the curve Γ is called the *exercise boundary*. We have seen in Theorem 6.3 that Γ does not intersect the axis $S = 0$.

Note also that from Theorem 6.1, the slope of $P(S,t)$ as a function of S is continuous across $S = \gamma(t)$. Only the second derivative of P with respect to S may jump at $S = \gamma(t)$.

Another interesting question (of practical interest) is the behavior of γ for t close to 0, i.e., near the maturity. The answer was given by Barles et al. [11], and Lamberton [86] for constant volatility. For local volatilities, a comparison argument can be applied, and the results cited above can be generalized as follows.

Proposition 6.8. *We have, for t close to* 0,

$$K - \gamma(t) \sim \sqrt{t \log\left(\frac{1}{t}\right)}. \tag{6.23}$$

Therefore, the function γ is not even Hölder continuous with exponent $1/2$.

6.4 Discrete Approximations to the Variational Inequality

We choose to discretize the problem with finite elements because they enable to discretize directly (6.14). A finite difference method is quite possible too, but one has to depart from (6.4).

We localize the problem on $(0, \bar{S})$ as usual, so V becomes

$$V = \left\{ v \in L^2((0,\bar{S})); \, S\frac{\partial v}{\partial S} \in L^2((0,\bar{S}); \, v(\bar{S})) = 0 \right\}$$

(where $\bar{S}$ is large enough so that $P_\circ(\bar{S}) = 0$), and $\mathcal{K} = \{v \in V, v \geq P_\circ\}$. The variational inequality is (6.14) with new meanings for V, $\mathcal{K}$, and a_t.

Moreover, if $\gamma_0 \in (0, K)$ as in (6.20) is known, one can focus on the smaller interval $[\underline{S}, \bar{S}]$ with $0 \leq \underline{S} < \gamma_0$ and obtain the equivalent weak formulation:

Find $P \in L^2((0,T),\mathcal{K}) \cap C^0([0,T]; L^2(\Omega))$, with $\frac{\partial P}{\partial t} \in L^2(0,T;V')$, such that $P(t=0) = P_\circ$ and (6.14) holds for all $v \in \mathcal{K}$, with the new definition of the closed set $\mathcal{K}$:

$$\mathcal{K} = \{v \in V, v \geq P_\circ \text{ in } (0,\bar{S}], \, P = P_\circ \text{ in } (0,\underline{S}]\}. \tag{6.24}$$

We introduce a partition of the interval $[0,T]$ into subintervals $[t_{n-1}, t_n]$, $1 \leq n \leq N$, with $\Delta t_i = t_i - t_{i-1}$, $\Delta t = \max_i \Delta t_i$ and a partition of the interval $[0,\bar{S}]$ into subintervals $\omega_i = [S_{i-1}, S_i]$, $1 \leq i \leq N_h + 1$, such that $0 = S_0 < S_1 < \cdots < S_{N_h} < S_{N_h+1} = \bar{S}$. The size of the interval ω_i is called h_i and we set $h = \max_{i=1,\dots,N_h+1} h_i$. The mesh $\mathcal{T}_h$ of $[0,\bar{S}]$ is the set $\{\omega_1, \dots, \omega_{N_h+1}\}$. In what follows, we will assume that both the strike K and the real number $\underline{S}$ coincide with nodes of $\mathcal{T}_h$: there exist $\alpha < \kappa$, $0 \leq \alpha < \kappa < N_h + 1$, such that $S_\kappa = K$ and $S_{\alpha-1} = \underline{S}$. We define the discrete space V_h by

$$V_h = \left\{ v_h \in V, \; \forall \omega \in \mathcal{T}_h, \, v_{h|\omega} \in \mathcal{P}_1(\omega) \right\}, \tag{6.25}$$

where $\mathcal{P}_1(\omega)$ is the space of linear functions on ω.

Since K is a node of $\mathcal{T}_h$, $P_\circ \in V_h$, and since $\underline{S}$ is also a node of $\mathcal{T}_h$, we can define the closed subset $\mathcal{K}_h$ of V_h by

$$\begin{aligned} \mathcal{K}_h &= \{v \in V_h, \; v \geq P_\circ \text{ in } [0,\bar{S}), \, v = P_\circ \text{ in } [0,\underline{S}]\} \\ &= \{v \in V_h, \; v(S_i) \geq P_\circ(S_i), \; i = 0,\dots,N_h+1, \; v(S_i) = P_\circ(S_i), \; i < \alpha\}. \end{aligned} \tag{6.26}$$

The discrete problem arising from an implicit Euler scheme is as follows:
Find $(P^n)_{0\le n\le N} \in \mathcal{K}_h$ satisfying

$$P^0 = P_\circ, \tag{6.27}$$

and for all n, $1 \le n \le N$,

$$\forall v \in \mathcal{K}_h, \quad \left(P^n - P^{n-1}, v - P^n\right) + \Delta t_n a_{t_n}(P^n, v - P^n) \ge 0. \tag{6.28}$$

Exercise 6.1. *Write the discrete problem arising from the Crank–Nicolson scheme.*

Let $(w^i)_{i=0,\dots N_h}$ be the nodal basis of V_h, and let $\mathbf{M}$ and $\mathbf{A}^m$ in $\mathbb{R}^{(N_h+1)\times(N_h+1)}$ be the mass and stiffness matrices defined by

$$\mathbf{M}_{i,j} = (w^i, w^j), \quad \mathbf{A}^m_{i,j} = a_{t_m}(w^j, w^i), \quad 0 \le i, j \le N_h.$$

These matrices are exactly those described in §4.3. Denoting

$$U^n = (P^n(S_0), \dots, P^n(S_{N_h}))^T \quad \text{and} \quad U^0 = (P_\circ(S_0), \dots, P_\circ(S_{N_h}))^T,$$

(6.28) is equivalent to

$$\begin{aligned} (\mathbf{M}(U^n - U^{n-1}) + \Delta t_n \mathbf{A}^n U^n)_i &\ge 0 \quad \text{for } i \ge \alpha, \\ U^n_i &= U^0_i \quad \text{for } i < \alpha, \\ U^n &\ge U^0, \\ (U^n - U^0)^T (\mathbf{M}(U^n - U^{n-1}) + \Delta t_n \mathbf{A}^n U^n) &= 0. \end{aligned} \tag{6.29}$$

We call $\mathbf{M}_\alpha$ (resp., $\mathbf{A}^n_\alpha$) the block of $\mathbf{M}$ (resp., $\mathbf{A}^n$) corresponding to $\alpha \le i, j \le N_h$.

6.4.1 Existence and Uniqueness of (6.28)

Theorem 6.9. *Consider λ such that Gårding's inequality* (2.25) *holds, and take $\Delta t < \frac{1}{\lambda}$; there exists a unique P^n satisfying* (6.28).

Proof. We reproduce the arguments of Stampacchia (the proof holds for infinite-dimensional Hilbert spaces assuming that the bilinear form $a(\cdot,\cdot)$ is continuous and satisfies a Gårding's inequality).

There exists a unique $Q^n \in V_h$ such that

$$(Q^n, v)_V = \left(P^n - P^{n-1}, v\right) + \Delta t_n a_{t_n}(P^n, v) \quad \forall v \in V_h.$$

We introduce a positive parameter ρ, which will be chosen later. It is straightforward to see that (6.28) is equivalent to

$$\forall v \in \mathcal{K}_h, \quad (P^n, v - P^n)_V - \rho(Q^n, v - P^n)_V \le (P^n, v - P^n)_V, \tag{6.30}$$

which amounts to saying that P^n is the projection of $P^n - \rho Q^n$ on $\mathcal{K}_h$, for the scalar product $(\cdot,\cdot)_V$:

$$P^n = \Pi_{\mathcal{K}_h}(P^n - \rho Q^n), \tag{6.31}$$

where $\Pi_{\mathcal{K}_h}$ is the projector on $\mathcal{K}_h$ with the scalar product $(\cdot,\cdot)_V$. To summarize, we have proven so far that (6.28) is equivalent to the fixed point problem (6.31). To use the Banach–Picard fixed point theorem, we need to find ρ such that the operator $P \mapsto \Pi_{\mathcal{K}_h}(P - \rho Q)$ is a contraction in the norm $\|\cdot\|_V$, where Q is given by $(Q, v)_V = \left(P - P^{n-1}, v\right) + \Delta t_n a_{t_n}(P, v)$ for all $v \in V_h$. Since the projector $\Pi_{\mathcal{K}_h}$ is Lipschitz continuous with the Lipschitz constant 1, it is enough to prove that for a positive constant $\chi < 1$, $\|P - \rho Q - P' - \rho Q'\|_V \le \chi \|P - P'\|_V$ for all $P, P' \in V_h$, and obvious meanings for Q and Q'. This amounts to proving that $\|P - \rho \tilde{Q}\|_V \le \chi \|P\|_V$, where $\tilde{Q}$ is the unique function in V_h such that $(\tilde{Q}, v)_V = (P, v) + \Delta t_n a_{t_n}(P, v)$ for all $v \in V_h$.

From the continuity of a_{t_n}, there exists a positive constant C such that $\|\tilde{Q}\|_V \le C\|P\|_V$.

Consider now λ such that Gårding's inequality (2.25) holds, and take $\Delta t < \frac{1}{\lambda}$. We have

$$\begin{aligned}
\|P - \rho\tilde{Q}\|_V^2 &= \|P\|_V^2 - 2\rho(\tilde{Q}, P)_V + \rho^2\|\tilde{Q}\|_V^2 \\
&= \|P\|_V^2 - 2\rho\left((P, P) + \Delta t_n a_{t_n}(P, P)\right) + \rho^2\|\tilde{Q}\|_V^2 \\
&\le (1 - 2\rho(1 - \lambda\Delta t_n))\|P\|_2^2 + \left(1 - \rho\frac{\sigma^2}{2}\right)|P|_V^2 + \rho^2 C^2 \|P\|_V^2.
\end{aligned}$$

For $\Delta t < \frac{1}{\lambda}$ and ρ small enough (the condition on ρ depends on Δt), we have proved that the mapping $P \mapsto P - \rho\tilde{Q}$ is a contraction in V_h, and we can apply the fixed point theorem: there exists a unique P^n satisfying (6.31), and therefore (6.28). □

6.4.2 Stability

Taking $v = P_\circ$ in (6.28),

$$(P^n - P^{n-1}, P^n - P_\circ) + \Delta t_n a_{t_n}(P^n, P^n - P_\circ) \le 0,$$

which implies from (2.22) and (2.25) that

$$\begin{aligned}
&\|P^n - P_\circ\|^2 + \|P^n - P^{n-1}\|^2 + \frac{\sigma^2}{2}\Delta t_n |P^n|_V^2 - 2\lambda\Delta t_n \|P^n\|^2 \\
&\le \|P^{n-1} - P_\circ\|^2 + 2\mu\Delta t_n |P^n|_V |P_\circ|_V.
\end{aligned}$$

Hence, assuming now that $2\lambda\Delta t < 1$,

$$(1 - 2\lambda\Delta t)\|P^n - P_\circ\|^2 + \frac{\sigma^2}{4}\Delta t_n |P^n|_V^2 \le \|P^{n-1} - P_\circ\|^2 + \Delta t\left(2\lambda\|P_\circ\|^2 + \frac{4\mu^2}{\sigma^2}|P_\circ|_V^2\right).$$

For $m \le N$, multiplying this by $(1 - 2\lambda\Delta t)^{m-n-1}$ and summing over n from $n = 1$ to $n = m$ yield

$$\begin{aligned}
&\|P^m - P_\circ\|^2 + \frac{\underline{\sigma}^2}{4} \sum_{n=1}^{m} (1 - 2\lambda\Delta t)^{m-n-1} \Delta t_n |P^n|_V^2 \\
&\le \Delta t \sum_{n=1}^{m} (1 - 2\lambda\Delta t)^{m-n-1} \left(2\lambda\|P_\circ\|^2 + \frac{4\mu^2}{\underline{\sigma}^2} |P_\circ|_V^2 \right) \\
&= \frac{1 - (1 - 2\lambda\Delta t)^{m+1}}{1 - 2\lambda\Delta t} \left(\|P_\circ\|^2 + \frac{2\mu^2}{\lambda\underline{\sigma}^2} |P_\circ|_V^2 \right).
\end{aligned}$$

Calling $P_{\Delta t}$ the piecewise affine function in time such that $P_{\Delta t}(t_n) = P^n$, the previous estimate implies that for any $c < 1$, there exists a constant C such that for all Δt with $2\lambda\Delta t < c$,

$$\sup_{0 \le t \le T} \|P_{\Delta t}(t)\|^2 + \int_0^T |P_{\Delta t}(t)|_V^2 dt \le C\|P_\circ\|_V^2. \tag{6.32}$$

6.4.3 Convergence

Only for simplicity, we assume hereafter that the grid in t is uniform: $\Delta t_n = \Delta t$ for all n.

Lemma 6.10. *Let P be the solution to* (6.10), (6.14) *with $\mathcal{K}$ given by* (6.24). *If $\tilde{P}_{\Delta t}$ is the piecewise bilinear (in the variables t and S) Lagrange interpolate of P on the (S, t) mesh, we have that $\tilde{P}_{\Delta t} \ge P_\circ$,*

$$\lim_{h,\Delta t \to 0} \|P - \tilde{P}_{\Delta t}\|_{L^2(0,T;V)} + \|P - \tilde{P}_{\Delta t}\|_{L^\infty(0,T;L^2(0,\bar{S}))} = 0, \tag{6.33}$$

and

$$\lim_{h,\Delta t \to 0} \left\| \frac{\partial P}{\partial t} - \frac{\partial \tilde{P}_{\Delta t}}{\partial t} \right\|_{L^2(0,T;V')} = 0. \tag{6.34}$$

Proof. It is possible to interpolate P at the mesh nodes since P is continuous. Then it is clear that the interpolate $\tilde{P}_{\Delta t}$ is a piecewise affine function of time with values in $\mathcal{K}_h$ (with $\tilde{P}_{\Delta t}(t_n) = \tilde{P}^n \in \mathcal{K}_h$). The asymptotics on the error in (6.33) and (6.34) follow from the regularity results in Theorem 6.1. □

Theorem 6.11. *Assume that the coefficients σ and r are smooth enough so that*

$$\lim_{\Delta t \to 0} \sup_{n=1,\dots,N} \sup_{t \in [t_{n-1}, t_n]} \sup_{v,w \in V} \frac{|(a_{t_n} - a_t)(v, w)|}{\|v\|_V \|w\|_V} = 0; \tag{6.35}$$

then

$$\lim_{h,\Delta t \to 0} \|P - P_{\Delta t}\|_{L^2(0,T;V)} + \|P - P_{\Delta t}\|_{L^\infty(0,T;L^2(0,\bar{S}))} = 0. \tag{6.36}$$

Proof. Let us define $R = P - P_\circ \in \mathcal{C}^0([0,T];\mathcal{K}_\circ)$ and $R_{\Delta t} = P_{\Delta t} - P_\circ \in \mathcal{C}^0([0,T];\mathcal{K}_\circ)$, as well as $r_h^m = R_{\Delta t}(t_m) = P_h^m - P_\circ \in \mathcal{K}_\circ \cap V_h$, where

$$\mathcal{K}_\circ = \{v \in V, v \geq 0 \text{ in } (0,\bar{S}), v = 0 \text{ in } (0,\underline{S}]\},$$

and where V_h is defined in (6.25). We have

1. $r_h^0 = 0$,

2. $$r_h^n \in \mathcal{K}_\circ \cap V_h \quad \forall v_h^n \in \mathcal{K}_\circ \cap V_h,$$
$$(r_h^n - r_h^{n-1}, v_h^n - r_h^n) + \Delta t a_{t_n}(r_h^n, v_h^n - r_h^n) \geq -rK\Delta t \int_0^K (v_h^n - r_h^n). \tag{6.37}$$

With λ in the Gårding's inequality (2.25), the inequality in (6.37) can be written as

$$\begin{aligned}&(r_h^n - r_h^{n-1}, v_h^n - r_h^n) - \lambda\Delta t(r_h^n, v_h^n - r_h^n)\\ &+\Delta t\left(a_{t_n}(r_h^n, v_h^n - r_h^n) + \lambda(r_h^n, v_h^n - r_h^n)\right) \geq -rK\Delta t \int_0^K (v_h^n - r_h^n).\end{aligned} \tag{6.38}$$

But

$$\begin{aligned}(r_h^n - r_h^{n-1}, v_h^n - r_h^n) = {}& (v_h^n - v_h^{n-1}, v_h^n - r_h^n) - \frac{1}{2}\|v_h^n - r_h^n\|^2\\ &+\frac{1}{2}\|v_h^{n-1} - r_h^{n-1}\|^2 - \frac{1}{2}\|v_h^n - v_h^{n-1} - r_h^n + r_h^{n-1}\|^2\end{aligned}$$

and

$$(r_h^n, v_h^n - r_h^n) = (v_h^n, v_h^n - r_h^n) - \|v_h^n - r_h^n\|^2.$$

Therefore,

$$\begin{aligned}&(r_h^n - r_h^{n-1} - \lambda\Delta t r_h^n, v_h^n - r_h^n) = (v_h^n - v_h^{n-1} - \lambda\Delta t v_h^n, v_h^n - r_h^n)\\ &-\frac{1}{2}(1 - 2\lambda\Delta t)\|v_h^n - r_h^n\|^2 + \frac{1}{2}\|v_h^{n-1} - r_h^{n-1}\|^2 - \frac{1}{2}\|v_h^n - v_h^{n-1} - r_h^n + r_h^{n-1}\|^2,\end{aligned} \tag{6.39}$$

and (6.38) becomes

$$\begin{aligned}&(v_h^n - v_h^{n-1} - \lambda\Delta t v_h^n, v_h^n - r_h^n) - \frac{1}{2}(1 - 2\lambda\Delta t)\|v_h^n - r_h^n\|^2 + \frac{1}{2}\|v_h^{n-1} - r_h^{n-1}\|^2\\ &-\frac{1}{2}\|v_h^n - v_h^{n-1} - r_h^n + r_h^{n-1}\|^2 + \Delta t\left(a_{t_n}(r_h^n, v_h^n - r_h^n) + \lambda(r_h^n, v_h^n - r_h^n)\right)\\ &\geq -rK\Delta t \int_0^K (v_h^n - r_h^n).\end{aligned} \tag{6.40}$$

Multiplying (6.40) by $(1-2\lambda\Delta t)^{n-N-1}$, taking $v_h^m = \tilde{r}_h^m = \tilde{P}^m - P_\circ$, and summing over n, we get (using the fact that $r_h^0 = \tilde{r}_h^0 = 0$)

$$
\begin{aligned}
&-\sum_{n=1}^{N}(1-2\lambda\Delta t)^{n-N-1}(\tilde{r}_h^n - \tilde{r}_h^{n-1} - \lambda\Delta t \tilde{r}_h^n, \tilde{r}_h^n - r_h^n)\\
&+\frac{1}{2}\|\tilde{r}_h^N - r_h^N\|^2 + \frac{1}{2}\sum_{n=1}^{N}(1-2\lambda\Delta t)^{n-N-1}\|\tilde{r}_h^n - \tilde{r}_h^{n-1} - r_h^n + r_h^{n-1}\|^2\\
&+\Delta t\sum_{n=1}^{N}(1-2\lambda\Delta t)^{n-N-1}\left(a_{t_n}(r_h^n, r_h^n - \tilde{r}_h^n) + \lambda(r_h^n, r_h^n - \tilde{r}_h^n)\right)\\
&\qquad\qquad \le -rK\Delta t\sum_{n=1}^{N}(1-2\lambda\Delta t)^{n-N-1}\int_0^K (r_h^n - \tilde{r}_h^n).
\end{aligned}
\tag{6.41}
$$

On the other hand, calling ϕ the piecewise constant function in $(0, T]$ defined by $\phi(t) = (1-2\lambda\Delta t)^{n-N-1}$ if $t_{n-1} < t \le t_n$, we know from (6.12) that, for every function $Q \in L^2(0, T; \mathcal{K}_\circ)$,

$$
\begin{aligned}
&\int_0^T \phi\left(\frac{\partial R}{\partial t} - \lambda R, Q - R\right) + \int_0^T \phi(\lambda(R, Q-R) + a_t(R, Q-R))\\
&\ge -rK\int_0^T\int_0^K \phi(Q-R).
\end{aligned}
\tag{6.42}
$$

Taking for Q the piecewise constant function with value in V_h such that $Q(t) = r_h^n$, for $t \in (t_{n-1}, t_n]$, we get that

$$
\begin{aligned}
&\sum_{n=1}^{N}(1-2\lambda\Delta t)^{n-N-1}\int_{t_{n-1}}^{t_n}\left(\frac{\partial R}{\partial t} - \lambda R, r_h^n - R\right)\\
&+\sum_{n=1}^{N}(1-2\lambda\Delta t)^{n-N-1}\int_{t_{n-1}}^{t_n}\left(\lambda(R, r_h^n - R) + a_t(R, r_h^n - R)\right)\\
&\ge -rK\sum_{n=1}^{N}(1-2\lambda\Delta t)^{n-N-1}\int_{t_{n-1}}^{t_n}\int_0^K (r_h^n - R)dS,
\end{aligned}
\tag{6.43}
$$

and using the stability estimate (6.32), the asymptotics on $R - \tilde{R}$, (6.33), (6.34), as well as (6.35), we obtain that

$$
\limsup_{h,\Delta t\to 0}\left(
\begin{aligned}
&\sum_{n=1}^{N}(1-2\lambda\Delta t)^{n-N-1}(\tilde{r}_h^n - \tilde{r}_h^{n-1} - \lambda\Delta t\tilde{r}_h^n, \tilde{r}_h^n - r_h^n)\\
&+\Delta t\sum_{n=1}^{N}(1-2\lambda\Delta t)^{n-N-1}\left(\lambda(\tilde{r}_h^n, \tilde{r}_h^n - r_h^n) + a_{t_n}(\tilde{r}_h^n, \tilde{r}_h^n - r_h^n)\right)\\
&-rK\Delta t\sum_{n=1}^{N}(1-2\lambda\Delta t)^{n-N-1}\int_0^K (r_h^n - \tilde{r}_h^n)dS
\end{aligned}
\right) \le 0.
\tag{6.44}
$$

Together with (6.41) and Gårding's inequality (2.25), this implies that

$$\lim_{h,\Delta t\to 0}\left(\begin{array}{l}\frac{1}{2}\|\tilde{r}_h^N - r_h^N\|^2 \\ +\frac{1}{2}\sum_{n=1}^{N}(1-2\lambda\Delta t)^{n-N-1}\|\tilde{r}_h^n - \tilde{r}_h^{n-1} - r_h^n + r_h^{n-1}\|^2 \\ +\Delta t\frac{\sigma^2}{4}\sum_{n=1}^{N}(1-2\lambda\Delta t)^{n-N-1}|r_h^n - \tilde{r}_h^n|_V^2\end{array}\right) = 0. \tag{6.45}$$

Then, realizing that in (6.45), N could be replaced by any n, $1 \le n \le N$, and combining (6.45) with (6.33), (6.34) yields (6.36). □

6.4.4 The Discrete Exercise Boundary

One may ask if there is a well-defined exercise boundary $t \to \gamma_h(t)$ also in the discrete problem. A positive answer has been given by Jaillet, Lamberton, and Lapeyre [78] in the case of a constant volatility, an implicit Euler scheme, and a uniform mesh in the logarithmic variable. The main argument of the proof lies in the fact that the solution to the discrete problem is nondecreasing with respect to the variable t. With a local volatility, this may not hold (see the numerical example below). The result of Jaillet, Lamberton, and Lapeyre has been generalized to a local volatility in [4], in the special case when the mesh is uniform in the variable S: here too, the discrete problem has a free boundary. The proof no longer relies on the monotonic character of the discrete solution with respect to t but on the discrete analogue of the bounds (6.17), i.e., $-1 \le \frac{\partial P}{\partial S} \le 0$. This is proved by studying a penalized problem for (6.28) (the discrete analogue to (6.15)) and by using a discrete maximum principle on the partial derivative with respect to S (for this reason, a uniform mesh is needed). We can summarize this in the following theorem.

Theorem 6.12. *Assume that the grid $\mathcal{T}_h$ is uniform and that $\underline{S} > 0$. Assume also that the parameters h and $\frac{h^2}{\Delta t}$ are small enough so that the matrices $\mathbf{A}_\alpha^n$ and $\mathbf{M}_\alpha + \Delta t_n \mathbf{A}_\alpha^n$ are tridiagonal irreducible M-matrices for all n, $1 \le n \le N$.*

Then there exist N real numbers γ_h^n, $1 \le n \le N$, such that

$$\begin{array}{l}\underline{S} \le \gamma_h^n < K, \\ \gamma_h^n \text{ is a node of } \mathcal{T}_h, \\ \forall i, 0 \le i \le N_h, \quad P^n(S_i) = P_\circ(S_i) \Leftrightarrow S_i \le \gamma_h^n.\end{array} \tag{6.46}$$

We believe that this may be extended to somewhat more general meshes.

6.5 Solution Procedures

We propose hereafter four algorithms for solving the discrete variational inequalities arising from an implicit time-stepping procedure. Other methods based on penalization can be found, for example, in [50].

6.5.1 The Projected SOR Algorithm

At each time step, we have to solve (6.32), which belongs to the following class of problems: Consider a matrix $\mathbf{A} \in \mathbb{R}^{M\times M}$, two vectors F and B in $\mathbb{R}^M$, and the variational inequality: find $U \in \mathbb{R}^M$ such that

$$\begin{aligned} &(V-U)^T(\mathbf{A}U - F) \geq 0 \quad \forall V \in \mathcal{K}, \\ &U \in \mathcal{K}, \end{aligned} \tag{6.47}$$

where

$$\mathcal{K} = \{V \in \mathbb{R}^M \ : \ V \geq B\}.$$

We assume that for two positive constants α and β, $\mathbf{A}$ satisfies

$$\alpha |V|^2 \leq V^T\mathbf{A}V \leq \beta|V|^2 \quad \forall V \in \mathbb{R}^M, \tag{6.48}$$

so (6.47) has a unique solution.

Let ω be a real number, $0 < \omega \leq 1$. The idea is to construct a sequence of vectors $U^{(k)}$ which hopefully converges to U by using a one-step recursion formula (starting from an initial guess $U^{(0)}$), $U^{(k+1)} = \psi(U^{(k)})$, where ψ is the nonlinear mapping in $\mathbb{R}^m$:

$$\begin{aligned} &\psi : X \mapsto Z = \psi(X) : \\ &\forall i = 1, \dots, M, \quad Z_i = \max(Y_i, B_i), \quad \text{and } Y_i \text{ is given by} \\ &\frac{1}{\omega}\mathbf{A}_{ii}Y_i + \sum_{j<i}\mathbf{A}_{ij}Z_j = F_i + \left(\frac{1}{\omega} - 1\right)\mathbf{A}_{ii}X_i - \sum_{j>i}\mathbf{A}_{ij}X_j. \end{aligned} \tag{6.49}$$

This construction is a modification of the so-called successive overrelaxation (SOR) method used to solve iteratively systems of linear equation; see [8, 62, 106] for iterative methods for systems of linear equations. For solving approximately the system $\mathbf{A}X = F$, the SOR methods constructs the sequence $(X^{(k)})_k$ (starting from an initial guess $X^{(0)}$) by the recursion

$$\forall i = 1, \dots, M, \quad \frac{1}{\omega}\mathbf{A}_{ii}X_i^{(k+1)} + \sum_{j<i}\mathbf{A}_{ij}X_j^{(k+1)} = F_i + \left(\frac{1}{\omega} - 1\right)\mathbf{A}_{ii}X_i^{(k)} - \sum_{j>i}\mathbf{A}_{ij}X_j^{(k)}.$$

Lemma 6.13. *If $\mathbf{A}$ is a diagonal dominant matrix and if $0 < \omega \leq 1$, then the mapping ψ defined in* (6.49) *is a contraction in $\mathbb{R}^M$ of the norm $\|\cdot\|_\infty$ ($\|V\|_\infty = \max_{1\leq i\leq M}|V_i|$).*

Proof. Denote $Z = \psi(X)$, $Z' = \psi(X')$. Obviously,

$$\forall j, \ 1 \leq j \leq M, \quad |Z_j - Z'_j| \leq |Y_j - Y'_j|, \tag{6.50}$$

where

$$\frac{1}{\omega}\mathbf{A}_{jj}Y_j + \sum_{k<j}\mathbf{A}_{jk}Z_k = F_j + \left(\frac{1}{\omega} - 1\right)\mathbf{A}_{jj}X_j - \sum_{k>j}\mathbf{A}_{jk}X_k$$

and

$$\frac{1}{\omega}\mathbf{A}_{jj}Y'_j + \sum_{k<j}\mathbf{A}_{jk}Z'_k = F_j + \left(\frac{1}{\omega} - 1\right)\mathbf{A}_{jj}X'_j - \sum_{k>j}\mathbf{A}_{jk}X'_k.$$

Therefore, denoting $\delta X = X - X'$, $\delta Y = Y - Y'$, and $\delta Z = Z - Z'$, we have

$$\frac{1}{\omega}\mathbf{A}_{ii}\delta Y_i + \sum_{j<i}\mathbf{A}_{ij}\delta Z_j = \left(\frac{1}{\omega} - 1\right)\mathbf{A}_{ii}\delta X_i - \sum_{j>i}\mathbf{A}_{ij}\delta X_j. \tag{6.51}$$

In view of (6.50), denoting by i, $1 \le i \le M$, an index such that $\|\delta Y\|_\infty = |\delta Y_i|$, it is enough to prove that for a constant ρ, $0 \le \rho < 1$, $|\delta Y_i| \le \rho\|\delta X\|_\infty$. But (6.50), (6.51), and $0 < \omega \le 1$ imply

$$\left(\frac{1}{\omega}|\mathbf{A}_{ii}| - \sum_{j<i}|\mathbf{A}_{ij}|\right)|\delta Y_i| \le \left(\left(\frac{1}{\omega} - 1\right)|\mathbf{A}_{ii}| + \sum_{j>i}|\mathbf{A}_{ij}|\right)\|\delta X\|_\infty \tag{6.52}$$

and

$$\|\delta Y\|_\infty \le \frac{(\frac{1}{\omega} - 1)|\mathbf{A}_{ii}| + \sum_{j>i}|\mathbf{A}_{ij}|}{\frac{1}{\omega}|\mathbf{A}_{ii}| - \sum_{j<i}|\mathbf{A}_{ij}|}\|\delta X\|_\infty,$$

for both the denominator and the numerator are positive, due to the diagonal dominance of $\mathbf{A}$ and the choice of ω. Calling

$$\rho = \max_{1\le k\le M}\frac{(\frac{1}{\omega} - 1)|\mathbf{A}_{kk}| + \sum_{j>k}|\mathbf{A}_{ij}|}{\frac{1}{\omega}|\mathbf{A}_{kk}| - \sum_{j<k}|\mathbf{A}_{kj}|},$$

we have $0 \le \rho < 1$ and $\|\psi(X) - \psi(X')\|_\infty \le \rho\|\delta X\|_\infty$. □

As a consequence of Lemma 6.13, the sequence $(U^{(k)})_k$ converges to the unique fixed point of ψ. There remains to prove the following lemma.

Lemma 6.14. *With the assumptions of Lemma* 6.13, *if U is the solution to* (6.47), *then U is the unique fixed point of ψ defined in* (6.49).

Proof. We know that U satisfies

$$U_i - B_i \ge 0, \quad \sum_j \mathbf{A}_{ij}U_j - F_i \ge 0, \quad \left(\sum_j \mathbf{A}_{ij}U_j - F_i\right)(U_i - B_i) = 0.$$

Denote $Z = \psi(U)$. Let us prove by induction on the indices that $Z = U$: if $U_1 > B_1$, then

$$\frac{1}{\omega}\mathbf{A}_{11}U_1 = F_1 + \left(\frac{1}{\omega} - 1\right)\mathbf{A}_{11}U_1 - \sum_{j>1}\mathbf{A}_{1j}U_j;$$

therefore with Y defined as above, $Y_1 = U_1$ and $Z_1 = U_1$. If, on the contrary, $U_1 = B_1$, we have that $Y_1 \le U_1$, and therefore $Z_1 = B_1$.

Suppose now that $Z_j = U_j$ for $j < i$. If $U_i > B_i$, we have that

$$\frac{1}{\omega}\mathbf{A}_{ii}U_i = F_i + \left(\frac{1}{\omega} - 1\right)\mathbf{A}_{ii}U_i - \sum_{j>i}\mathbf{A}_{ij}U_j - \sum_{j<i}\mathbf{A}_{ij}U_j \quad \text{and} \quad Y_i = U_i, \text{ i.e., } Z_i = U_i.$$

If, on the contrary, $U_i = B_i$, then $Y_i \le U_i$ and $Z_i = B_i$. □

Thanks to Lemmas 6.13 and 6.14, we have proved the following result.

Theorem 6.15. *With the assumptions of Lemma* 6.13, *the projected SOR method converges to the solution of* (6.47).

6.5.2 The Brennan and Schwartz Algorithm

The Brennan and Schwartz algorithm [19] is an algorithm which works under rather restrictive assumptions. It has been studied by Jaillet, Lamberton, and Lapeyre [78]. The algorithm is a modification of the Gaussian elimination algorithm, and is based on the factorization of $\mathbf{A}$: $\mathbf{A} = \mathbf{UL}$, where $\mathbf{U}$ is an upper triangular matrix whose diagonal coefficients are all 1 and $\mathbf{L}$ is a lower triangular matrix. In fact, the bandwidths of both $\mathbf{U}$ and $\mathbf{L}$ are 2, so solving a system with $\mathbf{U}$ or $\mathbf{L}$ is very easy, and can be done with a computational cost linear with respect to M. The complexity for computing $\mathbf{L}$ and $\mathbf{U}$ is linear too with respect to M. Then, assuming that $\mathbf{L}$ and $\mathbf{U}$ are computed, the algorithm for solving (6.47) is as follows:

- Solve $\mathbf{U}Y = F$.
- For $i = 1$ to M, do
  $U_i = \max((Y_i - \mathbf{L}_{i,i-1}U_{i-1})/\mathbf{L}_{i,i}, B_i)$.

In [78], Jaillet, Lamberton, and Lapeyre studied this method for the American put in the logarithmic variable, for a uniform grid in x. They prove that if the step size is small enough, then the Brennan and Schwartz algorithm yields the solution to (6.47): the proof relies on the fact that the discrete problem has a free boundary and on the fact that the matrix is an M-matrix.

This algorithm is really fast (its complexity grows linearly with M). However, it may lack robustness when the mesh is highly nonuniform.

6.5.3 A Front-Tracking Algorithm

Here, we propose an algorithm for computing the solution of (6.28), assuming that the free boundary is the graph of a function. In our experience, this algorithm, based on tracking the free boundary, is more robust (and slightly more expensive) than the Brennan and Schwartz algorithm (see [78]). Since the free boundary is the graph of a function, the idea is to look for γ_h^n by doing the following:

- Start from $\gamma_h^n = \gamma_h^{n-1}$.
- Solve the discrete problem corresponding to

$$\frac{P^n - P^{n-1}}{\Delta t_n} - \frac{\sigma^2(S,t_n)S^2}{2}\frac{\partial^2 P^n}{\partial S^2} - rS\frac{\partial P^n}{\partial S} + rP^n = 0 \quad \text{for } \gamma_h^n < S < \bar{S},$$
$$P^n = P_\circ \quad \text{for } 0 \le S \le \gamma_h^n,$$

  and $P^n(\bar{S}) = 0$.

- If P^n satisfies (6.28) and $P^n \geq P_\circ$, stop; else shift the point γ_h^n to the next node on the mesh left/right according to which constraint is violated by P^n.

With the notation introduced above, the algorithm for computing P_h^n is as follows.

Algorithm.
Choose k such that $\gamma_h^{n-1} = S_k$; set found=false;
while(**not** found)
.. solve

$$\begin{aligned}(\mathbf{M}(U^n - U^{n-1}) + \Delta t_n \mathbf{A}^n U^n)_i &= 0 \quad \text{for } i \geq k,\\ U_i^n &= U_i^0 \quad \text{for } i < k.\end{aligned} \tag{6.53}$$

.. **if** $((U^n - U^0)_k < 0)$
.. found=false; $k = k + 1$;
.. **else**
.. compute $a = (\mathbf{M}(U^n - U^{n-1}) + \Delta t_n \mathbf{A}^n U^n)_{k-1}$;
.. **if** $(a < 0)$
.. found=false; $k = k - 1$;
.. **else** found=true.

In our tests, we have computed the average (over the time steps) number of iterations to obtain the position of the free boundary: it was found that (with a rather fine time-mesh), this number is smaller than 2.

A Program in C++ for the American Put. Here we give a program for computing an American put. The mesh in S can vary in time exactly as in Chapter 5, so mesh adaption can be performed. For simplicity, the time scheme is Euler's implicit scheme. We write first a function for a single time step.

ALGORITHM 6.1. Time step.

```
int American_Euler_Scheme::Time_Step(int it,  vector< KN<double> >& P, const
double K, const int free_bdry_guess_p)
{
  int i,n;
  double dt,t,S,h_p,h_n,r;
  double a,b,c,d,e;

  n=S_steps[it].size();
  MatriceProfile<double> A(n,2);
  KN<double> y(n);
  KN<double> rhs(n);                                       //   right-hand side
  KN<double> ob(n);                                              //   obstacle

  t=grid_t[it];                                               //   current time
  dt=t-grid_t[it-1];                                              //   time step
  r=rate(t);
  e=0.5*dt;

                                                        //   assemble the matrix
  h_n=S_steps[it][0];
```

```
  A(0,0)=e*r*h_n+ h_n/3;
  A(0,1)=h_n/6;
  for(i=1;i< n-1;i++)
    {
      h_p=h_n;
      S=S_nodes[it][i];
      h_n=S_steps[it][i];
      a=pow(S*vol(t,S),2);
      b=a/h_p;
      c=a/h_n;
      d=r*S;

      A(i,i)=e*(b+c+r*(h_p+h_n))+ (h_p+h_n)/3;
      A(i,i-1)=e*(-b+d)+h_p/6;
      A(i,i+1)=e*(-c-d)+h_n/6;
    }
  h_p=h_n;
  S=S_nodes[it][i];
  h_n=S_steps[it][i];
  a=pow(S*vol(t,S),2);
  b=a/h_p;
  c=a/h_n;
  d=r*S;
  A(i,i)=e*(b+c+r*(h_p+h_n))+ (h_p+h_n)/3;
  A(i,i-1)=e*(-b+d)+h_p/6;


                                         //    assemble the right-hand side;
  if (change_grid[it])
    {
      build_rhs(rhs,P[it-1],S_steps[it-1],S_nodes[it-1],S_steps[it],
                S_nodes[it]);
    }
  else
    build_rhs(rhs,P[it-1],S_steps[it-1],S_nodes[it-1],S_steps[it],
              S_nodes[it]);
                                                //   the obstacle function
  int iK=0;
  while (S_nodes[it][iK]<K)
    iK++;
  ob=0;
  for (i=0;i<iK;i++)
    ob(i)=K-S_nodes[it][i];
              //    first guess for the position of the exercise boundary
  int free_bdry_guess=0;
  while  (S_nodes[it][free_bdry_guess]< S_nodes[it-1][free_bdry_guess_p])
    free_bdry_guess++;
                                                  //    solves the problem
  return  free_bdry<double>(A, P[it], y, rhs, ob, free_bdry_guess);
}
```

The program for the free boundary tracking is as follows.

ALGORITHM 6.2. Free boundary localization.

```
int
free_bdry(const  MatriceProfile<double> & A, KN<double> & x,   KN<double> & y,
          const  KN<double> & b, const  KN<double> & ob, int  free_bdry_guess)
  /* ob : a vector describing the obstacle function */
  /* x : the unknown function */
  /* y : auxiliary function */
  /* A the matrix of the problem */
  /* b the right-hand side */
  /* free_bdry_guess : guess for the position of the free boundary*/
  /* it should come from the previous time step */
{
  int found =0;
  int iterations=0;
  int fbpos= free_bdry_guess;
  int sense_of_motion=0;
  int prev_sense_of_motion=0;
  int not_infinite_loop=1;

  /* y contains b-A * ob */
  /* recall that the constraint b-A*x <= 0*/
  /* is to be satisfied */
  /* so the contact zone is a subset of the region b-A*ob <=0 */
  y=A*ob;
  y-=b;
  y*=-1.;
  int fst_ineq_threshold=0;
  while(y(fst_ineq_threshold+1)<0)
    fst_ineq_threshold++;
  /*  fst_ineq_threshold is the extremal point of the zone  b-A*ob < 0 */
  while((!found) && abs(fbpos-free_bdry_guess)<150 && iterations <150)
    {
      iterations++;
      prev_sense_of_motion=sense_of_motion;
      int matsize=x.size()-fbpos;    //   we shall solve a Dirichlet problem
                                                //    in the zone i>= fbpos

      /* fills the matrix and RHS*/
      KN<double> xaux(matsize);
      MatriceProfile<double>  auxmat(matsize,2);
      for (int i=0; i<matsize;i++)
        {
          auxmat(i,i)=A(i+fbpos,i+fbpos);
          xaux(i)=b(i+fbpos);
        }
      auxmat(0,0)=1.;
      for (int i=0; i<matsize-1;i++)
        {
          auxmat(i,i+1)=A(i+fbpos,i+fbpos+1);
          auxmat(i+1,i)=A(i+fbpos+1,i+fbpos);
        }
      auxmat(0,1)=0.;
      xaux(0)=ob(fbpos);
      auxmat.LU();
```

```
      auxmat.Solve(xaux,xaux);                        //   solves the system
      /*checks if the guess for the free boundary is correct*/
      /* if not, proposes a new guess */
      found=1;
      if (xaux(1)<ob(fbpos+1))              //   checks the inequality b>= ob
        {
          fbpos++;
          found=0;
          sense_of_motion=1;
        }
      else                                  //   check the inequality A*x>= b
        {
          double aux= A(fbpos,fbpos)*ob(fbpos)
            +A(fbpos,fbpos-1)*ob(fbpos-1)
            +A(fbpos,fbpos+1)*xaux(1);
          if (aux<b(fbpos))
            {
              found=0;
              fbpos--;
              sense_of_motion=-1;
            }
          else
            if(fbpos-1>fst_ineq_threshold)
              {
                found=0;
                fbpos--;
                sense_of_motion=-1;
              }
        }
      not_infinite_loop=sense_of_motion*prev_sense_of_motion;

      if (not_infinite_loop==-1)
        cout <<" enters an infinite loop"<<endl;

      if (found==1)     //   the guess is correct, saves the solution in the
                                                            //    vector x
        {
          for (int i=0; i< fbpos;i++)
            x(i)=ob(i);
          for (int i=fbpos; i<ob.size();i++)
            x(i)=xaux(i-fbpos);

        }
    }
  if (abs(fbpos-free_bdry_guess)<150&&iterations<150)
    return fbpos;
  else
    return -10;
}
```

6.5.4 A Regularized Active Set Strategy

The algorithm above is not easy to generalize in higher dimensions. For an algorithm based on active sets and generalizable in any dimension, we have to regularize first the problem.

Following [76], we first go back to the semidiscrete problem: find $P^n \in \mathcal{K}$ such that

$$\forall v \in \mathcal{K}, \quad (P^n - P^{n-1}, v - P^n) + \Delta t_n a_{t_n}(P^n, v - P^n) \geq 0.$$

For any positive constant c, this is equivalent to finding $P^n \in V$ and a Lagrange multiplier $\mu \in V'$ such that

$$\begin{aligned} &\forall v \in V, \quad \left(\frac{P^n - P^{n-1}}{\Delta t_n}, v\right) + a_{t_n}(P^n, v) - \langle \mu, v \rangle = 0, \\ &\mu = \max(0, \mu - c(P^n - P^0)). \end{aligned} \tag{6.54}$$

When using an iterative method for solving (6.54), i.e., when constructing a sequence $(P^{n,m}, \mu^m)$ for approximating (P^n, μ), the Lagrange multiplier μ^m may not be a function if the derivative of the $P^{n,m}$ jumps, whereas μ is generally a function. Therefore, a dual method (i.e., an iterative method for computing μ) may be difficult to use. As a remedy, Ito and Kunisch [76] considered a one-parameter family of regularized problems based on smoothing the equation for μ by

$$\mu = \alpha \max(0, \mu - c(P^n - P^0)) \tag{6.55}$$

for $0 < \alpha < 1$, which is equivalent to

$$\mu = \max(0, -\chi(P^n - P^0)) \tag{6.56}$$

for $\chi = c\alpha/(1 - \alpha) \in (0, +\infty)$. We may consider a generalized version of (6.56):

$$\mu = \max(0, \bar{\mu} - \chi(P^n - P^0)), \tag{6.57}$$

where $\bar{\mu}$ is a fixed function. This turns out to be useful when the complementarity condition is not strict.

It is now possible to study the full regularized problem

$$\begin{aligned} &\forall v \in V, \quad \left(\frac{P^n - P^{n-1}}{\Delta t_n}, v\right) + a_{t_n}(P^n, v) - \langle \mu, v \rangle = 0, \\ &\mu = \max(0, \bar{\mu} - \chi(P^n - P^0)), \end{aligned} \tag{6.58}$$

and prove that it has a unique solution, with μ a square integrable function. A primal-dual active set algorithm for solving (6.58) is the following.

ALGORITHM 6.3. Primal-dual active set algorithm.

1. Choose $P^{n,0}$, set $k = 0$.
2. Loop
   (a) Set
   $$\mathcal{A}^{-,k+1} = \{S : \bar{\mu}^k(S) - \chi(P^{n,k}(S) - P^0(S)) > 0\}$$
   and $\mathcal{A}^{+,k+1} = (0, \bar{S}) \backslash \mathcal{A}^{-,k+1}$.
   (b) Solve for $P^{n,k+1} \in V$: $\forall v \in V$,
   $$\left(\frac{P^{n,k+1} - P^{n-1}}{\Delta t_n}, v\right) + a_{t_n}(P^{n,k+1}, v) - (\bar{\mu} - \chi(P^{n,k+1} - P^0), 1_{\mathcal{A}^{-,k+1}} v) = 0. \tag{6.59}$$
   (c) Set
   $$\mu^{k+1} = \begin{cases} 0 & \text{on } \mathcal{A}^{+,k+1}, \\ \bar{\mu} - \chi(P^{n,k+1} - P^0) & \text{on } \mathcal{A}^{-,k+1}. \end{cases} \tag{6.60}$$
   (d) Set $k = k + 1$.

Denoting by A_n the operator from V to V': $\langle A_n v, w\rangle = (\frac{v}{\Delta t_n}, w) + a_{t_n}(v, w)$ and $F : V \times L^2(\mathbb{R}_+) \to V' \times L^2(\mathbb{R}_+)$,

$$F(v, \mu) = \begin{pmatrix} A_n v + \mu - \frac{P^{n-1}}{\Delta t_n} \\ \mu - \max(0, \bar{\mu} - \chi(v - P^0)) \end{pmatrix},$$

it is proved in [76] that $G(v, \mu) : V \times L^2(\mathbb{R}_+) \to V' \times L^2(\mathbb{R}_+)$, defined by

$$G(v, \mu)h = \begin{pmatrix} A_n h_1 + h_2 \\ h_2 - \chi 1_{\{\bar{\mu} - \chi(v - P^0) > 0\}} h_1 \end{pmatrix},$$

is a generalized derivative of F in the sense that

$$\lim_{\|h\| \to 0} \frac{\|F(v + h_1, \mu + h_2) - F(v, \mu) - G(v + h_1, \mu + h_2)h\|}{\|h\|} = 0;$$

this is seen from the fact that, for any $\beta \in \mathbb{R}$, the function $f : \mathbb{R} \to \mathbb{R},\ y \mapsto \max(y, 0)$ admits g as a generalized derivative where

$$g(y) = \frac{\max(0, y)}{y} \quad \text{for } y \neq 0; \quad g(0) = \beta,$$

and by taking $\beta = 0$. Note that

$$G(P^{n,k}, \mu^k)h = \begin{pmatrix} A_n h_1 + h_2 \\ h_2 - \chi 1_{\mathcal{A}^{-,k+1}} h_1 \end{pmatrix}.$$

Thus the primal-dual active set algorithm above can be seen as a semismooth Newton method applied to F, i.e.,

$$(P^{n,k+1}, \mu^{k+1}) = (P^{n,k}, \mu^k) + G^{-1}(P^{n,k}, \mu^k)F(P^{n,k}, \mu^k). \tag{6.61}$$

Indeed, calling $(\delta P^n, \delta\mu) = (P^{n,k+1} - P^{n,k}, \mu^{k+1} - \mu^k)$, it is straightforward to see that in the primal-dual active set algorithm, we have

$$\begin{aligned} A_n \delta P^n + \delta\mu &= -A_n P^{n,k} - \mu^k + \frac{P^{n-1}}{\Delta t_n}, \\ \delta\mu &= -\mu^k \text{ on } \mathcal{A}^{+,k+1}, \\ \delta\mu - \chi\delta P^n &= -\mu^k + \bar{\mu} - \chi(P^{n,k} - P^0) \text{ on } \mathcal{A}^{-,k+1}, \end{aligned}$$

which is precisely (6.61).

In [76], Ito and Kunish, by using the results proved in [72], established that the primal-dual active set algorithm converges from any initial guess, and that if the initial guess is sufficiently close to the solution of (6.58), then the convergence is superlinear.

To compute numerically the solution of (6.54), it is possible to compute successively the solutions $(P^n(\chi_\ell), \mu(\chi_\ell))$ of (6.58) for a sequence of parameters (χ_ℓ) converging to $+\infty$: to compute $(P^n(\chi_{\ell+1}), \mu(\chi_{\ell+1}))$, one uses the primal-dual active set algorithm with initial guess $(P^n(\chi_\ell), \mu(\chi_\ell))$.

Notice that it is possible to use the same algorithm for the fully discrete problem. Convergence results hold in the discrete case if there is a discrete maximum principle. The algorithm amounts to solving a sequence of systems of linear equations, and the matrix of the system varies at each iteration.

Exercise 6.2. *Write a program in order to apply the primal-dual active set algorithm to the pricing of an American option (use Euler's implicit scheme or Crank–Nicolson scheme). Vary the parameter χ and compare the active sets as $\chi \to \infty$.*

6.6 Results

6.6.1 Uniform Coefficients

We consider an American put with $\sigma = 0.2$, $r = 0.04$, and $T = 1$. In this case, the price of the option is an increasing function of the time to maturity, and therefore, the function γ introduced in Lemma 6.3 is decreasing.

We discretize the problem with the method described in §6.4, except that we allow the mesh in the variable S to vary in time. Doing so, we can adapt the mesh in S locally in time: mesh adaption is important in this case, because the solution is always singular (i.e., not twice differentiable in S) at the exercise boundary (which is unknown), so the mesh should be refined in this region. The routine for computing the matrix $\mathbf{A}$ and the right-hand side of the inequalities are similar to those presented in Chapter 5.

The evaluation of the error indicators follows along the same lines as in Chapter 5, except that the indicators are set to 0 inside the region where $P = P_\circ$. Purposely, we omit the code for the error indicators.

In Figure 6.3, we plot the function $P - P_\circ$ as a function of S and t for several meshes. Three successive meshes are plotted in Figure 6.4: we see that the mesh is refined near $t = 0$ and also on the last mesh at least that the mesh refinement follows the free boundary; this will be more visible in the next test case with a local volatility. In Figure 6.5, we plot the exercise boundary for several mesh refinements: when the mesh is properly tuned, we see that we obtain a smooth curve which is tangent to the S axis at $t = 0$. Finally, we plot in Figure 6.6 the error indicators with respect to S.

6.6.2 Local Volatility

Here, we consider an academic example, chosen to illustrate the power of the adaptive strategy. We still consider an American put, with strike $K = 100$. The interest rate is 0.04 as above, but the volatility is local and we choose

$$\sigma(S,t) = 0.1 + 0.1 * 1_{100(t-0.5)^2 + \frac{(S-90)^2}{100} < 2}(S,t),$$

so the volatility is piecewise constant and takes the value 0.2 in an ellipse and 0.1 outside. With such a choice, the exercise boundary is expected to change slope as it enters and comes out of the region where $\sigma = 0.2$. Note that this case is not covered by the theory above, because assumption (2.21) does not hold. In Figure 6.7, we plot the volatility surface as a function of S and t. In Figure 6.8, we plot the function $P - P_\circ$ versus S and t for two different meshes, and the exercise boundary is displayed in Figure 6.9: we see that the free boundary does change slope when the volatility jumps. We see also that refinement is crucial in order to catch properly the exercise boundary. Note that the function γ is not monotone. In Figure 6.10, two meshes are displayed: we see that the refinement follows the free boundary. In Figure 6.11, the error indicators with respect to S are plotted: here again, we see that the error indicators are large near the free boundary, where the function P is singular.

6.7 More Complex American Options

It is possible to consider American Options on baskets: the price of the option is then found by solving a parabolic variational inequality in dimension d (where d is the size of the basket).

Iterative algorithms like projected SOR in §6.5.1 can be used (see [61] for other iterative algorithms), but their convergence is generally slow. The two algorithms proposed in §6.5.2 and §6.5.3 cannot be applied without modification. The algorithm proposed in §6.5.4 may be applied.

Exercise 6.3. *Write a program in order to apply the primal-dual active set algorithm in §6.5.4 to the pricing of an American put option on a basket of two assets (use Euler's implicit scheme or Crank–Nicolson scheme). Vary the parameter χ and compare the active sets as $\chi \to \infty$.*

Designing a very efficient method for American options on baskets is still an interesting open problem.

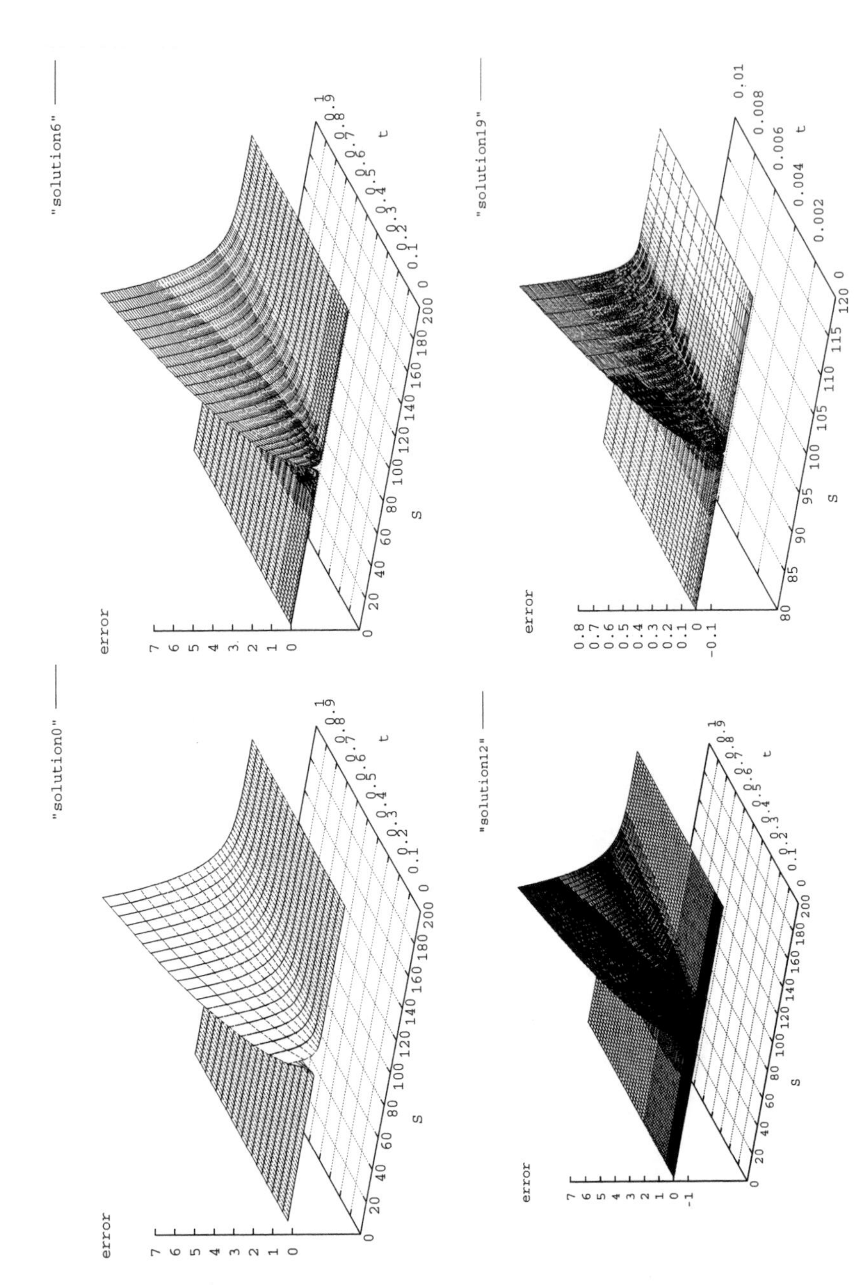

Figure 6.3. *The function $P - P_\circ$ with the adaptive strategy: the bottom right figure is a zoom.*

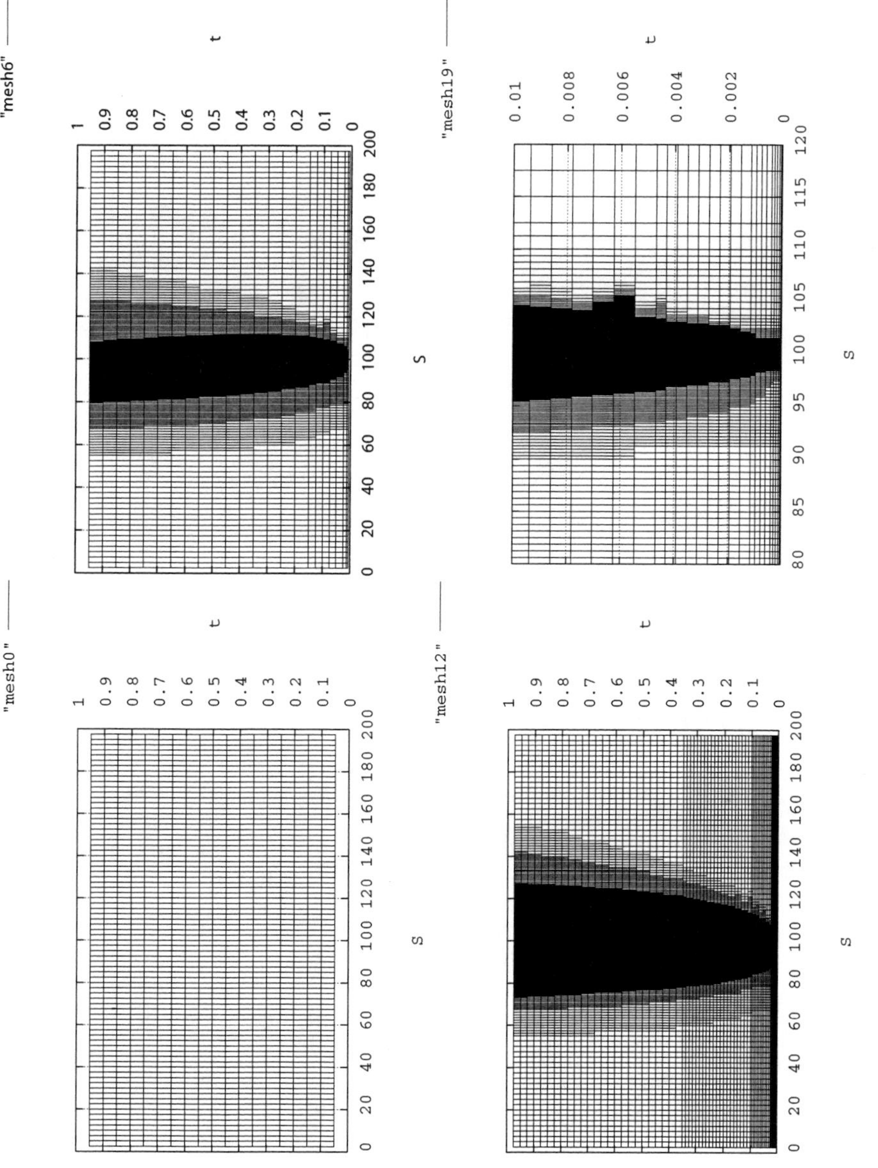

Figure 6.4. *Four successive mesh refinements: the bottom right figure is a zoom of the more refined mesh near the singularity.*

It is also possible to study American options with stochastic volatility: we refer the reader to [119], where the variational inequality is treated through a penalty method. American options on Lévy driven assets have been studied by Matache, Nitsche, and Schwab [93]: for the solution procedure, they used a wavelet basis for which the matrix $\mathbf{A}^n$ is well conditioned but then the constraint becomes difficult to handle.

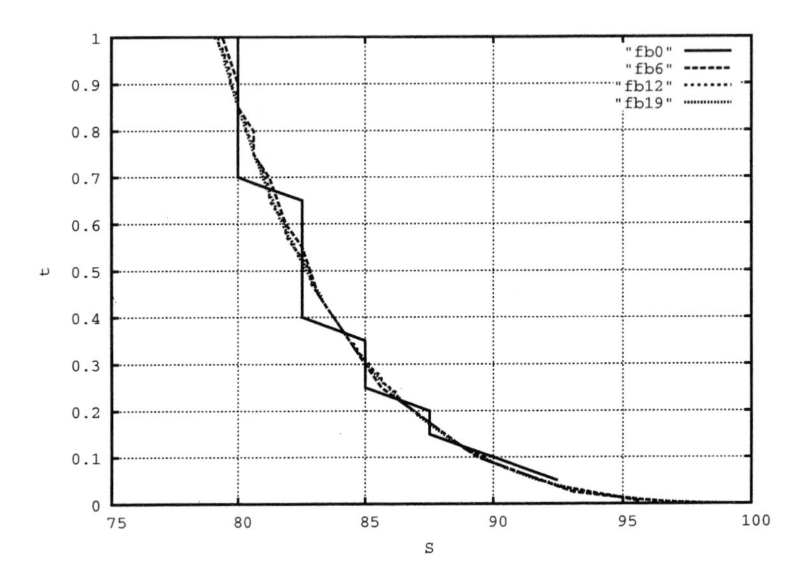

Figure 6.5. *The exercise boundaries for different mesh refinements.*

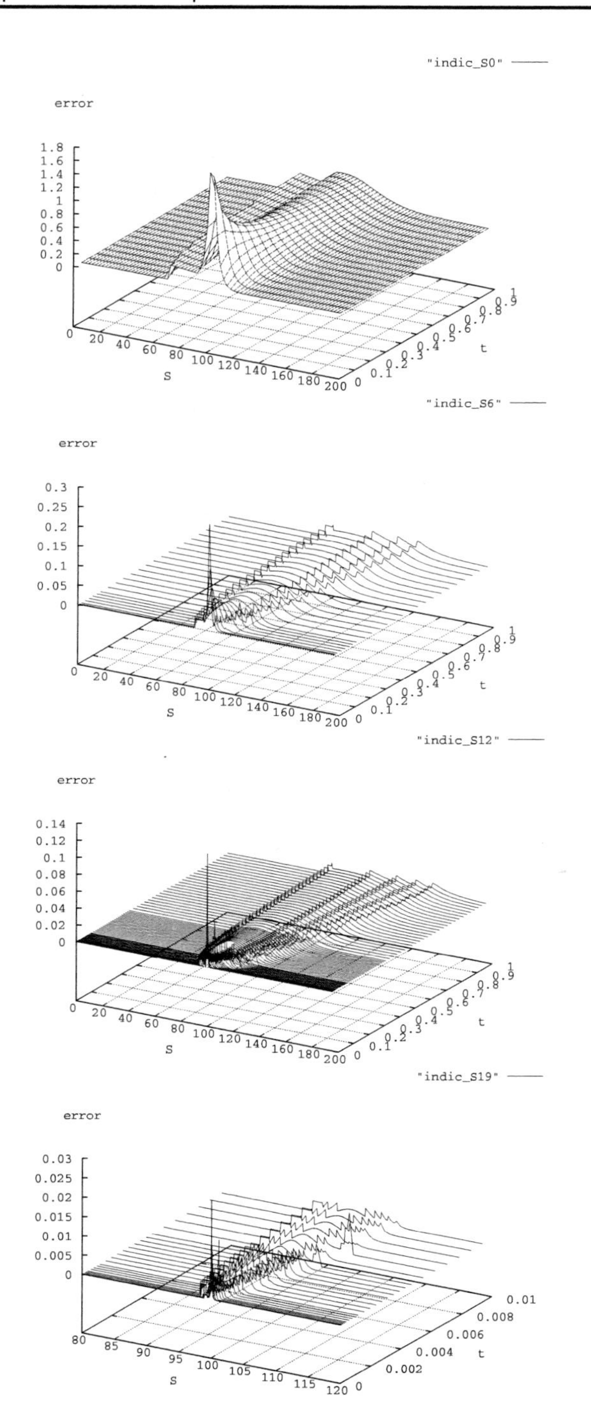

Figure 6.6. *Error indicators $\eta_{m,\omega}$: the last figure is a zoom.*

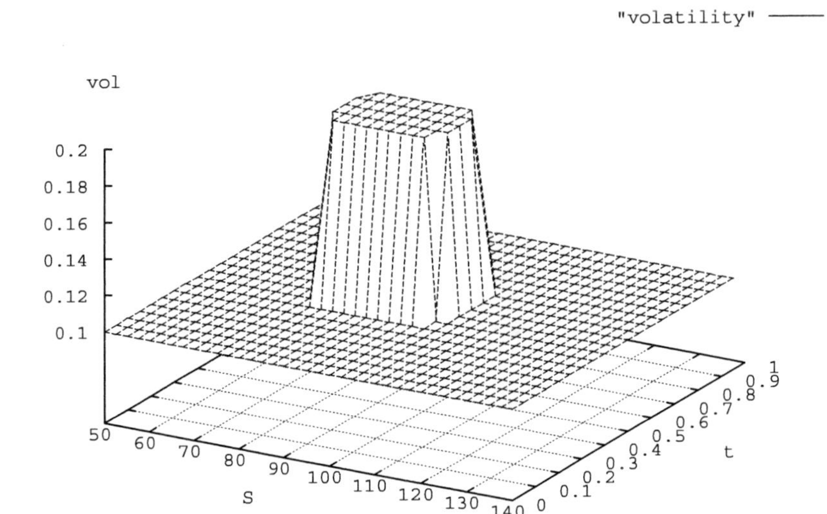

Figure 6.7. *The local volatility surface.*

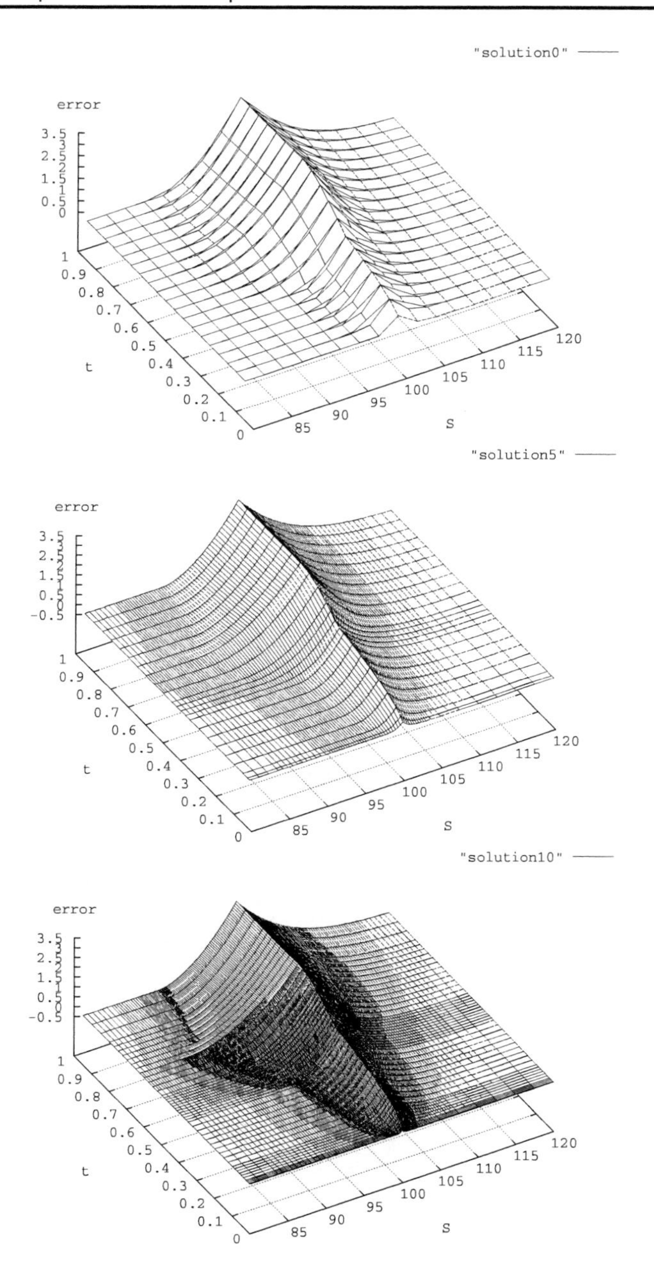

Figure 6.8. *The function $P - P_o$ with the adaptive strategy.*

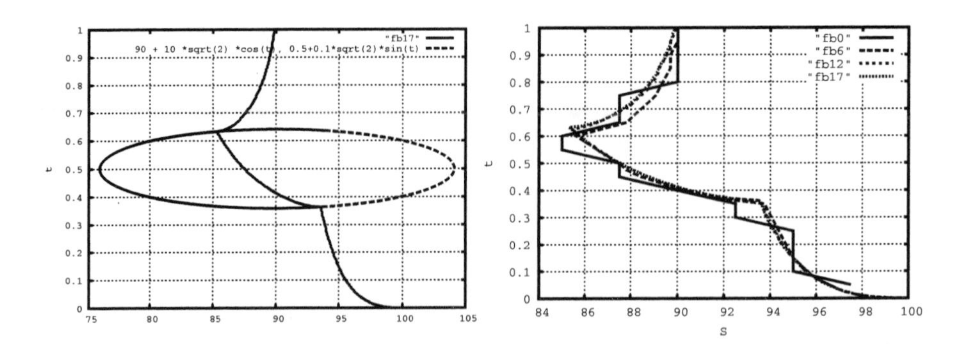

Figure 6.9. *Left: the exercise boundary for the final mesh and the ellipse where the volatility jumps: there are two singularities corresponding to the jumps of volatility. Right: the exercise boundaries for different mesh refinements.*

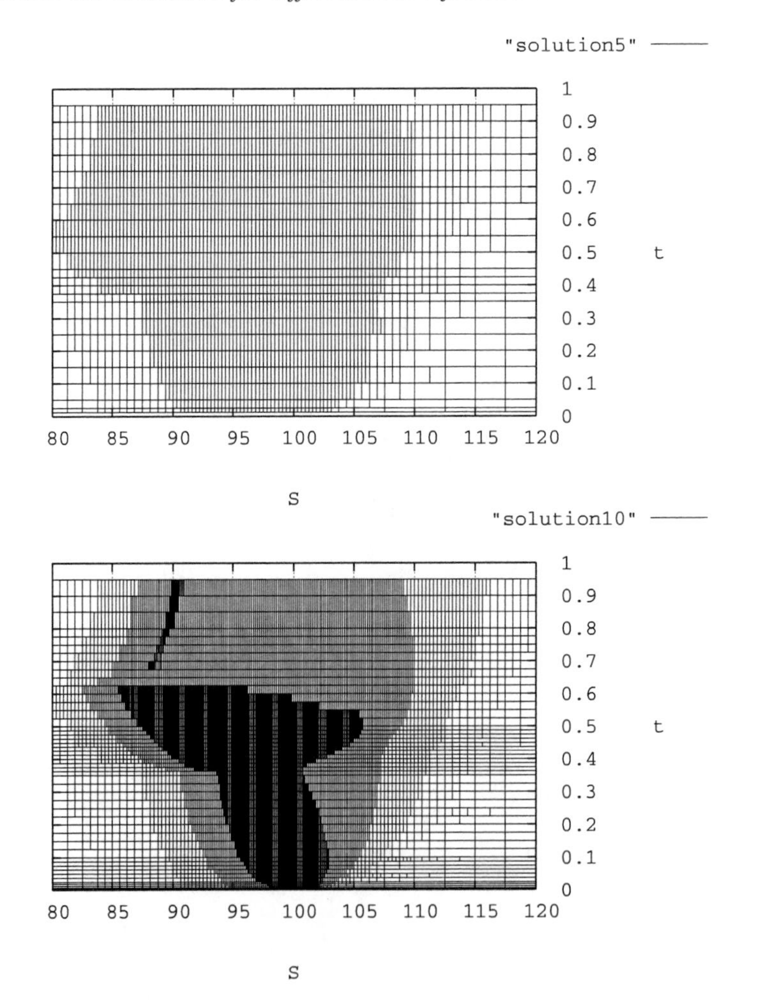

Figure 6.10. *Two successive mesh refinements: the mesh is refined along the exercise boundary; see Figure* 6.9.

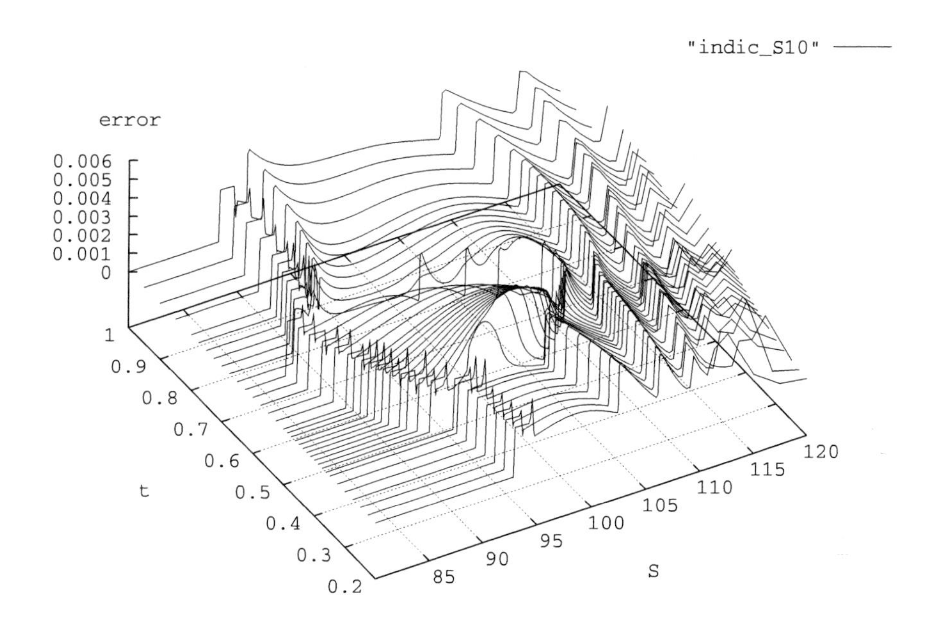

Figure 6.11. *Error indicators* $\eta_{m,\omega}$.

Chapter 7
Sensitivities and Calibration

7.1 Introduction

It is important in finance to compute the sensitivity of options with respect to parameters such as the price of the underlying asset or the volatility. In mathematical terms, $\partial_\sigma C$, for instance, can be obtained by differentiating the Black–Scholes equation and its boundary conditions. Automatic differentiation of a computer code for pricing provides a way to do that efficiently and automatically.

Equally important is the problem of adjusting the volatility to fit the observations. It is a difficult task but it can be formulated via least squares as a differentiable optimization problem.

Fast algorithms for optimization problems require derivatives of the cost functions, and here again when the number of variables is not too large, automatic differentiation is of great help.

In this chapter we present an introduction to automatic differentiation, an implementation of the forward mode by operator overloading in C++, and two applications: one to the computation of Greeks and another to simple calibration when the number of parameters is small.

This chapter is also an introduction to the last two chapters on calibration; to this end a short presentation of the conjugate gradient algorithm for optimization is given.

7.2 Automatic Differentiation of Computer Programs

Derivatives of functions defined by their computer implementations can be calculated automatically and exactly. Several techniques are available and we will present only one of them—the *forward mode* [66]. The basic idea is that each line of a computer program can be differentiated automatically, except perhaps branching statements, but since there are only a finite number of them in a computer program, differentiability will be obtained almost everywhere at worst.

The Forward Mode. Derivatives of a function can be computed from its differential form. This observation is easy to understand from the following example.

Let $J(u) = |u - u_d|^2$; then its differential is

$$\delta J = 2(u - u_d)(\delta u - \delta u_d) \tag{7.1}$$

and obviously the derivative of J with respect to u is obtained by putting $\delta u = 1, \delta u_d = 0$ in (7.1):

$$\frac{\partial J}{\partial u} = 2(u - u_d)(1.0 - 0.0).$$

Now suppose that J is programmed in C/C++ by

```
double J(double u, double u_d){
    double z = u-u_d;
    z = z*(u-u_d);
    return z;
}

int main(){
    double u=2,u_d = 0.1;
    cout << J(u,u_d) << endl;

}
```

A program which computes J and its differential can be obtained by writing above each differentiable line its differentiated form.

ALGORITHM 7.1. Automatic differentiation.

```
double JandDJ(double u, double u_d, double du,
                              double du_d, double *pdz)
{   double dz = du - du_d;
    double z = u-u_d;
    double dJ = dz*(u-u_d) + z*(du - du_d);
    z = z*(u-u_d);
    *pdz = dz;
    return z;
}
int main()
    {       double u=2,u_d = 0.1;
            double dJ;
            cout << J(u,u_d,1,0,&dJ) << endl;
    }
```

Except for the embarrassing problem of returning both `z,dz` instead of `z`, the procedure is fairly automatic. It can be automatized more systematically by introducing a structured type of differentiable variable to hold the value of the variable and the value of its derivative,

```
struct {double val[2];} ddouble;
```

and rewriting the above as follows.

ALGORITHM 7.2. AD.

```
ddouble JandDJ(ddouble u, ddouble u_d)
{
    ddouble z;
    z.val[1] = u.val[1]-u_d.val[1];
    z.val[0] = u.val[0]-u_d.val[0];
    z.val[1] = z.val[1]*(uval[0]-u_d.val[0])
                + z.val[0]*(uval[1]-u_d.val[1]);
    z.val[0] = z.val[0]*(uval[0]-u_d.val[0]);
    return z;
}
int main()
{
    ddouble u;
    u.val[0]=2; u_d.val[0] = 0.1; u.val[1]=1; u_d.val[1] = 0.;
    ddouble dJ;
    cout  << J(u,u_d).val[0]<<'\t'<< J(u,u_d,1,0).val[1]<< endl;
}
```

In C++ the program can be simplified further by redefining the operators $=$, $-$, and $*$. Then a class has to be used instead of a `struct` as follows.

ALGORITHM 7.3. AD++.

```
class ddouble{ public:
    double val[2];
ddouble(double a, double b=0)
        { v[0] = a; v[1]=b;}                          //    constructor
ddouble operator=(const ddouble& a)
  {
    val[1] = a.val[1]; val[0]=a.val[0];
    return *this;
  }
friend dfloat operator - (const dfloat& a, const dfloat& b)
        {    dfloat c;
             c.v[1] = a.v[1] - b.v[1];                //    (a-b)'=a'-b'
             c.v[0] = a.v[0] - b.v[0];
             return c;
        }
friend dfloat operator * (const dfloat& a, const dfloat& b)
        {    dfloat c;
             c.v[1] = a.v[1]*b.v[0] + a.v[0]* b.v[1];
             c.v[0] = a.v[0] * b.v[0];
             return c;
        }
};
```

As before a differentiable variable has two data fields: its value and the value of its derivative. Then we need a constructor to initialize such a variable and also the operator "=" to assign them to another one, so that `u=v` triggers `u.val[1]=v.val[1]` and `u.val[0]=v.val[0]`. The operator "−" does the usual minus operation on the value of the variables and also on the value of their differentials. For the product the rule for the differentiation of products is used. Finally the function and its calling program are

```
ddouble JandDJ(ddouble u, ddouble u_d) {
    ddouble z= u-u_d
    z = z*(u-u_d);
    return z;
}
int main()
{
    ddouble u(2,1), u_d=0.1;
    cout  << J(u,u_d).val[0]<<'\t'<< J(u,u_d,1,0).val[1]<< endl;

}
```

Note that << is an operator which can be redefined also inside the class `ddouble`.

The conclusion is that a C program can be differentiated simply by replacing the keyword `double` *by* `ddouble`*.*

Of course C programs are not only assignments, and it remains to check that branching statements, loops, function calls, etc., have the same property.

Branching statements like

```
A; if(B) C; else D; E;
```

are in fact a compound of two programs:

```
A;C;E  or, if not B  A;D;E
```

The method of differentiating each line works for both programs, giving

```
A';A;C';C;E';E  or, if not B  A';A;D';D;E';E
```

then we can compound these two programs into

```
A';A; if(B){C';C;} else {D';D;}E'; E;
```

But notice that we cannot account for `B'` that way, and so the differential may be wrong at the points where `B` changes from true to false; this does not occur if `B` does not depend upon the variable with respect to which the differentiation is done.

For loops it is the same; for instance,

```
for(int i=0;i<3;i++)A;
```

means `i=0;A;i=1;A;i=2;A`, which will be differentiated as

```
i=0; A';A;i=1;A';A;i=2;A';A,
```

which in turn is

```
for(int i=0;i<3;i++){A';A;}
```

Finally each function must be redefined; for instance,

```
ddouble sqrt (const ddouble& x)
{ ddouble r;r.val[0] = sqrt(x.val[0]);
  r.val[1] = 0.5*x.val[1]/(1e-30+r.val[0]);return r;}
```

The complete implementation of the class `ddouble` is given in the appendix (§7.7).

Notice that $\sqrt{x}$ is not differentiable at $x = 0$ and so a program differentiated automatically is never guaranteed to work. However, a test can be added to the function to avoid such trouble.

7.3 Computation of Greeks

7.3.1 Greeks for European Options

The sensitivities of an option with respect to the relevant parameters are called the *Greeks*: let C be the price of a vanilla European call:

- the δ (*delta*) is its derivative with respect to the stock price S: $\partial_S C$;
- the Θ or *time-decay* is its derivative with respect to time: $\partial_t C$;
- the *vega* κ is its derivative with respect to the volatility σ;
- the *rho* ρ is its derivative with respect to the interest rate, $\partial_r C$;
- η is its derivative with respect to the strike K;
- finally, the *gamma* is the rate of change of its delta, $\partial_{SS} C$.

Equations can be derived for these directly by differentiating the partial differential equation and the boundary conditions which define C, i.e., when r and σ are uniform,

$$\begin{aligned}
&\partial_t \delta + \frac{\sigma^2}{2} S^2 \partial_{SS} \delta + r S \partial_S \delta - r\delta = -2\sigma S \partial_{SS} C - r \partial_S C, \\
&\delta(S, T) = 1_{\{S>K\}}, \quad \delta(+\infty, t) \approx 1, \\
&\partial_t \kappa + \frac{\sigma^2}{2} S^2 \partial_{SS} \kappa + r S \partial_S \kappa - r\kappa = -\sigma S^2 \partial_{SS} C, \\
&\kappa(x, T) = 0, \quad \kappa(+\infty, t) \approx 0. \\
&\partial_t \eta + \frac{\sigma^2}{2} S^2 \partial_{SS} \eta + r S \partial_S \eta - r\eta = 0, \\
&\eta(x, T) = -1_{\{S>K\}}, \quad \eta(+\infty, t) \approx -e^{-r(T-t)}.
\end{aligned} \tag{7.2}$$

Note that for δ and κ, a sharp evaluation of the right-hand side $\partial_{SS} C$ can be tricky; automatic differentiation provides an approximation consistent with the numerical scheme used for C. For American options, it is not so easy to derive equations for the Greeks.

Instead of writing a new program to solve the above equations, it suffices to use Algorithm 4.4 written in C/C++ and to change all `double` or `float` to `ddouble`.

In the case of $\partial_K C$, for instance, here is the list of the necessary modifications:

- add the link to the ddouble library (`# include"ddouble.hpp"`);
- change the definition of `Real`: `typedef ddouble Real;`
- indicate with respect to which variable the differentiation is done, i.e., replace in the main function, `K=80` by `K(80,1);`
- for graphic display use `u[i].val[1]` instead of `u[i]`.

Automatic differentiation of computer codes can even be used to compute $\delta = \partial_S C$, starting from a code for pricing the option for $S \in [0, \bar{S}]$ with a uniform grid in S, thanks to the following trick: we observe that any function $f(S)$ defined for $S \in (0, \bar{S})$ satisfies the identity $\frac{\partial}{\partial \lambda} f(\lambda S) = S \frac{\partial f}{\partial S}(S)$ at $\lambda = 1$. The method then is to scale the array of grid points $S_i \to \lambda S_i$ and differentiate with respect to the scaling factor. As the Black–Scholes equation is integrated on $(0, \bar{S})$, and the grid points are uniformly distributed here, $S_i = i\bar{S}/N + 1$, $i = 0, \ldots, N + 1$, we can differentiate with respect to $\bar{S}$ the solution of the Black–Scholes partial differential equation in $(0, \bar{S})$. This gives

$$\frac{\partial C}{\partial S} = \frac{\bar{S}}{S} \frac{\partial C}{\partial \bar{S}}.$$

Figure 7.1 shows $\partial_\sigma P$, $\partial_K P$, $\partial_{\bar{S}} P$, $\partial_r P$ obtained by automatic differentiation of Algorithm 4.1.

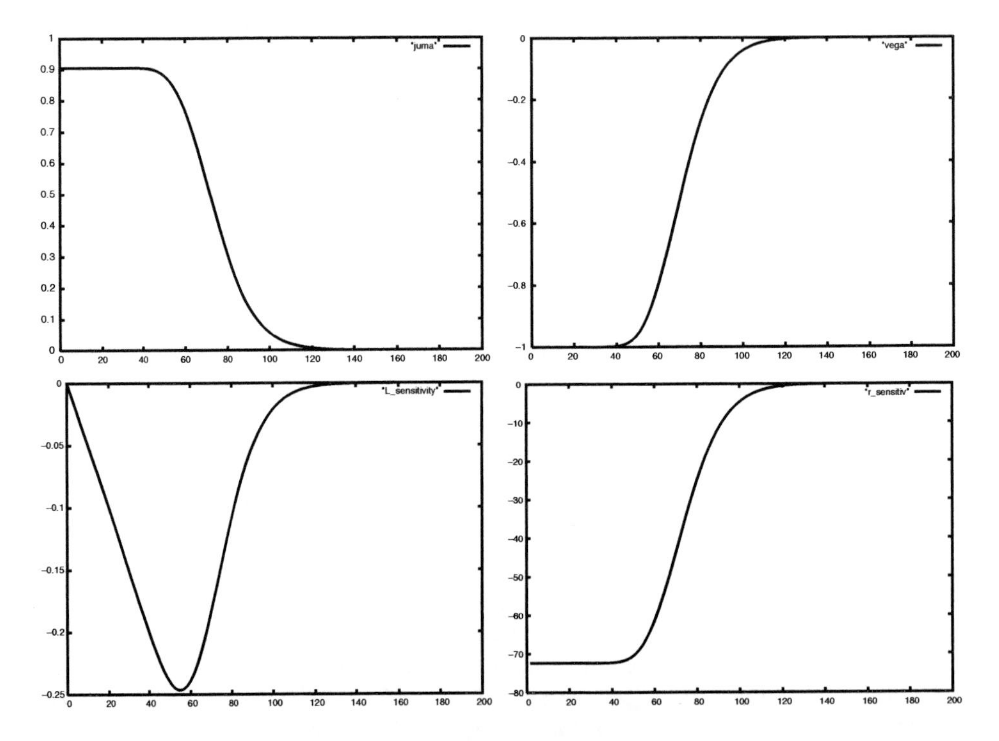

Figure 7.1. *Sensitivity of the European put one year to maturity with respect to K, S (vega), $\bar{S}$, and r obtained by automatic differentiation of Algorithm* 4.4.

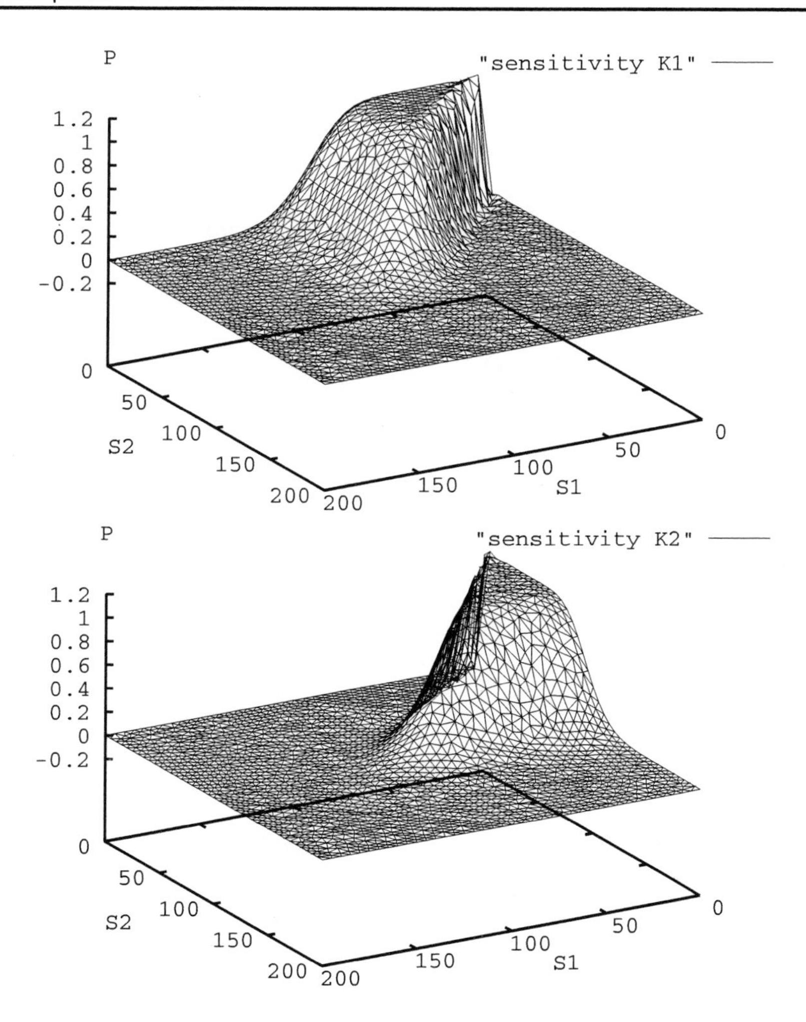

Figure 7.2. *Sensitivity with respect to* K_1 *and* K_2 *obtained by automatic differentiation of Algorithm* 4.10 *for the basket option with payoff* $P_0(S_1, S_2) = \min((K_1 - S_1)_+, (K_2 - S_2)_+)$ *computed for* $K_1 = K_2 = 100$.

Automatic differentiation can be applied to computationally more intensive programs: for example, we have used it for a basket put option with payoff $P_0(S_1, S_2) = \min((K_1 - S_1)_+, (K_2 - S_2)_+)$. In Figure 7.2, we have plotted the sensitivities with respect to K_1 and K_2 for $K_1 = K_2 = 100$.

7.3.2 Malliavin Calculus for Greeks

We use the notation of Chapter 1 and consider an option with payoff Q°. We assume that the spot price S_t satisfies (1.2) under the risk neutral probability $\mathbb{P}^*$.

For clarity denote by x the initial value $S_0 = x$ of the stochastic process S_t. To make explicit the fact that S_t depends on x, we denote it by $S_t(x)$. The value of the option at $t = 0$ is $P(x) = e^{-rT}\mathbb{E}^*(Q^\circ(S_T))$.

To compute $\partial_x \mathbb{E}^* \left(Q^\circ(S_T(x))\right)$, one could use a finite difference approximation in the Monte-Carlo method:

$$\partial_x \mathbb{E}^* \left(Q^\circ(S_T)\right) \approx \frac{1}{2\epsilon N} \sum_{n=1}^{N} \left(Q^\circ(S_T^n(x+\epsilon)) - Q^\circ(S_T^n(x-\epsilon))\right).$$

However, this is not precise and sometimes also unstable.

A better way to compute $\partial_x \mathbb{E}^* \left(Q^\circ(S_T)\right)$ comes from the remark that if $\partial_x S_T(x)$ had a meaning, we would have

$$\partial_x \mathbb{E}^* \left(Q^\circ(S_T)\right) = \mathbb{E}^* \left(Q^{\circ\prime}(S_T)\partial_x S_T\right).$$

Note that if the probability density $p(S_T, x)$ of S_T is known (see Broadie and Glasserman [22]), the differentiation with respect to x could be done by an integration by parts:

$$\partial_x \mathbb{E}^* \left(Q^\circ(S_T)\right) = \partial_x \int Q^\circ(s) p(s,x) ds = \int Q^\circ(s) \partial_x \log(p) p ds = \mathbb{E}^* \left(Q^\circ(S_T) \partial_x \log(p)\right).$$

More generally, the basic idea of Malliavin calculus is to use an integration by parts formula on the space of paths of the stochastic process in order to express the sensitivity of an expectation with respect to some parameter θ as the expectation of a new random variable, which can be computed by the Monte-Carlo method:

$$\frac{\partial}{\partial\theta} \mathbb{E}^*(f(X)) = \mathbb{E}^*(g(X)).$$

For two integrable random variables F and G (see [9]) an integration by parts is said to hold if there exists an integrable random variable $H(F;G)$ such that, for all smooth functions Φ with compact support,

$$\mathbb{E}^* \left(\Phi'(F)G\right) = \mathbb{E}^* \left(\Phi(F)H(F;G)\right).$$

Malliavin calculus provides a way to find the random variable $H(S_T; \partial_x S_T)$. Applied to the Greeks it gives (see [53, 52])

$$\begin{aligned}
\delta &= \mathbb{E}^* \left(e^{-rT} Q^\circ(S_T) \frac{W_T}{x\sigma T}\right), \\
\gamma &= \mathbb{E}^* \left(e^{-rT} Q^\circ(S_T) \frac{1}{x^2 T\sigma} \left(\frac{W_T^2}{\sigma T} - W_T - \frac{1}{\sigma}\right)\right), \\
\kappa &= \mathbb{E}^* \left(e^{-rT} Q^\circ(S_T) \left(\frac{W_T^2}{\sigma T} - W_T - \frac{1}{\sigma}\right)\right), \\
\rho &= \mathbb{E}^* \left(e^{-rT} Q^\circ(S_T) \left(\frac{W_T}{\sigma} - T\right)\right).
\end{aligned}$$

Figure 7.3 shows the performance of the formula for δ and gives a comparison with automatic differentiation applied to the partial differential equation. The computing time for δ is of the same order as the one for S_t. Automatic differentiation in the partial differential equation is much faster.

For a generalization to American options, see [10].

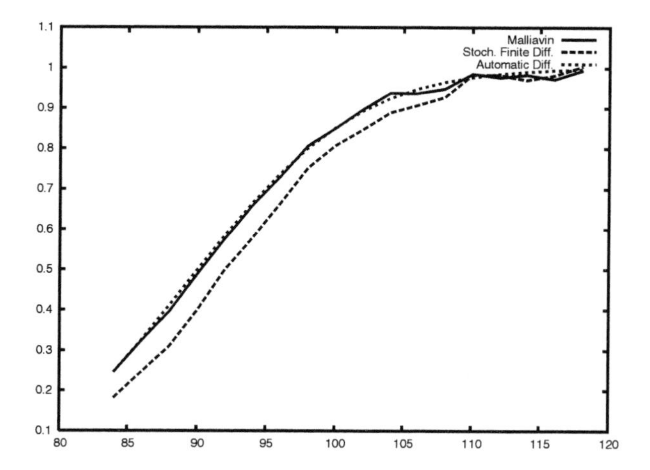

Figure 7.3. *The δ (derivative with respect to S_0) of a call with $\sigma = 0.1$, $r = 0.1$, $T = 1$, $K = 100$ for $S_0 \in (84, 118)$. Comparison between stochastic centered finite differences, the Malliavin formula, and automatic differentiation in the partial differential equation. The number of trials for the stochastic processes is* 50000 *and the number of time steps is* 100.

7.3.3 Greeks for American Options

We end this section on automatic differentiation with an application to American options. By the same trick, i.e., replacing `double` by `ddouble`, we can differentiate the program written in §6.5.3 and compute the sensitivity of a put with respect to the strike, for instance. The result is shown in Figure 7.4.

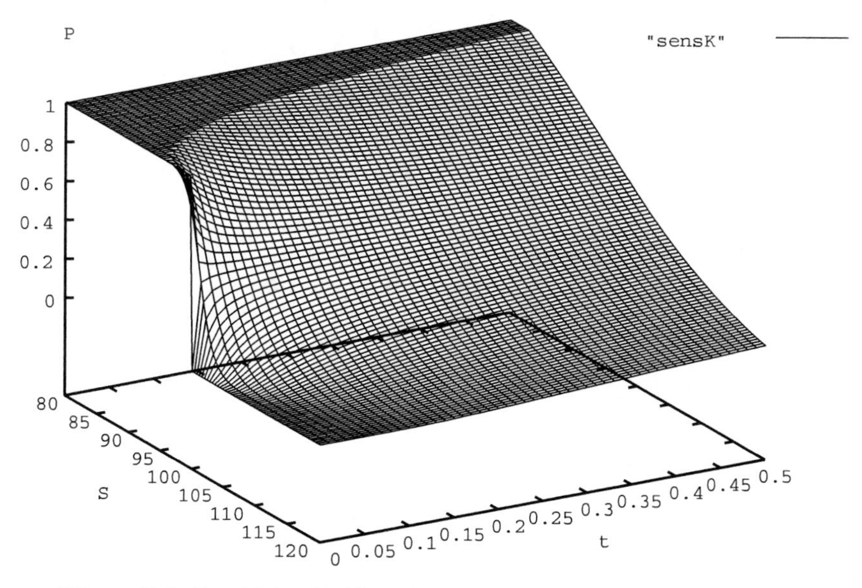

Figure 7.4. *Sensitivity dP/dK of an American put P with respect to its strike K as a function of asset price and time to maturity.*

7.4 An Introduction to the Calibration of Volatility

In practice financial models are not perfect and observed data differ from models predictions. A pragmatic approach to this problem is to relax the models by introducing parameters and calibrating these parameters with the observations.

The simplest in this line of thoughts is the concept of *historical volatility* for the stock price and *implied volatility* for European vanilla options.

- By observing the price S_t for various t one can find σ, the *historical* volatility, from the stochastic differential equation for S_t.

- Similarly, if one observes today ($t = 0$) the price $u_{d\,K,T}(S, 0)$ of an option with strike K at maturity T while the underlying asset is worth $S_\circ$, then one finds which *implied* volatility σ in the Black–Scholes analytical formulas (1.17) or (1.18) gives $u = u_d$.

If this is done for several strikes K, in general one does not find the same volatility but rather a volatility surface $\sigma(K, T)$ (called a "smile" because it is usually convex).

Exercise 7.1. *Write a C/C++ program to compute the implied volatility. One may use the Newton algorithm and to do so compute the derivative of the Black–Scholes formulas* (1.17) *or* (1.18) *with respect to* σ.

When the Black–Scholes equation with constant coefficients fails to match the observed prices, the simplest remedy is to use a *local volatility*, i.e., a carefully chosen function of S and t: $\sigma = \sigma(S, t)$. A possible way is to find σ among a family of surfaces usually defined by a few parameters, which fits the data u_d in the sense of least squares. The problem is of the type

$$\min_{a\in A}\left\{\sum_{i=1}^{n}\omega_i|u_{K_i,T_i}(S_\circ,0)-u_{di}|^2\ :\ \mathcal{L}_{\sigma(a)}u_{K_i,T_i}=0,\quad u_{K_i,T_i}(S,T_i)=(K_i-S)_-\right\},\tag{7.3}$$

where $\mathcal{L}_\sigma$ is the Black–Scholes partial differential operator with local volatility σ. The ω_i are suitably chosen nonnegative weights. These parameters are important in practice, because the prices of the options out the money can be very small. A common way to choose the ω_i is to relate them to the vega of the option; see [31].

More precisely (7.3) means that n calls on the same asset with different strikes and maturities are observed on the market to be worth $\{u_{di}\}_1^n$ today when the underlying asset is worth $S_\circ$; then one seeks for the best volatility surface $\sigma(a)$ parameterized by $a \in A \subset \mathbb{R}^p$ to fit the data.

7.4.1 Formulation with Dupire's Equation

A direct solution of (7.3) is possible but each evaluation of the cost function requires n solutions of the Black–Scholes partial differential equation (2.12). We have seen in Chapter 2 that fixing $t = 0$ and $S = S_\circ$, the price of a vanilla call with maturity τ and strike K,

$u(S_\circ, 0, K, \tau)$, as a function of τ and K satisfies Dupire's equation:

$$\mathcal{L}_\sigma^D(u) := \partial_\tau u - \frac{1}{2}\sigma^2(K,\tau)K^2\frac{\partial^2 u}{\partial K^2} + (r-q)K\frac{\partial u}{\partial K} + qu = 0, \quad u(K,0) = (K - S_\circ)_-, \tag{7.4}$$

where q is the dividend of the asset and r is the interest rate of the risk-free asset. If the options for all strikes and maturities were on the market, the local volatility would be

$$\sigma^2(K,\tau) = 2\frac{\partial_\tau u(K,\tau) + (r-q)K\frac{\partial u}{\partial K}(K,\tau) + qu(K,\tau)}{K^2\frac{\partial^2 u}{\partial K^2}(K,\tau)}. \tag{7.5}$$

It is not the case, so one may try to find σ as a minimizer of

$$\min_{a\in A}\left\{\sum_{i=1}^n \omega_i |v(K_i,\tau_i) - u_{di}|^2 \; : \; \mathcal{L}_{\sigma(a)}^D v = 0, \quad v(K,0) = (K - S_\circ)_-\right\}. \tag{7.6}$$

In contrast to the previous minimization problem (7.3), the evaluation of the cost function requires solving only one boundary value problem. The function v is uniquely defined and the evaluation of the cost function can be written as

$$a \to \sigma \to v \to J(a) := \sum_{i=1}^n \omega_i |v(K_i,\tau_i) - u_{di}|^2,$$

and the problem is to minimize $J(a)$ subject to $a \in A$.

This problem is usually ill-posed and very unstable; i.e., small variations of J can be obtained from very large variations of a. A remedy is to add to J a Tychonoff regularization functional, which ensures at least that J is coercive, i.e., $\lim_{a\in A, |a|\to\infty} J(a) = +\infty$.

7.4.2 First Numerical Tests

Consider the following volatility space:

$$\sigma^2(a) = a_0 + \sum_i \mathrm{Re}(a_i e^{2\mathbf{i}\pi t}) e^{-\lambda(K-K_i)^2}, \tag{7.7}$$

where a_0, λ, K_i are given constants. We take $K_i = 75+50(i/I)$ and $\lambda = 100$; the advantage is that $\sigma^2(a)$ is equal to a_0 for large and small K and never too small or too big in between when the real and imaginary parts of a_i are not too large; to ensure this, we add a penalty term to the criteria,

$$J = J(a) + J_p, \qquad J_p = \sum_{i=1}^I ((|a_i|^2 - 0.008)_+)^2.$$

In the following numerical test, a target put option is computed first by solving Dupire's equation with a nonconstant target volatility surface given by (7.7) with $a_j = (j - 1 + \mathbf{i}j)/100$, $j = 1, 2, 3, 4$. There are 5 observation points, at $K = 70, 80, 90, 11, 120$ made alternatively at $T = 1$ and 1.8.

Then the optimization program described below is run starting with $a_i = 0$, $i = 1, \ldots, 4$. After a few iterations the cost function is reduced to 10^{-11} and yet there is a

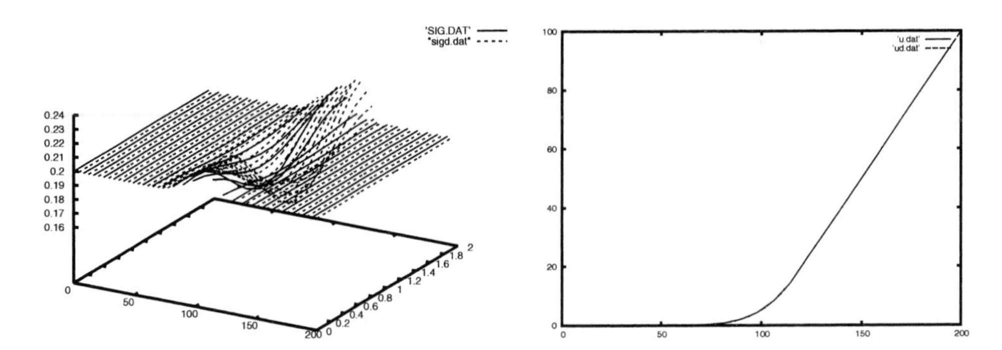

Figure 7.5. *Results of a calibration with* 8 *parameters and* 5 *observations. Left: the volatility surface.*

visible difference between the target and the computed volatility surface; see Figure 7.5. The problem is stiff in that in some domain of the parameters a large variation of σ makes no noticeable change in u while in other domains it is the opposite (for example, if at some places σ becomes very small). Note also that the amount of data is too small for practical use. The next section explains how the minimization of J was performed.

7.5 Finite-Dimensional Differentiable Optimization

Consider the minimization of the functional $J : A \subset \mathbb{R}^N \to \mathbb{R}$. When J is differentiable, *gradient* methods are applicable, and when J is twice differentiable, Newton and quasi-Newton methods can be used (see Polak [101], for example). Here, we cover briefly the steepest descent method and the conjugate gradient method for unconstrained problems.

Consider the unconstrained case, $A = \mathbb{R}^N$: if J is lower semicontinuous, bounded from below, and coercive, i.e., $\lim_{x\to\infty} J(x) = +\infty$, then there is at least one solution of the minimization problem, and the solution is unique if in addition J is strictly convex.

Furthermore if J is differentiable, the gradient of J satisfies $\nabla J(a^*) = 0$ at a solution a^*.

7.5.1 Gradient Methods

Gradient methods stem from the fact that at each point a, $-\nabla J(a)$ is a descent direction for J: indeed, if $\rho > 0$ is small enough,

$$J(a - \rho\nabla J(a)) < J(a),$$

because

$$J(a - \rho\nabla J(a)) = J(a) - \rho|\nabla J(a)|^2 + o(\rho).$$

The gradient methods construct sequences by the induction relation

$$a^{m+1} = a^m - \rho_m \nabla J(a^m),$$

where the *step size* $\rho_m > 0$ is chosen so that $J(a^{m+1}) < J(a^m)$.

One possibility is to take ρ_m constant: $\rho_m = \rho$, with ρ small enough. For convex functions with a positive convexity modulus α and with a Lipschitz continuous gradient (with Lipschitz constant Λ), fixed step gradient methods converge (geometrically) if $\rho < \frac{2\alpha}{\Lambda^2}$. In the *steepest descent method*, the step size ρ_m is chosen as the solution of the one-dimensional minimization of $\rho \mapsto f(\rho) := J(a^m - \rho\nabla J(a^m))$. In practice, it may not be possible to compute exactly this minimum, and line search methods are available to compute efficiently approximations of the minimum. Later, we discuss one of them—Armijo's rule. It can be proved that for convex functions with a positive convexity modulus α and with a uniformly Lipschitz continuous gradient, the steepest descent method converges linearly (i.e., the error tends to 0 like a geometric progression). However, if the minimum is in a very narrow valley (i.e., for C^2 functionals, the Hessian matrix is very ill conditioned near the minimum), the steepest descent method produces a lot of oscillations which slow down the convergence. The conjugate gradient method (see below) was developed to avoid these oscillations.

7.5.2 The Conjugate Gradient Method

The conjugate gradient method (Hestenes and Stiefel [69]) is a descent method $a^{m+1} = a^m + \rho_m d^m$ where the descent direction d^m is a linear combination of $\nabla J(a^m)$ and d^{m-1}: $d^m = -\nabla J(a^m) + \gamma_m d^{m-1}$ and where the step size is the minimum ρ_m of the real variable function $\rho \mapsto f(\rho) := J(a^m + \rho d^m)$: $\rho_m = \operatorname{argmin} f(\rho)$. For a convex quadratic function J, the linear combination d^m is chosen in order to minimize $J(a^m + \rho_m d^m)$, and the method converges in a finite number of iterations (assuming there are no round-off errors). In this case, there are several ways to write the best coefficients γ_m. These formulas can also be used when J is not quadratic, but then they are not equivalent: the three best known formulas are

- the Fletcher–Reeves formula:

$$\gamma_m = \frac{|\nabla J(a^m)|^2}{|\nabla J(a^{m-1})|^2};$$

- the Polak–Ribière formula:

$$\gamma_m = \frac{\nabla J(a^m) \cdot (\nabla J(a^m) - \nabla J(a^{m-1}))}{|\nabla J(a^{m-1})|^2};$$

- the Hestenes–Stiefel formula:

$$\gamma_m = \frac{\nabla J(a^m) \cdot (\nabla J(a^m) - \nabla J(a^{m-1}))}{d^{m-1} \cdot (\nabla J(a^m) - \nabla J(a^{m-1}))}.$$

7.5.3 Line Search by Armijo's Rule

In the gradient and conjugate gradient methods, the search for an optimal step size can be replaced by Armijo's rule.

Let us denote

$$E(\rho) = J(u + \rho h) - J(u).$$

The following will find k such that $\rho = \rho_0 2^k$ satisfies

$$E(\rho) < \frac{\rho E'(0)}{2}, \qquad 2\rho\frac{E'(0)}{2} \leq E(2\rho).$$

Convergence of the gradient and conjugate gradient methods can be shown with such a step size while preserving their linear and superlinear rate.

Armijo's rule is implemented as follows:

- choose ρ_0, set $\rho = \rho_0$;
- if $(E(\rho) < \rho\frac{E'(0)}{2})$, then while $(E(\rho) < \rho\frac{E'(0)}{2})$ set $\rho := 2\rho$;
- otherwise while $(E(\rho) \geq \rho\frac{E'(0)}{2})$ set $\rho := \frac{\rho}{2}$.

7.5.4 An Implementation in C++

Thus it is possible to write a small program which can solve any differentiable optimization problem and for which only the cost function is problem-dependent.

A class is defined to encapsulate the data of a minimization problem as follows.

ALGORITHM 7.4. A class for an unconstrained minimization problem.

```
class Minproblem
{   public:
        int pmax;                                    //    Number of iterations
        bool beginning, finishing;
        int na;                                             //    nb of unknowns
        ddouble* a;                                      //    array of unknowns
        double rho;                                             //    step size

        ddouble E();                                          //    cost function
        void dE(double* grad);                       //    returns the gradient
        Minproblem (double rho1, int pmax1, const int na1,
                     rho(rho1), pmax(pmax1), na(na1),
          {
            beginning=true; finishing=false;
            a = new ddouble[nu];
          };
        void descent();                  //    the conjugate gradient algo.
        double e(double rho, double E0, double* h);
        double armijorule(double rho,double dE0, double alpha, double E0,
double* h);
  };
```

Thanks to automatic differentiation the function de is implemented by the following program.

ALGORITHM 7.5. Computation of the gradient by automatic differentiation.

```
  void Minproblem::dE(double* grad)
 {
    for(int i =0; i<na; i++)
    {
```

```
        for(int j =0; j<na; j++)
            a[j].val[1]= 0;
        a[i].val[1]=1;
        grad[i] = E().val[1];
    }
}
```

Armijo's rule is programmed as follows.

ALGORITHM 7.6. Armijo's rule.

```
double Minproblem::armijorule(double rho,double dE0, double alpha, double E0,
double* h)
{
  double  E1=e(rho, E0, h);  int i = 0;
  if(E1 < alpha * rho * dE0)
  {
    do rho = 2*rho;
    while( e(rho, E0, h) < alpha * rho * dE0 && i++ < 100);
    return rho/2;
  }
  else
  {
    do rho = rho/2;
    while( e(rho, E0, h) > alpha * rho * dE0 && i++ < 100);
    return rho;
  }
}
```

The following is Armijo's rule with the auxiliary function.

ALGORITHM 7.7. Auxiliary function for Armijo's rule.

```
double Minproblem::e(double rho, double E0, double* h)
{
   for(int i = 0; i < nu; i++)
      u[i] += rho*h[i];
   double aux = E().val[0] - E0;
   for(int i = 0; i < nu; i++)
      u[i] -= rho*h[i];
   return aux;
}
```

Finally the conjugate gradient with Armijo's rule is given below.

ALGORITHM 7.8. Conjugate gradient with Armijo's rule.

```
void Minproblem::descent()
{
  double normg2old = 1e60;
  for(int i =0; i<nu; i++)
    h[i]=0;
  for(int p = 0; p < pmax; p++)
  {
```

```
        double E0 = E().val[0], normg2 =0;
        dE();
        for(int i =0; i<nu; i++)
           normg2 += grad[i]*grad[i];
        double gam = normg2/normg2old;
        for(int i =0; i<nu; i++)
           h[i] = -grad[i] + gam*h[i];
        normg2old = normg2;
        double dE2 = 0;
        for(int i =0; i<nu; i++)
           dE2 += grad[i]*h[i];
        for(int i =0; i<nu; i++)
           u[i] += rho*h[i];
        double E1 = E().val[0];
        if(fabs(E1-E0)<1e-30)
           cout<<"exit: no variation of Cost"<<endl;
        if(fabs(E1-E0)<1e-30)
           break;
        double rhom =  armijorule(rho,dE2,0.3,E0,h);
        for(int i =0; i<nu; i++)
           u[i] += rhom*h[i];
        cout<<p<<'\t'<<E0<<'  '<<normg2<<'\t'<<rhom<<'\t'<<gam<<endl;
        E1 = E().val[0];
        if(E1>E0)
           cout<<E1<<" cost grows, wrong gradient"<<endl;
        if(normg2<eps*eps)
          cout<< "optimization done\n";
        if(E1>E0 || normg2<eps*eps)
          break;
      }
}
```

7.6 Application: Calibration on a Basis of Solutions

7.6.1 The Method

In the Black–Scholes model for an option of maturity T and strike K, the calibration of the volatility so as to fit observable data is difficult because of stability issues. In the remaining chapters, we will give a general inverse method based on Tychonoff regularizations.

Let us go back to the least squares problem (7.6). From formula (7.5), we see that the knowledge of the pricing function $(K,\tau)\mapsto v(K,\tau)$ implies the local volatility: therefore, a possibility is to look for a pricing function which fits the observed prices, then deduce the volatility. Instead of solving (7.4) by a finite element method, let us look for v in the form

$$v(K,\tau)=\sum_1^n a_j v^j(K,\tau), \tag{7.8}$$

where the v^j are linearly independent solutions to Dupire's equation (7.4) with a given volatility field $\sigma^j(K,\tau)$ and such that $v^j(K,0)=(K-S)_+$, with the natural conditions $v^j(\tau,0)=0$, $\lim_{K\to\infty}(v^j(\tau,K)-K)=S$. Once the nonnegative parameters a_j are found such that the function v fits the observed prices, the local volatility will be given by (7.5). In that context, it seems important to obtain as many analytical solutions v^j of Dupire's equations as possible.

For example, assume that the pricing functions v^j correspond to constant volatilities $\sigma^j(K,\tau) = \sigma^j$ for $j = 1, \ldots, n$; then the parametrization $(a_j)_{j=1,\ldots,n} \mapsto \sigma$ is

$$\sigma^2(K,\tau) = \frac{\sum_1^n a_j \sigma_j^2 K^2 \partial_{KK} v^j}{\sum_1^n a_j K^2 \partial_{KK} v^j}. \tag{7.9}$$

This approach is very simple and very fast, and all equations are satisfied; however, the stability is ensured only if the basis $\{\sigma_j\}$ is cleverly chosen (this is an open problem).

7.6.2 Numerical Results

The numerical procedure of §7.6 was tested with the idea of recovering the target volatility

$$\sigma(x,\tau) = \sum_1^3 \frac{k}{100} \cos(2\pi\tau) \exp\left(-\frac{(75 + 16.66k - x)^2}{100}\right). \tag{7.10}$$

Several put options on the same underlying asset S on day 0 with strike K_i and maturity T_i are observed (their price are u_{di}). We took five different options, which is a very small number. The results of the calibration procedure with $n = 5$ and $n = 20$ are given in Figure 7.6. This method seems to have a good stability and a fair accuracy. It is clear that for practical use, more data are necessary.

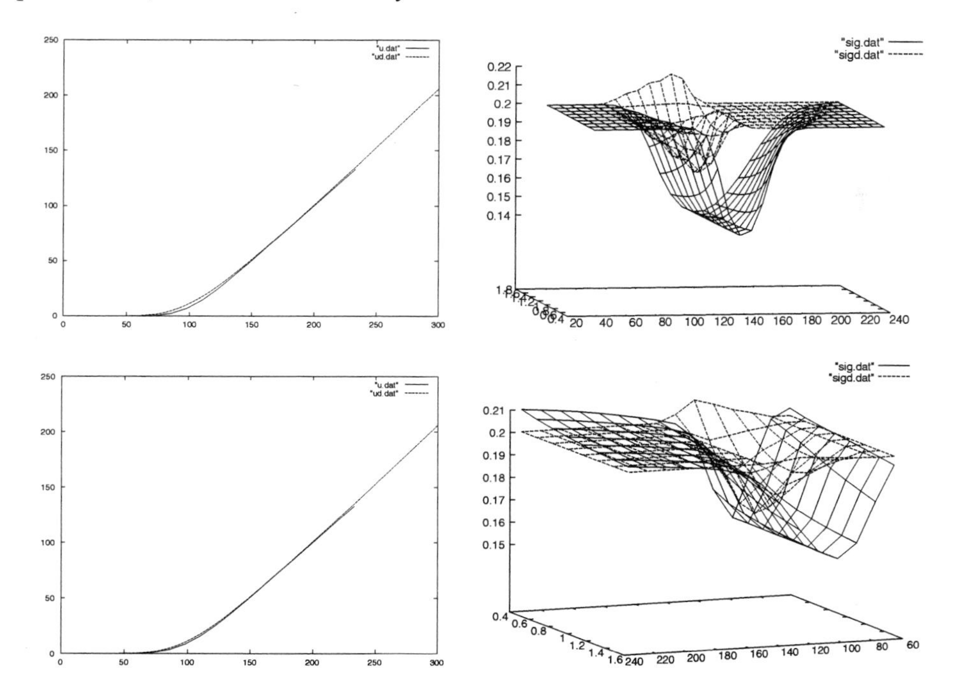

Figure 7.6. *Result of the procedure in §7.6 with parameterization (7.9). Top: 5 basis functions have been used. Bottom: 20 basis functions have been used (note that the centering and the zoom are different from above).*

Project 7.1. *Build the program that produced the numerical results above and replace the optimization module by the BFGS optimization function of the GSL:* `gsl_multimin_fdfminimizer_vector_bfgs`*. Then propose some other parameterization for* σ *(such as B-splines) and rerun the calibration.*

7.7 Appendix

The complete implementation of the class `ddouble` is given below. It can also be found online at www.ann.jussieu.fr/pironneau.

ALGORITHM 7.9. The ddouble library for automatic differentiation.

```
        //    file ddouble.h, for automatic differentiation w/r one variable
                                              //    adapted from M. Grundmann's MIPAD
#ifndef _DDOUBLE__H_
#define _DDOUBLE__H_

#include <stdlib.h>
#include <math.h>
#include <iostream.h>

class ddouble
{
  public:   double val[2];              //    val[0]=value, val[1]= derivative

 ddouble() { val[0] = 0; val[1] = 0;}
 ddouble(const ddouble& a){ val[0] = a.val[0]; val[1] = a.val[1]; }
 ddouble(double a, double b=0){ val[0]=a; val[1]=b;}

 ddouble& operator= (double a)
    { val[0] = a; val[1] = 0.0; return *this;}
 ddouble& operator= (const ddouble& a)
    { val[0] = a.val[0]; val[1] = a.val[1]; return *this;}

 double& operator[] (const int ii){ return this->val[ii];}
 double operator[] (const int ii) const { return this->val[ii];}

  ddouble& operator  + (){return *this;};
  ddouble& operator += (double);
  ddouble& operator += (const ddouble&);
  ddouble& operator -= (double );
  ddouble& operator -= (const ddouble&);
  ddouble& operator *= (double);
  ddouble& operator *= (const ddouble&);
  ddouble& operator /= (double) ;
  ddouble& operator /= (const ddouble&) ;

  ddouble operator++(int);
  ddouble operator--(int);
  ddouble&  operator++();
  ddouble&  operator--();

  friend ostream& operator << (ostream&, const ddouble&);
  friend ddouble& operator << (ddouble&,double);
```

```
   friend ddouble parameter(double);

   friend int operator != (const ddouble&,const ddouble&);
   friend int operator != (double,const ddouble&);
   friend int operator != (const ddouble&,double);
   friend int operator == (const ddouble&,const ddouble&);
   friend int operator == (double,const ddouble&);
   friend int operator == (const ddouble&,double);
   friend int operator >= (const ddouble&,const ddouble&);
   friend int operator >= (double,const ddouble&);
   friend int operator >= (const ddouble&,double);
   friend int operator <= (const ddouble&,const ddouble&);
   friend int operator <= (double,const ddouble&);
   friend int operator <= (const ddouble&,double);
   friend int operator > (const ddouble&,const ddouble&);
   friend int operator > (double,const ddouble&);
   friend int operator > (const ddouble&,double);
   friend int operator < (const ddouble&,const ddouble&);
   friend int operator < (double,const ddouble&);
   friend int operator < (const ddouble&,double);

   friend ddouble operator + (const ddouble& x);
   friend ddouble operator + (const ddouble&,const ddouble&);
   friend ddouble operator + (double, const ddouble&);
   friend ddouble operator + (const ddouble&, double);
   friend ddouble operator - (const ddouble& x ,double y);
   friend ddouble operator - (const ddouble&,const ddouble&);
   friend ddouble operator - (double, const ddouble&);
   friend ddouble operator - ( const ddouble& );
   friend ddouble operator * (const ddouble&,const ddouble&);
   friend ddouble operator * (double, const ddouble& );
   friend ddouble operator * (const ddouble& x, double y);
   friend ddouble operator / (const ddouble& x, double y);
   friend ddouble operator / (const ddouble&,const ddouble&);
   friend ddouble operator / (double,const ddouble&) ;
   friend ddouble exp (const ddouble&);
   friend ddouble log (const ddouble&) ;
   friend ddouble sqrt (const ddouble&) ;
   friend ddouble sin (const ddouble&);
   friend ddouble cos (const ddouble&);
   friend ddouble tan (const ddouble&);
   friend ddouble pow (const ddouble&,double);
   friend ddouble pow (const ddouble&,const ddouble&);
   friend ddouble pow (const ddouble&, const int);
   friend ddouble abs (const ddouble&) ;

};

inline double sign(const ddouble& x)
{ return ( x < 0.0 ? -1.0 : 1.0); }

inline double sign(const ddouble& x, double y)
{ return ( x < 0.0 ? -fabs(y) : fabs(y)); }

                                                          //    used with f2c

inline ddouble d_abs(ddouble * x){ return abs(*x);  }
```

```
inline ddouble d_cos(ddouble * x){ return cos(*x);  }
inline ddouble d_sin(ddouble * x){ return sin(*x);  }
inline ddouble d_tan(ddouble * x){ return tan(*x);  }
inline ddouble d_exp(ddouble * x){ return exp(*x);  }
inline ddouble d_log(ddouble * x){ return log(*x);  }

inline ddouble d_sign(ddouble * x){ return sign(*x);    }
inline ddouble d_sign(ddouble * x,double*y){ return sign(*x,*y);    }

inline ddouble d_sqrt(ddouble * x){ return sqrt(*x);    }

inline ddouble pow_dd(ddouble * x,ddouble*y)    { return pow(*x,*y);    }
inline ddouble pow_dd(double * x,ddouble*y)     { return pow(*x,*y);    }
inline ddouble pow_dd(ddouble * x,double*y)     { return pow(*x,*y);    }
inline ddouble pow_di(ddouble * x,int*y)    { return pow(*x,*y);    }

#endif
```

ALGORITHM 7.10. The file ddouble.cpp.

```
                                //    file ddouble.cpp, for AD with one variable
#include "ddouble.h"

const double eps = 1.0e-50;                 //    avoids NaN in (sqrt(y))'at 0

ostream& operator<<(ostream& f, const ddouble& a)
{ f << "[" << a[0] << ','<< a[1] << "]"; return f;}

ddouble ddouble::operator++(int)
{   ddouble r=(*this);  r[0]++; return r;}

ddouble ddouble::operator--(int)
{   ddouble r=(*this);  r[0]--; return r;}

ddouble& ddouble::operator++(){  (*this)[0]++; return *this;}

ddouble& ddouble::operator--(){  (*this)[0]--; return *this;}

ddouble& ddouble::operator += (double y)
{  (*this)[0] += y; return *this; }

ddouble operator - (const ddouble& a)
{ ddouble r; r[0] = -a[0]; r[1] = -a[1];  return r;}

ddouble& ddouble::operator -= (double y)
{  (*this)[0]-=y; return *this;}

ddouble& ddouble::operator += (const ddouble& y)
{   (*this)[0]+=y[0];(*this)[1]+=y[1];  return *this; }

ddouble& ddouble::operator -= (const ddouble& y)
{   (*this)[0]-=y[0];(*this)[1]-=y[1];  return *this; }

ddouble& ddouble::operator *= (double y)
{ (*this)[0] *=y; (*this)[1] *=y;  return *this;}
```

```
ddouble& ddouble::operator *= (const ddouble& y)
{ return *this = *this * y;}

ddouble& ddouble::operator /= (const ddouble& y)
{  return *this = *this / y;}

ddouble& ddouble::operator /= (double y)
{ const double inv = 1.0 / y;
    (*this)[1]  *= inv; (*this)[1]  *= inv;
return *this;
}

int operator != (const ddouble& u,const ddouble& v)
{  return u[0] != v[0];}

int operator != (double u,const ddouble& v){  return u != v[0];}

int operator != (const ddouble& v,double u){  return v[0] != u;}

int operator == (const ddouble& u,const ddouble& v)
{  return u[0] == v[0];}

int operator == (double u,const ddouble& v){  return u == v[0];}

int operator == (const ddouble& v,double u){  return v[0] == u;}

int operator <= (const ddouble& u,const ddouble& v)
{  return u[0] <= v[0];}

int operator <= (double u,const ddouble& v){  return u <= v[0];}

int operator <= (const ddouble& v,double u){  return v[0] <= u;}

int operator >= (const ddouble& u,const ddouble& v)
{  return u[0] >= v[0];}

int operator >= (double u,const ddouble& v){  return u >= v[0];}

int operator >= (const ddouble& v,double u){  return v[0] >= u;}

int operator > (const ddouble& u,const ddouble& v)
{  return u[0] > v[0];}

int operator > (double u,const ddouble& v){  return u > v[0];}

int operator > (const ddouble& v,double u){  return v[0] > u;}

int operator < (const ddouble& u,const ddouble& v){  return u[0] < v[0];}

int operator < (double u,const ddouble& v){  return u < v[0];}

int operator < (const ddouble& v,double u){  return v[0] < u;}

ddouble operator + (const ddouble& x, const ddouble& y)
{ ddouble r;
    r[0]    = x[0] + y[0];r[1]  = x[1] + y[1];   return r;
}
```

```
ddouble operator + (double x, const ddouble& y)
{   ddouble r(y);  r[0] += x;  return r;}

ddouble operator + (const ddouble& y, double x)
{   ddouble r(y);  r[0] += x;  return r;}

ddouble operator - (const ddouble& x, const ddouble& y)
{   ddouble r;  r[0] = x[0] - y[0];r[1] = x[1] - y[1]; return r;}

ddouble operator - (double x, const ddouble& y)
{  ddouble r; r[1]  = - y[1];  r[0] = x - y[0];   return r;}

ddouble operator - (const ddouble& x, double y)
{  ddouble  r(x);  r[0] -= y;  return r;        }

ddouble operator * (const ddouble& x, const ddouble& y)
{ ddouble r; r[0] = x[0]*y[0]; r[1]=x[0]*y[1]+x[1]*y[0];return r;}

ddouble operator * (double x, const ddouble& y)
{ddouble r; r[0]    = x * y[0]; r[1]    = x * y[1]; return r;}

ddouble operator * ( const ddouble& y, double x)
{return x * y;}

ddouble operator / (const ddouble& x, const ddouble& y)
{  ddouble r; r[0] = x[0]/y[0]; r[1]=(x[1]-x[0]*y[1]/y[0])/y[0];return r;}

ddouble operator / (double x, const ddouble& y)
{ ddouble r; r[0] = x/y[0]; r[1]=-x*y[1]/y[0]/y[0];return r;}

ddouble operator/(const ddouble& x, double y)
{ddouble r; r[0] = x[0]/y; r[1]=x[1]/y;return r;}


ddouble exp (const ddouble& x)
{  ddouble r;r[0] = exp(x[0]); r[1] = x[1]*r[0];return r;}

ddouble log (const ddouble& x)
{ ddouble r;r[0] = log(x[0]);r[1]     = x[1]/x[0];return r;}

ddouble sqrt (const ddouble& x)
{ ddouble r;r[0] = sqrt(x[0]); r[1] = 0.5*x[1]/(eps+r[0]);return r;}

ddouble sin (const ddouble& x)
{   ddouble r; r[0]=sin(x[0]); r[1]=x[1]*cos(x[0]);  return r;}

ddouble cos (const ddouble& x)
{   ddouble r; r[0]=cos(x[0]); r[1]=-x[1]*sin(x[0]);  return r;}

ddouble tan (const ddouble& x)
{ return (sin(x) / cos(x));}

ddouble pow (const ddouble& x,double y)
{return exp(log(x) * y);}

ddouble pow (const ddouble& x,const int y)
{return exp(log(x) * (double)y);}
```

```
ddouble pow (const ddouble& x,const ddouble& y)
{return exp(log(x) * y);}

ddouble abs (const ddouble& x)
{ ddouble y;if(x[0] >= 0) y=x; else y = -x; return y;}
```

ALGORITHM 7.11. An example of automatic differentiation.

```
#include "ddouble.h"
void main(void)
{
ddouble  x(0,1);
ddouble a(1),b,c;

b = sin(x); c = (a+x) * b ; cout << "the derivative of(1+x)*
sin(x) at x=0 is "<<c << endl; }
```

Chapter 8

Calibration of Local Volatility with European Options

In this chapter, we describe a least squares algorithm to calibrate the local volatility by fitting the prices of a set of vanilla European calls available on the market. As seen in Chapter 7, using Dupire's equation saves a lot of work for evaluating the least squares functional. For computing the gradient of the cost function with respect to volatility, we propose an optimal control approach involving an adjoint state.

Coleman, Li, and Verma [29], Jackson, Süli, and Howison [77], and Lagnado and Osher [83, 84] use a rather similar method, but they do not rely on Dupire's equation, so they have to perform as many partial differential solves as observed prices. Also, their method differs in the computation of the gradient of the cost function: Coleman, Li, and Verma use automatic differentiation or finite differences to evaluate the gradient of the cost functional, and Jackson, Süli, and Howison use finite differences, whereas Lagnado and Osher derive a partial differential equation for the gradient of the option price with respect to σ. Let us also mention Andersen and Brotherton-Ratcliffe [6] for another method special to vanilla options using discrete Green's functions, and Boyle and Thangaraj [17] .

We will close the chapter by reviewing a different and attractive method proposed by Avellaneda et al. [7], using dynamic programming. It has the advantage of not relying on any parametrization of the volatility, and of leading to an unconstrained optimization problem, at the cost of solving nonlinear Hamilton–Jacobi–Bellman equations.

Finally, note that other models also need calibration: for example, Cont and Tankov [31] calibrate jump diffusion models.

8.1 The Least Squares Problem

Hereafter, we shall call η the square of the local volatility. We shall localize Dupire's equation by picking $\bar{K}$ and $\bar{\tau}$ large enough, and by considering Dupire's equation in the rectangle $Q = [0, \bar{K}] \times [0, \bar{\tau}]$. We must impose an artificial boundary condition on $K = \bar{K}$, so we take $C(K = \bar{K}, \tau) = 0$ for all $\tau \in (0, \bar{\tau}]$. Calling $S_\circ$ the spot price, the price

$C(K,\tau) := C(S_\circ, 0, K, \tau)$ is a solution to the boundary value problem

$$
\begin{aligned}
\partial_\tau C - \frac{\eta K^2}{2}\partial^2_{KK} C + (r-q)K\partial_K C + qC = 0, \quad (\tau, K) \in Q,\\
C(K,0) = (S_\circ - K)_+, \; K \in (0,\bar{K}),\\
C(0,\tau) = S_\circ e^{-q\tau}, \tau \in (0,\bar{\tau}],\\
C(\bar{K},\tau,) = 0, \quad \tau \in (0,\bar{\tau}].
\end{aligned}
\tag{8.1}
$$

Here, η is a function of τ and K. For simplicity only, we assume that the interest rate and the dividend yield are constant.

Remark 8.1. *Note that the boundary condition $C(\tau, 0) = S_\circ e^{-q\tau}$ need not be imposed, because it is implied by the other equations in* (8.1).

As in Chapter 2, the notion of weak solution to (8.1) is helpful: denote $V = \{v \in L^2((0,\bar{K})) : x\frac{dv}{dx} \in L^2((0,\bar{K})),\; v(\bar{K}) = 0\}$, and

$$
\begin{aligned}
a_\tau(v,w) = &\int_0^{\bar{K}} \frac{K^2\eta(K,\tau)}{2}\frac{\partial v}{\partial K}\frac{\partial w}{\partial K}\, dK\\
&+ \int_0^{\bar{K}} \left(r - q + \eta(K,\tau) + \frac{K}{2}\frac{\partial \eta}{\partial K}(K,\tau)\right) K\frac{\partial v}{\partial K} w\, dK\\
&+ q\int_0^{\bar{K}} vw\, dK.
\end{aligned}
\tag{8.2}
$$

There exists a unique function $C \in C^0([0,\bar{\tau}]; L^2((0,\bar{K}))) \cap L^2(0,\bar{\tau}; V)$ such that $C(0,K) = (S_\circ - K)_+$ and for a.e. $\tau \in (0,\bar{\tau})$,

$$
\forall v \in V, \quad \left(\frac{\partial C}{\partial \tau}(\tau), v\right) + a_\tau(C(\tau), v) = 0,
\tag{8.3}
$$

as soon as

- η is bounded;
- η is bounded from below by a positive constant $\underline{\eta}$;
- $|K\frac{\partial \eta}{\partial K}|$ is a bounded function.

Exercise 8.1. *Prove that since we work with the localized problem instead of the original one, the last condition above can be replaced by $\frac{\partial \eta}{\partial K} \in L^\infty((0,\bar{\tau}); L^2(0,\bar{K}))$ and $\|\frac{\partial \eta}{\partial K}\|_{L^\infty((0,\bar{\tau});L^2(0,\bar{K}))}$ small enough (compared to $\underline{\eta}$).*

The calibration problem consists in finding η from the observations of

- the spot price $S_\circ$ today;
- the prices $(c_i)_{i\in I}$ of a family of European vanilla calls with different maturities and different strikes $(\tau_i, K_i)_{i\in I}$.

We can choose, for example, $\bar{\tau} > \max_{i\in I} \tau_i$ and $\bar{K} \gg \max((K_i)_{i\in I}, S_\circ)$. We consider the least squares problem: find $\eta \in \mathcal{H}_h$ minimizing

$$J(\eta) + J_R(\eta), \qquad J(\eta) = \sum_{i\in I} |C(K_i, \tau_i) - c_i|^2,$$

where $\mathcal{H}_h$ is a suitable closed subset of a possibly infinite-dimensional function space, J_R is a suitable Tychonoff regularization functional, and C is the solution to (8.1). Of course, with straightforward modifications, all that follows would carry over to the functional $J(\eta) = \sum_{i\in I} w_i |C(K_i, \tau_i) - c_i|^2$, where w_i are nonnegative weights.

8.2 Which Space for η and Which Tychonoff Functional?

The squared volatility η will eventually belong to a finite-dimensional space (for example, a space made of piecewise polynomial functions), but since the dimension of this space may be large (especially if the degrees of the polynomials are low), it is safer to define J_R in an infinite-dimensional space Y, in which it is meaningful to impose bounds on η and its partial derivative with respect to K. Similarly, since $\bar{K}$ may be very large, it is safer to choose the functional J_R as if $\bar{K}$ was infinite.

We have seen above that we must impose constraints of the type $\underline{\eta} \le \eta \le \bar{\eta}$ and $|K\frac{\partial \eta}{\partial K}| \le M$. On the one hand, the space Y must be chosen in such a way that these constraints define a closed subset of Y for the minimization problem to have a solution. On the other hand, we wish Y to be a Hilbert space, because Hilbertian norms are easy to differentiate.

Let X be the space

$$X = \left\{ \eta \in L^2(\mathbb{R}_+);\ (K+1)\frac{\partial \eta}{\partial K} \in L^2(\mathbb{R}_+) \right\}, \tag{8.4}$$

which is a Hilbert space with the norm

$$\|v\|_X = \left(\|v\|^2_{L^2(\mathbb{R}_+)} + \left\| (K+1)\frac{\partial \eta}{\partial K} \right\|^2_{L^2(\mathbb{R}_+)} \right)^{\frac{1}{2}}. \tag{8.5}$$

We can check the Sobolev-type embedding $X \subset L^\infty(\mathbb{R}_+) \cap \mathcal{C}^0(\mathbb{R}_+)$ with continuous injection. This comes from the facts that $H^1(I) \subset \mathcal{C}^0(I)$ for any compact interval and that $K\eta(K) = \int_0^K \kappa \frac{\partial \eta}{\partial K}(\kappa) d\kappa - \int_0^K \eta(\kappa) d\kappa$, and the Cauchy–Schwarz inequality yields that for all $K > 0$, $|\eta(K)|^2 \le \frac{2}{K}\int_0^K \left(|\kappa \frac{\partial \eta}{\partial K}(\kappa)|^2 + \eta^2(\kappa) \right) d\kappa$. In fact, from this inequality, we deduce that the embedding $X \subset L^\infty(\mathbb{R}_+)$ is also compact, because it is possible to approximate a function $\eta \in X$ by means of a piecewise constant function η_N with

- $\eta_N(K) = 0$ for $K \ge N$;
- $\eta_N(K) = \frac{1}{N}\int_{i+\frac{j}{N}}^{i+\frac{j+1}{N}} \eta(\kappa) d\kappa$ for $i + \frac{j}{N} \le K < i + \frac{j+1}{N}$, $0 \le i, j \le N-1$,

and prove that there exists a positive constant C such that $\|\eta - \eta_N\|_{L^\infty(\mathbb{R}_+)} \le \frac{C}{\sqrt{N}}\|\eta\|_X$.

We denote by Y the space of functions from $\mathbb{R}_+ \times (0, \bar{\tau}) \to \mathbb{R}$ defined by

$$Y = \left\{ \eta : \eta, K\frac{\partial \eta}{\partial K} \in H^1((0, \bar{\tau}); X) \right\}, \tag{8.6}$$

which is a Hilbert space with the norm

$$\|v\|_Y = \left(\|v\|^2_{L^2((0,\bar{\tau});X)} + \left\|\frac{\partial \eta}{\partial \tau}\right\|^2_{L^2((0,\bar{\tau});X)} + \left\|K\frac{\partial v}{\partial K}\right\|^2_{L^2((0,\bar{\tau});X)} + \left\|K\frac{\partial^2 v}{\partial \tau \partial K}\right\|^2_{L^2((0,\bar{\tau});X)} \right)^{\frac{1}{2}}. \tag{8.7}$$

We have that

$$Y \subset \left\{ \eta : \eta, K\frac{\partial \eta}{\partial K} \in L^\infty(\mathbb{R}_+ \times (0, \bar{\tau})) \right\},$$

with a continuous and compact embedding. Therefore,

- the set $\mathcal{H} = \{\eta \in Y, \underline{0} < \eta \leq \eta \leq \bar{\eta}, |K\frac{\partial \eta}{\partial K}| \leq M\}$ is a closed subset of Y;
- from a bounded sequence (η_n) in Y, one can extract a subsequence $(\eta_{n'})$ that converges in $L^\infty(\mathbb{R}_+ \times (0, \bar{\tau}))$ and such that $K\frac{\partial \eta_{n'}}{\partial K}$ converges in $L^\infty(\mathbb{R}_+ \times (0, \bar{\tau}))$.

Now, we change the definitions of X and Y by replacing $\mathbb{R}_+$ by $(0, \bar{K})$ in (8.4), (8.5), (8.6). If the least squares minimization is set in $Y_h \cap \mathcal{H}$, where Y_h is a finite-dimensional subspace of Y, a good choice for the Tychonoff regularization functional is of the form

$$\begin{aligned} J_R(\eta) = a_0\|v\|^2_{L^2(Q)} + a_1\left\|\frac{\partial \eta}{\partial \tau}\right\|^2_{L^2(Q)} + a_2\left\|K\frac{\partial \eta}{\partial K}\right\|^2_{L^2(Q)} + a_3\left\|K\frac{\partial^2 v}{\partial \tau \partial K}\right\|^2_{L^2(Q)} \\ + a_4\left\|K^2\frac{\partial^2 v}{\partial K^2}\right\|^2_{L^2(Q)} + a_5\left\|K^2\frac{\partial^3 v}{\partial \tau \partial K^2}\right\|^2_{L^2(Q)}, \end{aligned} \tag{8.8}$$

with suitable positive parameters a_i, $i = 0, \ldots, 5$.

A possible example for $Y_h \subset Y$ is the space of bicubic splines whose control nodes lie on the nodes of a quadrangular mesh of Q; see [39] for a review on splines, and [77] for a detailed description of splines in calibration.

If the space Y_h is not a subspace of Y, then the definition of J_R has to be modified by using, for example, mesh dependent norms: for example, if Y_h is made of piecewise bilinear functions on a quadrangular mesh $\mathcal{T}_h$ of Q, then $\frac{\partial^2 v}{\partial K^2}$ is not a function, so it has to be replaced in each element by some differential quotient involving $\frac{\partial v}{\partial K}$ on the left and right adjacent elements.

Remark 8.2. *If the space where the minimization is carried out is of very small dimension with regular functions which are constant for K large enough, then almost any squared norm can be used for J_R.*

For simplicity, we shall assume hereafter that the least squares minimization is set in $\mathcal{H}_h = Y_h \cap \mathcal{H}$, where Y_h is a finite-dimensional subspace of Y, and that J_R is given by (8.8).

Remark 8.3. *It is possible to add to J a squared error between η and a function $\tilde{\eta}$ chosen a priori, often from historical considerations. For example, $\tilde{\eta}$ could be the result of calibration on a previous day. Adding such a term prevents the local volatility from varying too much from one day to another.*

8.3 Existence of a Minimizer

Lemma 8.1. *Let $\tilde{Q} = (0, \tilde{K}) \times (0, \bar{\tau})$, with $0 < \max(S_\circ, K_i) < \tilde{K} < \bar{K}$. Consider two functions η_1 and η_2 in $\mathcal{H}_h$, call $\delta\eta = \eta_1 - \eta_2$ and $C(\eta_1)$ (resp., $C(\eta_2)$) the weak solutions to* (8.1) *corresponding to $\eta = \eta_1$ (resp., $\eta = \eta_2$), and $\delta C = C(\eta_1) - C(\eta_2)$. Then, for a constant c,*

$$\|K\delta C\|_{L^\infty(\tilde{Q})} \leq c\|\delta\eta\|_{L^\infty(Q)}. \tag{8.9}$$

Proof. It is possible to prove an analogue to Proposition 2.13, i.e., denoting $\tilde{V} = \{v \in L^2((0, \tilde{K})) : K\frac{\mathrm{d}v}{\mathrm{d}K} \in L^2((0, \tilde{K}))\}$, that the weak solution to (8.1) belongs to $\mathcal{C}^0([0, \bar{\tau}]; \tilde{V})$, is such that $K^2\partial^2_{KK}C \in L^2(\tilde{Q})$, and $\|C(\eta)\|_{L^\infty((0,\bar{\tau});\tilde{V})}$ and $\|K^2\partial^2_{KK}C(\eta)\|_{L^2(\tilde{Q})}$ are bounded independently of η for $\eta \in \mathcal{H}_h$.

The function δC satisfies

$$\partial_\tau\delta C - \frac{\eta_1 K^2}{2}\partial^2_{KK}\delta C + (r-q)K\partial_K\delta C + q\delta C = \frac{\delta\eta K^2}{2}\partial^2_{KK}C(\eta_2). \tag{8.10}$$

Therefore, $\|\delta C\|_{L^\infty((0,\bar{\tau});L^2(0,\tilde{K}))} + \|\delta C\|_{L^2((0,\bar{\tau});V)} \leq c\|\delta\eta\|_{L^\infty(Q)}$. By using an argument close to the proof of Proposition 2.13, it is even possible to prove that $\|\delta C\|_{L^\infty((0,\bar{\tau});\tilde{V})} \leq c\|\delta\eta\|_{L^\infty(Q)}$. Then (8.9) follows from a Sobolev-type estimate. □

Corollary 8.2. *The functional J is continuous with respect to the norm $\|\cdot\|_{L^\infty(Q)}$.*

Proposition 8.3. *There exists a minimizer of $J + J_r$ in $\mathcal{H}_h$.*

Proof. A minimizing sequence (η_n) is bounded in the norm $\|\cdot\|_Y$ because J_R is coercive. It is possible to extract a subsequence $(\eta_{n'})$ which converges in Y_h to $\eta \in \mathcal{H}_h$. Therefore, from the continuity of J_R and from Corollary 8.2, $J(\eta_{n'}) + J_R(\eta_{n'})$ converges to $J(\eta) + J_R(\eta)$. □

Remark 8.4. *The conclusions of Proposition* 8.3 *hold if the minimization is performed in $\mathcal{H}$ because $\mathcal{H}$ is closed and $J + J_R$ is lower semicontinuous.*

8.4 The Gradient of J

Theorem 8.4. *The functional $\eta \mapsto J(\eta)$ is Fréchet differentiable in $\mathcal{H}_h$, and its differential is*

$$\begin{aligned} DJ :&Y \to \mathbb{R}, \\ &v \mapsto -\frac{1}{2}\int_Q K^2 P(\eta)\partial^2_{KK} C(\eta) v, \end{aligned} \tag{8.11}$$

where $P(\eta) \in L^2(Q)$ is the unique function such that, for any $v \in L^2((0,\bar{\tau}), V)$ with $\partial_\tau v \in L^2(Q)$ and $K^2 \partial^2_{KK} v \in L^2(Q)$,

$$\begin{aligned} &-\int_Q \left(\partial_\tau v - \frac{\eta K^2}{2}\partial^2_{KK} v + (r-q)K\partial_K v + qv\right) P(\eta) \\ &= 2\sum_{i \in I}(C(K_i,\tau_i) - c_i)v(K_i,\tau_i). \end{aligned} \tag{8.12}$$

The function $P(\eta)$ is called the adjoint state.

Proof. Consider $\eta \in \mathcal{H}_h$ and the variation of J produced by a small admissible variation of η: $\delta\eta$. The cost function J varies like

$$\delta J = 2\sum_{i \in I}(C(K_i,t_i) - c_i)\delta C(K_i,t_i) + \sum_{i \in I}\delta C(K_i,t_i)^2.$$

From Lemma 8.1, there exists a constant c such that $\sum_{i \in I}\delta C(K_i,t_i)^2 \le c\|\delta\eta\|^2_{L^\infty(Q)}$. The variation δC is a solution to

$$\begin{aligned} \partial_\tau \delta C - \frac{\eta K^2}{2}\partial^2_{KK}\delta C + (r-q)K\partial_K \delta C + q\delta C &= K^2\frac{\delta\eta}{2}\partial^2_{KK} C(\eta), \\ \delta C(K,0) &= 0, \\ \delta C(0,\tau) &= 0, \\ \delta C(\bar{K},\tau) &= 0. \end{aligned} \tag{8.13}$$

Of course, the regularity results for C hold for δC.

To express δJ explicitly in terms of $\delta\eta$, an adjoint state function P is introduced: consider the adjoint problem

$$\begin{aligned} \partial_\tau P + \partial^2_{KK}\left[\frac{\eta K^2}{2}P\right] - \partial_K[P(r-q)K] - qP &= 2\sum_{i \in I}(C(K_i,\tau_i) - c_i)\delta_{K_i,\tau_i}, \\ P(K,\bar{\tau}) &= 0, \\ P(\bar{K},\tau) &= 0, \end{aligned} \tag{8.14}$$

where on the right-hand side the $\delta_{K,\tau}$ denote Dirac functions in time and strike, at τ and K. The problem (8.14) is a backward parabolic problem in Q with singular data (singular in τ and K). Yet, it is possible to find very weak solutions to (8.14) by considering first the auxiliary boundary value problem

$$\begin{aligned}\partial_\tau W + \partial_K \left[\frac{\eta K^2}{2}\partial_K W\right] - [(r-q)K]\partial_K W - qW &= 2\sum_{i\in I}(C(K_i,\tau_i)-c_i)\delta_{\tau_i}1_{K>K_i},\\ W(K,\bar{\tau}) &= 0,\\ \partial_K W(\bar{K},\tau) &= 0.\end{aligned} \tag{8.15}$$

If $\eta \in \mathcal{H}_h$, (8.15) has a unique weak solution which is piecewise continuous in time with values in $L^2(0,\bar{K})$ and which belongs to $L^2((0,\bar{\tau}),\bar{V})$, where $\bar{V} = \{v \in L^2(0,\bar{K});\, K\partial_K v \in L^2(0,\bar{K})\}$. The function $P = \partial_K W$ is the unique solution in $L^2(Q)$ to the following problem: find $P \in L^2(Q)$, such that for any $v \in L^2((0,\bar{\tau}),V)$ with $\partial_\tau v \in L^2(Q)$ and $K^2\partial^2_{KK}v \in L^2(Q)$,

$$-\int_Q \left(\partial_\tau v - \frac{\eta K^2}{2}\partial^2_{KK}v + (r-q)K\partial_K v + qv\right)P = 2\sum_{i\in I}(C(K_i,\tau_i)-c_i)v(K_i,\tau_i),$$

which is a very weak solution to (8.14).

Then by multiplying (8.14) by δC and integration on Q, the following is found:

$$\begin{aligned}&2\left\langle \sum_{i\in I}(C(K_i,\tau_i)-c_i)\delta_{K_i,\tau_i},\, \delta C\right\rangle\\ &= -\int_Q\left\{\partial_\tau\delta C - \frac{\eta K^2}{2}\partial^2_{KK}\delta C + (r-q)K\partial_K\delta C + q\delta C\right\}P = -\int_Q \frac{\delta\eta}{2}K^2P\partial^2_{KK}C.\end{aligned}$$

Note that $K^2\partial^2_{KK}C \in L^2(Q)$, $P \in L^2(Q)$, and $\delta\eta \in L^\infty(Q)$, so the previous integral makes sense. Therefore,

$$\left|\delta J + \int_Q \frac{\delta\eta}{2}K^2P\partial^2_{KK}C(\eta)\right| \le c\|\delta\eta\|^2_{L^\infty(Q)}, \tag{8.16}$$

which implies that J is Fréchet differentiable, and its differential is given by (8.11). □

Proposition 8.5 (necessary optimality conditions). *A necessary condition for η in $\mathcal{H}_h$ to be a minimum of $J + J_R$ is that for any $\eta' \in \mathcal{H}_h$,*

$$\frac{1}{2}\int_Q K^2P(\eta)\partial^2_{KK}C(\eta)(\eta'-\eta) \le 0,$$

where $P(\eta)$ is the solution to (8.12).

8.5 The Discrete Problem

We introduce a partition of the interval $[0,\bar{\tau}]$ into subintervals $[t_{n-1},t_n]$, $1 \le n \le N$, with $\Delta t_i = t_i - t_{i-1}$, $\Delta t = \max_i \Delta t_i$, and we assume that all the maturities τ_i, $i \in I$, coincide with some grid node. We also introduce a partition of the interval $[0,\bar{K}]$ into subintervals $\omega_i = [\kappa_{i-1},\kappa_i]$, $1 \le i \le N_h+1$, such that $0 = \kappa_0 < \kappa_1 < \cdots < \kappa_{N_h} < \kappa_{N_h+1} = \bar{K}$. The size of the interval ω_i is called h_i and we set $h = \max_{i=1,\dots,N_h+1} h_i$. The mesh $\mathcal{T}_h$ of $[0, \bar{K}]$ is the set $\{\omega_1, \dots, \omega_{N_h+1}\}$. We assume that $S_\circ$ and the strikes K_i, $i \in I$, coincide with some mesh node.

We use piecewise linear finite elements for the discretization with respect to K: we define the discrete space V_h by

$$V_h = \left\{ v_h \in V,\ \forall \omega \in \mathcal{T}_h,\ v_{h|\omega} \in \mathcal{P}_1(\omega) \right\}, \tag{8.17}$$

where $\mathcal{P}_1(\omega)$ is the space of linear functions on ω.

The discrete problem arising from Euler's implicit scheme is as follows:

Find $(C^n)_{0 \le n \le N} \in V_h$ satisfying

$$C^0(K) = (K - S_\circ)_-, \tag{8.18}$$

and for all m, $1 \le m \le N$,

$$\forall v \in V_h, \quad \left(C^m - C^{m-1}, v\right) + \Delta t_m a_{t_m}(C^m, v) = 0, \tag{8.19}$$

where a_τ is defined in (8.2).

Let $(w^i)_{i=0,\dots,N_h}$ be the nodal basis of V_h, and let $\mathbf{M}$ and $\mathbf{A}^m$ in $\mathbb{R}^{(N_h+1)\times(N_h+1)}$ be the mass and stiffness matrices defined by $\mathbf{M}_{i,j} = (w^i, w^j)$, $\mathbf{A}^m_{i,j} = a_{t_m}(w^j, w^i)$, $0 \le i, j \le N_h$. The matrix $\mathbf{A}^m$ is tridiagonal and its entries are

$$\begin{aligned}
\mathbf{A}^m_{i,i-1} &= -\frac{\kappa_i^2 \eta(\kappa_i, t_m)}{2h_i} + \frac{(q-r)\kappa_i}{2} + r\frac{h_i}{6}, \quad 1 \le i \le N,\\
\mathbf{A}^m_{i,i} &= \frac{\kappa_i^2 \eta(\kappa_i, t_m)}{2}\left(\frac{1}{h_i} + \frac{1}{h_{i+1}}\right) + \left(\frac{q}{2} - \frac{r}{6}\right)(h_{i+1} + h_i), \quad 1 \le i \le N,\\
\mathbf{A}^m_{0,0} &= \left(\frac{q}{2} - \frac{r}{6}\right) h_1,\\
\mathbf{A}^m_{i,i+1} &= -\frac{\kappa_i^2 \eta(\kappa_i, t_m)}{2h_{i+1}} - \frac{(q-r)\kappa_i}{2} + r\frac{h_{i+1}}{6}, \quad 0 \le i \le N-1.
\end{aligned}$$

Denoting $\mathbf{C}^n = (C^n(\kappa_0), \dots, C^m(\kappa_{N_h}))^T$, (8.19) is equivalent to

$$\mathbf{M}(\mathbf{C}^m - \mathbf{C}^{m-1}) + \Delta t_m \mathbf{A}^m \mathbf{C}^m = 0. \tag{8.20}$$

The cost function is now

$$J(\eta) + J_R(\eta), \qquad J(\eta) = \sum_{i\in I} \sum_{m=0}^{N} \delta_{t_m = \tau_i} (C^m(K_i) - c_i)^2.$$

Note that a variation $\delta\eta$ of η produces a variation $\delta\mathbf{A}^m$ of $\mathbf{A}^m$:

$$\begin{aligned}
\delta\mathbf{A}^m_{i,i-1} &= -\frac{\kappa_i^2 \delta\eta(\kappa_i, t_m)}{2h_i}, \quad 1 \le i \le N, \qquad \delta\mathbf{A}^m_{0,0} = 0,\\
\delta\mathbf{A}^m_{i,i} &= \frac{\kappa_i^2 \delta\eta(\kappa_i, t_m)}{2}\left(\frac{1}{h_i} + \frac{1}{h_{i+1}}\right), \quad 1 \le i \le N,\\
\delta\mathbf{A}^m_{i,i+1} &= -\frac{\kappa_i^2 \delta\eta(\kappa_i, t_m)}{2h_{i+1}}, \quad 0 \le i \le N-1,
\end{aligned} \tag{8.21}$$

whereas the variation of $\mathbf{C}^m$ satisfies $\delta\mathbf{C}^0 = 0$ and

$$\mathbf{M}(\delta\mathbf{C}^m - \delta\mathbf{C}^{m-1}) + \Delta t_m \mathbf{A}^m \delta\mathbf{C}^m = -\Delta t_m \delta\mathbf{A}^m \mathbf{C}^m.$$

We have

$$\begin{aligned}\delta J(\eta) &= \sum_{i\in I}\sum_{m=0}^{N} \delta_{t_m=\tau_i}\left(2(C^m(K_i) - c_i)\delta C^m(K_i) + (\delta C^m(K_i))^2\right)\\ &= 2\sum_{i\in I}\sum_{m=0}^{N} \delta_{t_m=\tau_i}(C^m(K_i) - c_i)\delta C^m(K_i) + o(\delta\eta).\end{aligned}$$

The discrete adjoint state $(\mathbf{P}^m)_{0\le m\le N}$ is a solution to $\mathbf{P}^N = 0$, and

$$\begin{aligned}&\mathbf{M}(\mathbf{P}^m - \mathbf{P}^{m+1}) + \Delta t_m \mathbf{A}^{T,m}\mathbf{P}^m = 2\sum_{i\in I}\delta_{t_m=\tau_i}\mathbf{G}^m,\\ &\mathbf{G}^m = \left(0, \sum_{i\in I}\delta_{K_1=K_i}(C_1^m - c_i), \dots, \sum_{i\in I}\delta_{K_j=K_i}(C_j^m - c_i), \dots\right)^T,\end{aligned} \tag{8.22}$$

where $\mathbf{A}^{T,m}$ is the matrix obtained by transposing $\mathbf{A}^m$. Therefore, we have

$$\begin{aligned}&2\sum_{i\in I}\sum_{m=1}^{N-1}\delta_{t_m=\tau_i}(C^m(K_i) - c_i)\delta C^m(K_i)\\ &= \sum_{m=1}^{N-1}(\delta\mathbf{C}^m)^T\left(\mathbf{M}(\mathbf{P}^m - \mathbf{P}^{m+1}) + \Delta t_m \mathbf{A}^{T,m}\mathbf{P}^m\right)\\ &= \sum_{m=1}^{N-1}(\mathbf{P}^m)^T\left(\mathbf{M}(\delta\mathbf{C}^m - \delta\mathbf{C}^{m-1}) + \Delta t_m \mathbf{A}^m\delta\mathbf{C}^m\right)\\ &= -\sum_{m=1}^{N-1}\Delta t_m(\mathbf{P}^m)^T\delta\mathbf{A}^m\mathbf{C}^m.\end{aligned} \tag{8.23}$$

From this and from (8.21), it is easy to compute $grad(J(\eta))$.

Exercise 8.2. *Write down the adjoint problem when a Crank–Nicolson scheme is used for computing* $\mathbf{C}^m$.

Exercise 8.3. *Write down the adjoint problem when a transparent boundary condition is used for Dupire's equation along with an Euler or Crank–Nicolson scheme.*

8.6 A Taste of the Program

Here, we focus on the code for computing the gradient of the functional J, assuming that $(\mathbf{C}^m)_{m=0,\dots,M}$ has already been computed: the routine is as follows.

ALGORITHM 8.1. Gradient.

```
template <class Vecto, class Tab, class Mat>
void dupire_cost<Vecto,Tab,Mat>::comp_grad_J(const Tab &eta,  const  Tab  &c
, Tab & grad)
{
  for(int tit=0;tit<grad.size();tit++)
    grad[tit]=0;
  for(int j=0;j<p.size();j++)
    p[j]=0;
     //  constructs the right-hand side for the adjoint backward parabolic
                                                               //   problem
                    //  the observed prices are contained in the array c_ob
  for( int i_ob=0;i_ob<c_ob.size(); i_ob++)
    p[maturities[i_ob]][strikes[i_ob]] +=
      2*weights_[i_ob]*(c[maturities[i_ob]][strikes[i_ob]]-c_ob[i_ob]);
                               //  solves the adjoint backward parabolic problem
  scheme->backward_loop(eta, p);
                                              //  computes sum_m (P^m)^T δA^m U^m
  scheme->Diff_loop_CP(c,p,grad,eta);
}
```

The programs for the backward loop for computing the adjoint state and the program for (8.23) are as follows.

ALGORITHM 8.2. Backward loop: Adjoint problem.

```
template <class Vecto, class Tab, class Mat>
void  euler_scheme<Vecto,Tab, Mat>::backward_loop_iv(const int init_time,
const Tab& eta, Tab& y,  const Tab &p, const Vecto & ob,const int strike)
                                                  //  p :  price of the put
                                                    //  y :  adjoint state
                                                  //  ob:  obstacle (K-x)_+
                            //  strike:  the index of the strike in the mesh
{
  int ti=t.size()-1;
  dt=t[ti]-t[ti-1];
                                    //  builds the matrix A^T at final time
  disc->Build_Imp_Euler_LHS_Matrix_T(dt,eta[ti],LhsmatT);
                       //  modifies the adjoint system in the exercise zone
  for (int i=0;i<p[ti].size();i++)
    if(p[ti][i]==ob[i]&&  disc->meshpoints()[i]<2
       * disc->meshpoints()[strike])
      {
        LhsmatT(i,i)=1.;
        if (i< p[ti].size()-1)
          LhsmatT(i,i+1)=0.;
        if (i>0)
          LhsmatT(i,i-1)=0;
        y[ti][i]=0;
      }
                                 //  solves for adjoint state at final time
  gauss(LhsmatT,y[ti]);
                                                           //  loop on t
  for(int tit=t.size()-1;tit>init_time+1;tit--)
```

```
    {
      dt=t[tit-1]-t[tit-2];
      disc->rate()=rate[tit-1];
                                                      //   builds the matrix A^T
      disc->Build_Imp_Euler_LHS_Matrix_T(dt,eta[tit-1],LhsmatT);
                        //    modifies the adjoint system in the exercise zone
      for (int i=0;i<p[tit-1].size();i++)
        if(p[tit-1][i]==ob[i]&& disc->meshpoints()[i]<2
           *  disc->meshpoints()[strike])
          {
            LhsmatT(i,i)=1.;
            if (i< p[tit-1].size()-1)
              LhsmatT(i,i+1)=0.;
            if (i>0)
              LhsmatT(i,i-1)=0;
          }
                        //   builds the right-hand side of the adjoint problem
      for (int i=0; i<y[0].size();i++)
        y[tit-1][i]+=Massmat(i,i)*y[tit][i];
      for (int i=0; i<y[0].size()-1;i++)
        y[tit-1][i]+=Massmat(i,i+1)*y[tit][i+1];
      for (int i=1; i<y[0].size();i++)
        y[tit-1][i]+=Massmat(i,i-1)*y[tit][i-1];
                       //    modifies the right-hand side in the exercise zone
      for (int i=0;i<p[tit-1].size();i++)
        if(p[tit-1][i]==ob[i]&& disc->meshpoints()[i]<2
           *disc->meshpoints()[strike] )
          y[tit-1][i]=0;
                                              //    solves for the adjoint state
      gauss(LhsmatT,y[tit-1]);
    }
  y[init_time]=0;
}
```

It calls the following routine.

ALGORITHM 8.3. Formula (8.23).

```
template <class Vecto, class Mat>
void dupire_fem<Vecto,Mat>::_Diff_LHS_matrix_UP(const double dt,const  Vecto
& u,const  Vecto & p,Vecto & v,const Vecto & eta)
{
  v=0;
  int i;
  double xx,hp,hn;
  double a,b,c,d;
  double   e=-0.5*dt;
  hn=h[0];
  for( i=1;i<h.size()-1;i++)
    {
      hp=hn ;                              //    diameter of previous element
      hn=h[i];                             //     diameter of current element
      xx=x[i];                                                //    current node
      a=xx*xx;
```

```
        b=a/hp;
        c=a/hn;
        v[i] +=e*(b+c)*u[i]*p[i];
        v[i]-=e*b*u[i-1]*p[i];
        v[i]-=e*c*u[i+1]*p[i];
      }
    hp=hn;                              //  diameter of previous element
    hn=h[i];                              //  diameter of last element
    xx=x[i];                                         //   last node
    a=xx*xx;
    b=a/hp;
    c=a/hn;
    v[i] +=e*(b+c)*u[i]*p[i];
    v[i]-=e*b*u[i-1]*p[i];
}
```

8.7 Results

We consider a family of calls on the S&P 500 index. The spot price is 590 and the implied volatilities for the family of calls are given in Table 8.1.

The volatility is discretized by functions that are the sum of

- a piecewise affine function in the x-variable which is constant in the regions $K < 250$ and $x > 830$ and affine in the region $250 < x < 830$;
- a bicubic spline in the region $250 < x < 830$, $-0.1 < \tau < 5.5$, whose value and derivatives vanish on the boundary of this rectangle. The control points of the spline are plotted in Figure 8.1. We see that the control points are not uniformly distributed: the mesh is refined for small times τ and at the money region.

In the least squares functional, we have chosen to put a larger weight on the options with a strike close to the money region. Table 8.2 contains the relative errors between the actual observed prices $(c_i)_{i \in I}$ and the prices computed by the algorithm.

The local volatility surface obtained after calibration is plotted in Figure 8.2.

Table 8.1. *The implied volatility observed on call options on the S&P 500.*

$T\backslash\frac{K}{S_o}$	0.85	0.9	0.95	1	1.05	1.1	1.15	1.2	1.3	1.4
0.175	0.19	0.168	0.133	0.113	0.102	0.097	0.120	0.142	0.169	0.2
0.425	0.177	0.155	0.138	0.125	0.109	0.103	0.100	0.114	0.130	0.150
0.625	0.172	0.157	0.144	0.133	0.118	0.104	0.100	0.101	0.108	0.124
0.94	0.171	0.159	0.149	0.137	0.127	0.113	0.106	0.103	0.100	0.110
1	0.171	0.159	0.150	0.138	0.128	0.115	0.107	0.103	0.099	0.108
1.5	0.169	0.160	0.151	0.142	0.133	0.124	0.119	0.113	0.107	0.102
2	0.169	0.161	0.153	0.145	0.137	0.130	0.126	0.119	0.115	0.111
3	0.168	0.161	0.155	0.149	0.143	0.137	0.133	0.128	0.124	0.123
4	0.168	0.162	0.157	0.152	0.148	0.143	0.139	0.135	0.130	0.128
5	0.168	0.164	0.159	0.154	0.151	0.148	0.144	0.140	0.136	0.132

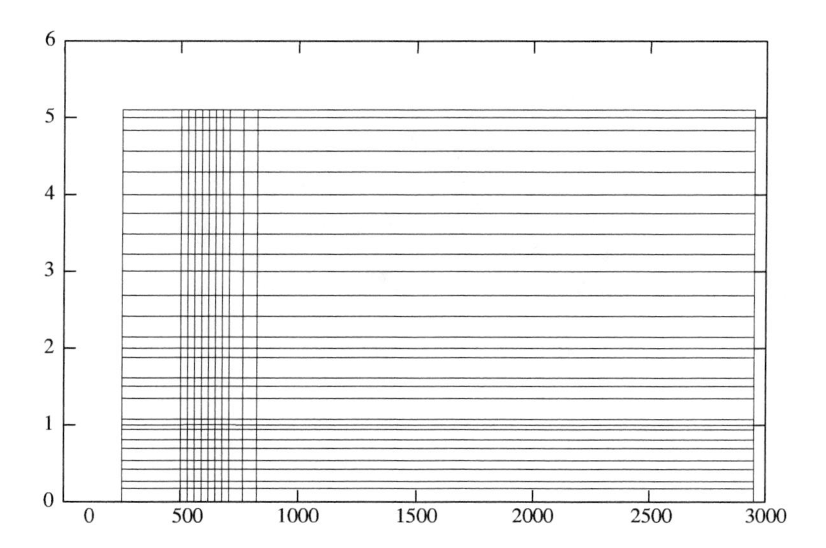

Figure 8.1. *The grid used for the local volatility.*

Table 8.2. *The relative error between the observed and computed prices.*

$T\backslash\frac{K}{S_\circ}$	0.85	0.9	0.95	1	1.05
0.175	−0.00050360	−0.00134184	0.00874306	−0.00011394	0.000961893
0.425	−0.006577	−0.011745	−0.022311	−0.000725	0.0053118
0.625	−0.005356	−0.004013	−0.004495	−0.000188	0.0007585
0.94	−0.004170	−0.003589	−0.011314	0.0008474	−0.005444
1	−0.00416	−0.00274	−0.01173	−0.00061	0.002959
1.5	−0.00222	−0.00124	−0.00422	1.76366e-5	−0.00026
2	−0.003215	−0.002036	−0.002488	0.0021965	0.0064208
3	−0.002092	0.0006371	0.0009923	0.0036407	0.0078316
4	−0.001912	0.0006624	0.0011179	0.0030971	0.0033606
5	−0.000979	−0.001172	0.0003663	0.0036968	0.0035785

$T\backslash K$	1.1	1.15	1.2	1.3	1.4
0.175	0.0119657	−0.023659	0.0035202	−0.411383	−0.975634
0.425	−0.047604	−0.068850	0.0445737	−0.722029	−0.925981
0.625	−0.027442	0.0196397	−0.051659	0.492077	−0.753774
0.94	−0.000799	0.0110919	0.0262597	0.66291	−0.118613
1	0.000552	0.027384	0.113846	0.752714	0.018957
1.5	0.000711	−0.00420	0.017756	0.190795	0.896564
2	0.0135983	0.0461694	0.172374	0.12785	0.0883746
3	0.015847	0.0208292	0.0444584	0.0886675	0.0162867
4	0.0094648	0.012883	0.0186873	0.0384731	0.0322194
5	0.0026771	0.0072061	0.0136582	0.0167708	0.0421486

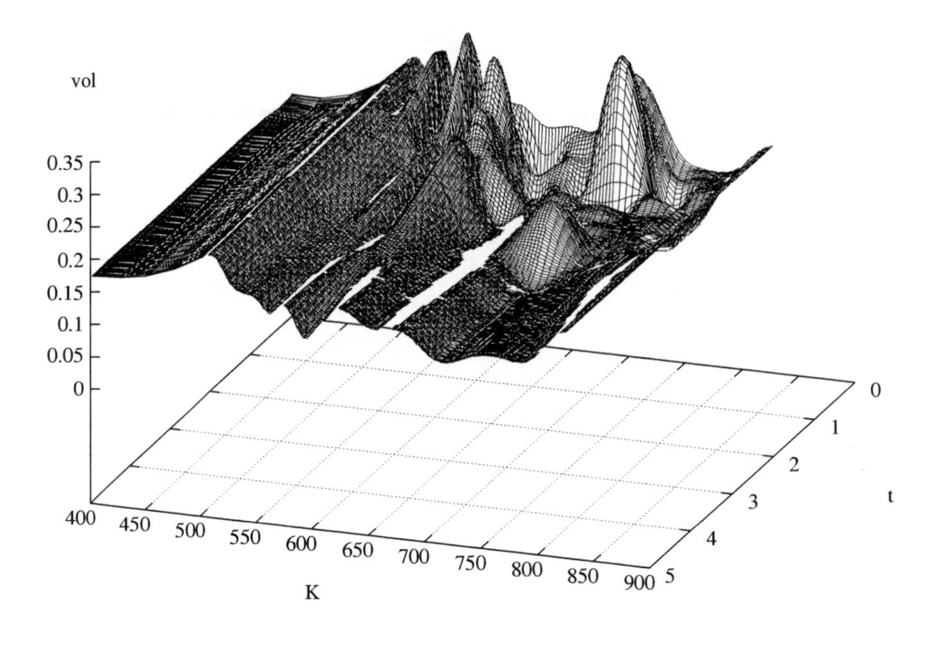

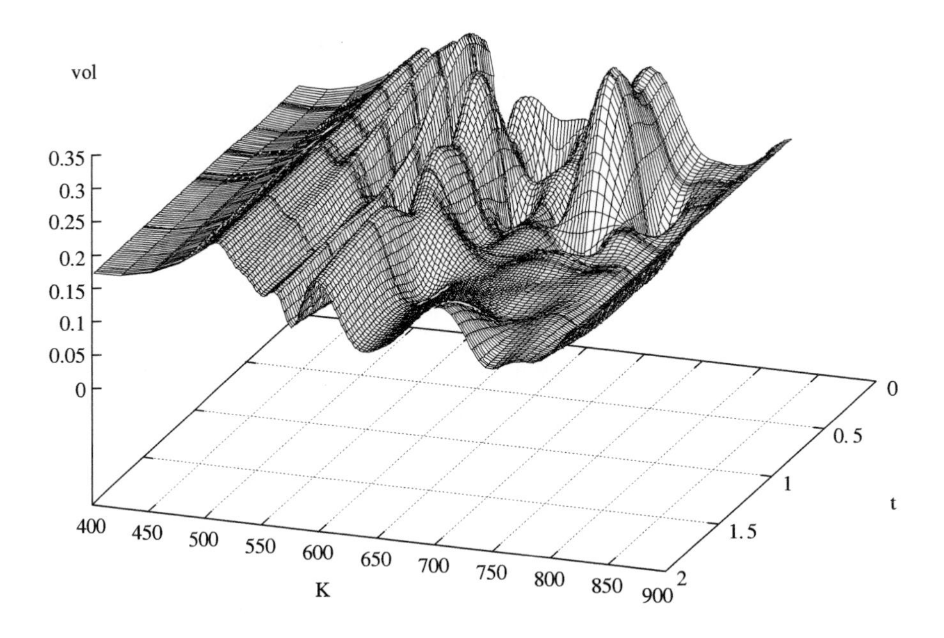

Figure 8.2. *Two views of the local volatility surface obtained by calibration.*

8.8 A Stochastic Control Approach

The method described below was proposed by Avellaneda et al. [7] and studied in a deeper way by Samperi [108]. It has the advantages of not relying on any parametrization of the volatility and of leading to an unconstrained optimization problem.

8.8.1 A Minimax Dual Program

Following Avellaneda et al. [7], we assume that the risk neutralized price of the asset is the process

$$dS_t = S_t(\sqrt{\eta_t}dB_t + rdt), \qquad S(t=0) = S_\circ, \tag{8.24}$$

where B_t is a standard Brownian motion and where r is the interest rate, assumed constant here. From the observations, we know that

$$\mathbb{E}^\eta(e^{-r\tau_i}(S_{\tau_i} - K_i)_+) = c_i, \quad i \in I, \tag{8.25}$$

where $\mathbb{E}^\eta$ is the expectation with respect to the measure on path space corresponding to a given choice of η_t.

Let $\mathcal{H}$ be the set of real-valued processes for $t \in [0, \bar{\tau}]$ that are progressively measurable with respect to some reference probability system and that satisfy $0 < \underline{\eta} \le \eta_t \le \bar{\eta}$.

Let $\eta_\circ$ be a given value such that $\underline{\eta} \le \eta_\circ \le \bar{\eta}$. This value is a prior squared volatility which may be fixed from historical considerations. Let H be a smooth and strictly convex real-valued function defined on $\mathbb{R}$, with minimal value 0 at $\eta_\circ$.

The approach to calibration proposed in [7] consists of finding η^* that solves

$$\sup_{\eta\in\mathcal{H}} -\mathbb{E}^\eta\left(\int_0^{\bar{\tau}} e^{-rt}H(\eta_t)dt\right), \tag{8.26}$$

subject to the constraints (8.25).

Note that $\mathbb{E}^\eta(\int_0^{\bar{\tau}} e^{-rt}H(\eta_t)dt)$ depends on η_t through $H(\eta_t)$ but also through the probability measure.

Remark 8.5. *It is possible to localize the problem for prices lying in* $[\underline{S}, \bar{S}]$. *Then the functional in* (8.26) *should be changed.*

Remark 8.6. *In* [7], *the function* H *is called a pseudoentropy, as a reference to relative entropies of probability measures.*

We define $f(\eta) = -\mathbb{E}^\eta(\int_0^{\bar{\tau}} e^{-rt}H(\eta_t)dt)$, and $g_i(\eta) = \mathbb{E}^\eta(e^{-r\tau_i}(S_{\tau_i} - K_i)_+)$.

Assuming that the data $(c_i)_{i\in I}$ and $S_\circ$ are such that the problem is solvable, and calling n_I the cardinal of I, then the problem is equivalent to finding $\eta^* \in \mathcal{H}$ that solves

$$\sup_{\eta\in\mathcal{H}} \inf_{\mu\in\mathbb{R}^{n_I}} f(\eta) + \sum_{i\in I} \mu_i(g_i(\eta) - c_i), \tag{8.27}$$

and it is clear that

$$\sup_{\eta\in\mathcal{H}} \inf_{\mu\in\mathbb{R}^{n_I}} f(\eta) + \sum_{i\in I} \mu_i(g_i(\eta) - c_i) \le \inf_{\mu\in\mathbb{R}^{n_I}} \sup_{\eta\in\mathcal{H}} f(\eta) + \sum_{i\in I} \mu_i(g_i(\eta) - c_i). \tag{8.28}$$

Denoting $U(\eta) = \inf_{\mu\in\mathbb{R}^{n_I}} f(\eta) + \sum_{i\in I} \mu_i(g_i(\eta) - c_i)$ and $V(\mu) = \sup_{\eta\in\mathcal{H}} f(\eta) + \sum_{i\in I} \mu_i g_i(\eta)$, the problem of maximizing U on $\mathcal{H}$ is called the primal program, whereas the problem of minimizing $W(\mu) = V(\mu) - \sum_{i\in I} \mu_i c_i$ on $\mathbb{R}^{n_I}$ is called the dual program. The gap between $\sup_{\eta\in\mathcal{H}} U(\eta)$ and $\inf_{\mu\in\mathbb{R}^{n_I}} W(\mu)$ is called the duality gap.

The following lemma can be found in Samperi [108].

Lemma 8.6. *Assume that there exists a convex open subset A of $\mathbb{R}^{n_I}$ such that*

1. *for all $\mu \in A$, there exists a unique $\eta(\mu) \in \mathcal{H}$ such that $\eta(\mu)$ solves $V(\mu) = f(\eta(\mu)) + \sum_{i\in I} \mu_i g_i(\eta(\mu))$;*
2. *the function V is differentiable in A and $\nabla V(\mu) = (g_1(\eta(\mu)), \dots, g_{n_I}(\eta(\mu)))^T$;*
3. *the function V is twice differentiable and strictly convex in A.*

Assume also that there exists $\mu^* \in A$ such that $\nabla V(\mu^*) = (c_i)^T_{i\in I}$. Then the problem (8.26) has a solution, and the corresponding Lagrange multiplier can be found as a local minimum of W.

Proof. We know that there exists μ^* such that $\nabla V(\mu^*) = (c_i)^T_{i\in I}$. This implies on the one hand that $\nabla W(\mu^*) = 0$, and since A is convex, and W is strictly convex, thanks to item 3, μ^* is the minimum of W on A. On the other hand, from item 2, $g_i(\eta(\mu^*)) = c_i, i \in I$.

Therefore, calling $\eta^* = \eta(\mu^*)$, we know that η^* satisfies the constraints (8.25). On the other hand, it is clear that

$$W(\mu^*) \geq \sup_{\eta\in\mathcal{H}} U(\eta),$$

but

$$W(\mu^*) = f(\eta^*) = U(\eta^*),$$

which implies that $U(\eta^*) = \sup_{\eta\in\mathcal{H}} U(\eta)$. ☐

In [108], Samperi proves that under some assumptions, the modified value function V corresponding to the regularized problem

$$\tilde{V}(\mu) = \sup_{\eta\in\mathcal{H}} -\mathbb{E}^{\eta}\left(\int_0^{\tau} e^{-rt} H(\eta_t)dt + \sum_{i\in I} \mu_i \int_0^{\tau} e^{-rt}\phi_i(S_t, t)dt\right), \qquad (8.29)$$
$$\tau = \min(\bar{\tau}, \inf\{t, S_t \notin (\underline{S}, \bar{S})\}),$$

where $\phi_i(S,t)$ are smooth approximations of $(S-K_i)_+\delta_{t=\tau_i}$ with compact support in $(\underline{S}, \bar{S})$, satisfies assumptions 1, 2, 3 of Lemma 8.6. To our knowledge, the theory for the genuine V is not complete. With this partial justification, the idea is to now to solve the dual program—minimize W on $\mathbb{R}^{n_I}$—rather than the primal one.

Following the theory for stochastic dynamic programming (see [48, 49] and §8.8.2 for a formal derivation), even though the regularity assumptions needed by the theory are not satisfied in the present case, one finds $V(\mu) = V(S_\circ, 0)$ by solving the Hamilton–Jacobi–Bellman (HJB) nonlinear partial differential equation:

$$\partial_t V + \frac{1}{2}\Phi\left(\frac{S^2}{2}\partial^2_{SS} V\right) + rS\partial_S V - rV = -\sum_{i\in I} \mu_i (S - K_i)_+ \delta_{t=\tau_i}, \qquad (8.30)$$

with the Cauchy condition $V(\cdot, \bar{\tau}) = 0$. Here

$$\Phi(x) = \sup_{\underline{\eta} \le \eta \le \bar{\eta}} x\eta - H(\eta) \tag{8.31}$$

is the Legendre dual of Φ; the function Φ is convex, and we have

$$H(\eta) = \sup_{x \in \mathbb{R}} (x\eta - \Phi(x)). \tag{8.32}$$

Furthermore, η such that $f(\eta) + \sum_{i \in I} \mu_i g_i(\eta) = V(\mu)$ is given by

$$\eta_t = \Phi'\left(\frac{S_t^2}{2} \partial_{SS}^2 V(S_t, t)\right). \tag{8.33}$$

The theory of HJB equations exceeds the scope of the book. The notion of viscosity solution [37, 49] was developed for that purpose.

For the well-posedness, the following lemma is useful.

Lemma 8.7. *The function Φ is C^1-regular and satisfies $\Phi(0) = 0$, $\Phi'(0) = \eta_\circ$, and*

$$0 < \underline{\eta} \le \Phi'(x) \le \bar{\eta},$$

$$0 < \underline{\eta} \le \frac{\Phi(x)}{x} \le \bar{\eta},$$

where $\frac{\Phi(x)}{x}$ is set to $\Phi'(0)$ for $x = 0$. Furthermore, Φ is smooth and strictly convex in the open region where $\underline{\eta} < \Phi'(x) < \bar{\eta}$. There exists $x_1 < 0 < x_2$ such that $\Phi'(x) = \underline{\eta} \Leftrightarrow x \le x_1$, and $\Phi'(x) = \bar{\eta} \Leftrightarrow x \ge x_2$, and Φ'' jumps at x_1 and x_2.

8.8.2 Formal Derivation of (8.30)

Let us replace the term $e^{-r\tau_i} \sum_{i \in I} \mu_i (S_{\tau_i} - K_i)_+ = \int_0^{\bar{\tau}} e^{-rv} \mu_i (S_v - K_i)_+ \delta_{v=\tau_i} dv$ in $\mu_i g_i(\eta)$ by $\int_0^{\bar{\tau}} e^{-rv} \mu_i \phi_i(S_v, v) dv$, where ϕ_i is a smooth function with compact support. The goal is to compute the function $\sup_{\eta \in \mathcal{H}} f(\eta) + E^\eta(\sum_{i \in I} \mu_i \int_0^{\bar{\tau}} e^{-rv} \phi_i(S_v, v) dv)$.

The principle of dynamic programming is as follows: Define

$$J(t, S, \eta) = -\mathbb{E}^\eta\left(\int_t^{\bar{\tau}} e^{-rv} H(\eta_v) dv | F_t\right) + \mathbb{E}^\eta\left(\int_t^{\bar{\tau}} e^{-rv} \sum_{i \in I} \mu_i \phi_i(S_v, v) dv | F_t\right),$$

where S_v is the process given by (8.24) with $S_t = S$ and where $\eta_v = \eta(S_v, v)$. Let η^* be an optimal volatility, and denote $\tilde{V}(S, t) = e^{rt} J(t, S, \eta^*)$.

Take the volatility η_1 to be

$$\eta_1(s, v) = \begin{cases} \eta^*(s, v) & \text{for } v > s, \\ \eta(s, v) & \text{for } t \le v \le s. \end{cases}$$

We have that

$$\begin{aligned} & J(t, S, \eta_1) \\ = & -\mathbb{E}^{\eta_1}\left(\int_t^s e^{-rv} H(\eta_v) dv + \int_t^s e^{-rv} \sum_{i \in I} \mu_i \phi_i(S_v, v) dv \,\middle|\, F_t\right) + \mathbb{E}^{\eta_1}(J(s, S_s, \eta^*) | F_t). \end{aligned}$$

But $e^{-rt}\tilde V(S,t) \geq J(t,S,\eta_1)$ and $e^{-rs}\tilde V(S_s,s) = J(s,S_s,\eta^*)$; therefore

$$\begin{aligned} e^{-rt}\tilde V(S,t) \geq & -\mathbb{E}^{\eta_1}\left(\int_t^s e^{-rv}H(\eta_v)dv + \int_t^s e^{-rv}\sum_{i\in I}\mu_i\phi_i(S_v,v)dv \,\middle|\, F_t\right) \\ & + \mathbb{E}^{\eta_1}(e^{-rs}\tilde V(S_s,s)|F_t). \end{aligned} \tag{8.34}$$

But Itô's formula yields that

$$\begin{aligned} & e^{-rt}\tilde V(S,t) \\ & = -\mathbb{E}^{\eta_1}\left(\int_t^s \partial_t V(S_v,v) + \eta_v \frac{S_v^2}{2}\partial_{SS}^2\tilde V(S_v,v) + rS_v\partial_S\tilde V(S_v,v)dv \,\middle|\, F_t\right) \\ & \quad - \mathbb{E}^{\eta}\left(\int_t^s re^{-rv}\tilde V(S_v,v)dv \,\middle|\, F_t\right) + \mathbb{E}^{\eta}(e^{-rs}\tilde V(S_s,s)|F_t). \end{aligned} \tag{8.35}$$

This is formal, because we do not check the assumptions required by Itô's lemma. Let us subtract (8.35) from (8.34), divide by $s-t$, and let s tend to t_+; this gives, formally,

$$0 \geq \partial_t\tilde V + \eta_v \frac{S^2}{2}\partial_{SS}^2\tilde V + rS\partial_S\tilde V - r\tilde V - H(\eta_t) + \sum_{i\in I}\mu_i\phi_i(S_t,t),$$

and equality holds if $\eta_t = \eta_t^*$: we have obtained formally the analogue of (8.30):

$$\partial_t\tilde V + \max_\eta\left(\eta\frac{S^2}{2}\partial_{SS}^2\tilde V - H(\eta)\right) + rS\partial_S\tilde V - r\tilde V = -\sum_{i\in I}\mu_i\phi_i(S,t). \tag{8.36}$$

8.8.3 Verification

To verify formally why (8.30) and (8.33) lead to the value function, let us assume that we are interested in computing $\tilde V(\mu)$, the value function of the regularized problem

$$\tilde V(\mu) = \sup_{\eta\in\mathcal{H}} -\mathbb{E}^{\eta}\left(\int_0^{\bar\tau} e^{-rt}H(\eta_t)dt + \sum_{i\in I}\mu_i\int_0^{\bar\tau} e^{-rt}\phi_i(S_t,t)dt\right), \tag{8.37}$$

where ϕ_i are smooth functions of S and t.

Let $\tilde V$ be a smooth enough solution to the HJB equation

$$\partial_t\tilde V + \frac{1}{2}\Phi\left(\frac{S^2}{2}\partial_{SS}^2\tilde V\right) + rS\partial_S\tilde V - r\tilde V = -\sum_{i\in I}\mu_i\phi_i(S_t,t), \quad 0 < S, \tag{8.38}$$

with the Cauchy condition $\tilde{V}(\cdot, \bar{\tau}) = 0$. From Itô's lemma, we know that

$$d(e^{-rt}\tilde{V}(S_t, t))$$
$$= e^{-rt}\left(\partial_S\tilde{V}(S_t, t)dS_t + \left(\partial_t\tilde{V}(S_t, t) + \eta\frac{S^2}{2}\partial_{SS}^2\tilde{V}(S_t, t) - r\tilde{V}(S_t, t)\right)dt\right)$$
$$\leq e^{-rt}\left(\begin{array}{l}\eta_t S_t\partial_S\tilde{V}(S_t, t)dB_t \\ +\left(\partial_t\tilde{V}(S_t, t) + \Phi\left(\frac{S^2}{2}\partial_{SS}^2\tilde{V}(S_t, t)\right) + rS_t\partial_S\tilde{V}(S_t, t) - r\tilde{V}(S_t, t)\right)dt \\ +H(\eta)dt\end{array}\right)$$
$$= e^{-rt}\left(\eta_t S_t\partial_S\tilde{V}(S_t, t)dB_t + \left(H(\eta_t) - \sum_{i\in I}\mu_i\phi_i(S_t, t)\right)dt\right),$$

where the inequality stems from the definition (8.31) of Φ. Integrating with respect to time and taking the expectation, we obtain that

$$\tilde{V}(S_\circ, 0) = -\mathbb{E}^\eta\left(e^{-r\bar{\tau}}\tilde{V}(S_{\bar{\tau}}, \bar{\tau})\right) + \tilde{V}(S_\circ, 0)$$
$$\geq \mathbb{E}^\eta\left(\int_0^{\bar{\tau}} e^{-rt}\left(\sum_{i\in I}\mu_i\phi_i(S_t, t) - H(\eta_t)\right)dt\right).$$

This proves that $\tilde{V}(S_\circ, 0)$ is greater than $\tilde{V}(\mu)$. Moreover, this inequality is an equality if η satisfies $H(\eta) = \eta\frac{S^2}{2}\partial_{SS}^2\tilde{V}(S_t, t) - \Phi(\frac{S^2}{2}\partial_{SS}^2\tilde{V}(S_t, t))$, and from (8.32) and the fact that Φ is C^1-regular, if and only if $\eta_t = \Phi'(\frac{S_t^2}{2}\partial_{SS}^2\tilde{V}(S_t, t))$. We have proved that if $\tilde{V}$ is a smooth solution to (8.38), then the value function is $\tilde{V}(S_\circ, 0)$ and that the minimum is reached for η_t given by $\eta_t = \Phi'(\frac{S^2}{2}\partial_{SS}^2\tilde{V}(S_t, t))$.

8.8.4 Algorithm

Knowing the function $V(S, t)$, the partial derivative $V_{,i} = \frac{\partial V}{\partial\mu_i}$ is computed by solving the linear parabolic equation

$$\partial_t V_{,i} + \frac{S^2}{4}\Phi'\left(\frac{S^2}{2}\partial_{SS}^2 V\right)\partial_{SS}^2 V_{,i} + rS\partial_S V_{,i} - rV_{,i} = -(S - K_i)_+\delta_{t=\tau_i}, \tag{8.39}$$

with the Cauchy condition $V_{,i}(\cdot, \bar{\tau}) = 0$.

Therefore, it is possible to use a gradient-based algorithm for solving the dual problem, and when the dual problem is solved, a tentative solution to the primal problem is given by (8.33).

The dual problem is unconstrained. The evaluation of $V(\mu)$ necessitates N_i solves of HJB equations, and $\nabla V(\mu)$ is computed by solving N_I linear parabolic problems.

Chapter 9

Calibration of Local Volatility with American Options

9.1 The Calibration Problem

In this chapter, we describe a least squares method to calibrate the local volatility σ by fitting the prices of a set of vanilla American puts available on the market. In contrast to calibration with European options, it is not possible to use Dupire's equation here, because of the nonlinear character of the American options pricing problem. Therefore, as we shall see below, calibration with American options necessitates many more computations.

The calibration problem consists in finding $\eta = \sigma^2$ from the observations of

- the spot price $S_\circ$ today;
- the prices $(\bar{P}_i)_{i\in I}$ of a family of American puts with different maturities and different strikes $(T_i, K_i)_{i\in I}$.

We call $T = \max_{i\in I} T_i$. We consider the following least squares problem: find $\eta \in \mathcal{H}$ minimizing

$$J(\eta) + J_R(\eta), \qquad J(\eta) = \sum_{i\in I} |P_i(S_\circ, T_i) - \bar{P}_i|^2,$$

where $\mathcal{H}$ is a suitable closed subset of a possibly infinite-dimensional function space, J_R is a suitable Tychonoff regularization functional, and

$$\left.\begin{aligned}
&\frac{\partial P_i}{\partial t} - \frac{\eta(S, T_i - t)S^2}{2}\frac{\partial^2 P_i}{\partial S^2} - rS\frac{\partial P_i}{\partial S} + rP_i \geq 0,\\
&P_i \geq (K_i - S)_+,\\
&\left(\frac{\partial P_i}{\partial t} - \frac{\eta(S, T_i - t)S^2}{2}\frac{\partial^2 P_i}{\partial S^2} - rS\frac{\partial P_i}{\partial S} + rP_i\right)(P_i - (K_i - S)_+) = 0,
\end{aligned}\right\} \quad \begin{aligned}0 &< t,\\ S &> 0,\end{aligned}$$

$$P_i(S, t = 0) = (K_i - S)_+, \qquad S > 0. \tag{9.1}$$

In [2], the inverse problem is studied and optimality conditions are given for suitable choices of $\mathcal{H}$ and J_R. Here, we are interested in the discrete counterpart of the previous least squares problem. As in Chapter 6, we localize the problem (i.e., we consider only $S \in (0, \bar{S})$ for $\bar{S} \gg \max(S_\circ, \max_{i\in I} K_i)$). We introduce a partition of the interval $[0, T]$ into subintervals $[t_{n-1}, t_n]$, $1 \le n \le N$, with $\Delta t_i = t_i - t_{i-1}$, $\Delta t = \max_i \Delta t_i$, and a partition of the interval $[0, \bar{S}]$ into subintervals $\omega_i = [S_{i-1}, S_i]$, $1 \le i \le N_h + 1$, such that $0 = S_0 < S_1 < \cdots < S_{N_h} < S_{N_h+1} = \bar{S}$. The size of the interval ω_i is called h_i and we set $h = \max_{i=1,\dots,N_h+1} h_i$. We assume that for any $i \in I$, the maturity T_i coincides with some node of the time grid, i.e., there exists $N_i \le N$ such that $t_{N_i} = T_i$. We also assume that for any $i \in I$, the strike K_i is a node of the S-grid, i.e., there exists $\kappa_i < N_h$ such that $K_i = S_{\kappa_i}$.

We introduce the function $P_{\circ,i}(S) = (K_i - S)_+$. We know from Theorem 6.3 that there exists a constant γ_0, $0 < \gamma_0 < \min_{i\in I} K_i$, independent of η in the class of functions satisfying (9.8) and (9.9) below, such that for all $i \in I$, $P_i(S, t)$ coincides with $P_{\circ,i}(S)$ in the rectangle $[0, 2\gamma_0] \times [0, T_i]$. We choose a mesh node $\underline{S}$ smaller than γ_0. Let $\rho < \min_{i\in I} \kappa_i$ be the index such that $S_\rho = \underline{S}$.

With $V = \{v \in L^2((0, \bar{S})) : S\frac{dv}{dS} \in L^2((0, \bar{S}))\ v(\bar{S}) = 0\}$, we introduce V_h as in (6.25), the nodal basis functions $(w^i)_{0\le i\le N_h}$, and the closed sets

$$\mathcal{K}_i = \{v \in V,\ v \ge P_{\circ,i},\ v = P_{\circ,i} \text{ in } (0, \underline{S})\}, \quad \mathcal{K}_{h,i} = V_h \cap \mathcal{K}_i. \tag{9.2}$$

With this notation, the new calibration problem is as follows:

$$\text{find } \eta \in \mathcal{H} \text{ minimizing} \quad J(\eta) + J_R(\eta), \qquad J(\eta) = \sum_{i\in I} |P_i^{N_i}(S_\circ) - \bar{P}_i|^2, \tag{9.3}$$

where $\mathcal{H}$ is a suitable closed subset of a finite-dimensional function space, J_R is a suitable Tychonoff regularization functional, and P_i^n is the solution of the following discrete problem:

Find $(P_i^n)_{0\le n\le N_i}$, $P_i^n \in \mathcal{K}_{h,i}$ satisfying

$$P_i^0 = P_{\circ,i}, \tag{9.4}$$

and for all n, $1 \le n \le N_i$,

$$\forall v \in \mathcal{K}_{h,i}, \quad \left(P_i^n - P_i^{n-1}, v - P_i^n\right) + \Delta t_n a_{T_i - t_n}(P_i^n, v - P_i^n) \ge 0, \tag{9.5}$$

with a_t given by

$$\begin{aligned} &a_t(v, w) \\ &= \left(\frac{\eta(S,t)}{2} S\frac{\partial v}{\partial S}, S\frac{\partial w}{\partial S}\right) + \left(\left(-r + \eta(S,t) + \frac{1}{2} S\frac{\partial \eta}{\partial S}(S,t)\right) S\frac{\partial v}{\partial S}, w\right) + r(v, w). \end{aligned} \tag{9.6}$$

Many choices are possible for $\mathcal{H}$. For example, let Y be a finite-dimensional vector space of C^1 and piecewise smooth real-valued functions on $[0, \bar{S}] \times [0, T]$ (Y may be a space of bicubic spline functions on $[0, \bar{S}] \times [0, T]$). Then it will be convenient for what follows to take

$$\mathcal{H} = \left\{\chi \in Y; \forall (S,t) \in [0, \bar{S}] \times [0, T], \underline{\eta} \le \chi(S,t) \le \overline{\eta}, \left|S\frac{\partial \chi}{\partial S}(S,t)\right| \le C_\eta\right\} \tag{9.7}$$

with $0 < \eta_{\min} < \underline{\eta} < \overline{\eta} < \eta_{\max}$ and $0 < C_\eta$.

The space Y can also be made of less regular functions such as continuous and piecewise bilinear functions of S and t on the elements of a Cartesian mesh. In any case, the functions of $\mathcal{H}$ must be continuous and satisfy, for fixed constants $\eta_{\min}$, $\eta_{\max}$, and C_η,

$$\eta_{\min} \le \eta(S,t) \le \eta_{\max} \quad \text{in } [0,\bar{S}] \times [0,T], \tag{9.8}$$

and

$$\left| S \frac{\partial \eta}{\partial S} \right| \le C_\eta \quad \forall t \in [0,T] \text{ a.e. in } (0,\bar{S}). \tag{9.9}$$

With (9.8) and (9.9), the bilinear form a_t is continuous on V and we have Gårding's inequality: there exists $\lambda > 0$ independent of η for η satisfying (9.8) and (9.9), such that

$$a_t(v,v) \ge \frac{\eta_{\min}}{4} |v|_V^2 - \lambda \|v\|^2. \tag{9.10}$$

The Tychonoff functional J_R is chosen as in Chapter 8.

9.2 A Penalized Problem

Orientation. We want to find necessary optimality conditions for the least squares problem (9.3). A straightforward derivation of these conditions does not seem possible. For this reason, we are going to prove that the solutions P to (9.4), (9.5) can be found as the limits of the solutions P_ϵ to suitably penalized nonlinear problems, when the penalty parameter ϵ tends to zero. This result, stated in Proposition 9.1 below, is the discrete counterpart of the result stated in Theorem 6.1. Then, in Proposition 9.2 below, we derive optimality conditions for a new least squares problem where the state functions P are replaced by P_ϵ. Finally, we will pass to the limit in the optimality conditions as $\epsilon \to 0$. The final result is stated in Theorem 9.6 below.

The Penalized Problem. We fix a strike K (assumed to be a mesh point) and a maturity T, we call $P_\circ(S) = (K-S)_+$, $\mathcal{K} = \{v \in V_h,\ v \ge P_\circ\ v = P_\circ \text{ in } (0,\underline{S})\}$, and we consider the following problem:

Find $(P^n)_{0\le n\le N}$, $P^n \in \mathcal{K}$ satisfying

$$P^0 = P_\circ, \tag{9.11}$$

and for all n, $1 \le n \le N$,

$$\forall v \in \mathcal{K}, \quad \left(P^n - P^{n-1}, v - P^n\right) + \Delta t_n a_{T-t_n}(P^n, v - P^n) \ge 0. \tag{9.12}$$

We consider a smooth nonincreasing convex function $\mathcal{V}$ satisfying assumptions (6.16) (see Figure 6.2), and we denote $\mathcal{V}_\epsilon(x) = \mathcal{V}(\frac{x}{\epsilon})$. We denote $\tilde{V}_h = \{v_h \in V_h,\ v_h(S_i) = 0\ \forall i < \rho\}$ and we define the discrete penalized problem as follows:

Find $(P_\epsilon^n)_{0\le n\le N}$, $P_\epsilon^n \in V_h$ satisfying

$$P_\epsilon^0 = P_\circ, \tag{9.13}$$

and for all n, $1 \le n \le N$, $P_\epsilon^n - P_\circ \in \tilde{V}_h$, and for any $v \in \tilde{V}_h$,

$$0 = \left(P_\epsilon^n - P_\epsilon^{n-1}, v\right) + \Delta t_n \left(a_{T-t_n}(P_\epsilon^n, v) - rK \sum_{\rho \le i \le \kappa} \frac{|\tilde{\Omega}_i|}{2} \mathcal{V}_\epsilon(P_\epsilon^n(S_i) - P_\circ(S_i)) v(S_i) \right), \tag{9.14}$$

where $\tilde{\Omega}_i = \Omega_i \cap (0, K)$ and $\Omega_i \subset (0, \bar{S})$ is the support of w^i, so $\frac{|\tilde{\Omega}_i|}{2} = \int_0^K w^i$, where w^i is the nodal basis function associated to S_i.

In matrix form, (9.14) becomes the following:

Find $\mathbf{P}_\epsilon^n \in \mathbb{R}^{N_h+1}$ such that

$$\begin{aligned} \left(\mathbf{M}(\mathbf{P}_\epsilon^n - \mathbf{P}_\epsilon^{n-1}) + \Delta t_n (\mathbf{A}^n \mathbf{P}_\epsilon^n - rK G_\epsilon(\mathbf{P}_\epsilon^n))\right)_i &= 0 \qquad \text{for } \rho \le i \le N_h, \\ (\mathbf{P}_\epsilon^n)_i &= P_\circ(S_i) \quad \text{for } i < \rho, \end{aligned} \tag{9.15}$$

where $\mathbf{M}$ and $\mathbf{A}^n$ are the mass and stiffness matrices introduced in §4.3 and where G_ϵ is the nonlinear function in $\mathbb{R}^{N_h+1}$ such that, with obvious notation, $(G_\epsilon(\mathbf{P}))_i = 1_{\rho \le i \le \kappa} \frac{|\tilde{\Omega}_i|}{2} \mathcal{V}_\epsilon(\mathbf{P}_i - P_\circ(S_i))$. The function G_ϵ is obviously monotone, i.e.,

$$(G_\epsilon(\mathbf{P}) - G_\epsilon(\mathbf{Q}), \mathbf{P} - \mathbf{Q}) \le 0. \tag{9.16}$$

Denoting by λ the constant in the Gårding's inequality (9.10), we obtain, by using the Lipschitz regularity, the monotonicity of $\mathcal{V}_\epsilon$, and a fixed point argument due to Stampacchia, that for $\Delta t < \frac{1}{2\lambda}$, (9.13), (9.14) has a unique solution.

It is also possible to prove a priori bounds on $(P_\epsilon^n)_{0 \le n \le N}$. For this, using the identity

$$a_{T-t_n}(P_\circ, v) - rK \sum_{\rho \le i \le \kappa} \frac{|\tilde{\Omega}_i|}{2} \mathcal{V}_\epsilon(P_\circ(S_i) - P_\circ(S_i)) v(S_i) = -\frac{\eta(T - t_n, K) K^2}{2} v(K)$$

for all $v \in V_h$, and defining $e_\epsilon^n = P_\epsilon^n - P_\circ$, we obtain that for any $v \in \tilde{V}_h$,

$$\begin{aligned} &\left(e_\epsilon^n - e_\epsilon^{n-1}, v\right) + \Delta t_n a_{T-t_n}(e_\epsilon^n, v) \\ &- rK \Delta t_n \sum_{\rho \le i \le \kappa} \frac{|\tilde{\Omega}_i|}{2} (\mathcal{V}_\epsilon(P_\epsilon^n(S_i) - P_\circ(S_i)) - \mathcal{V}_\epsilon(P_\circ(S_i) - P_\circ(S_i))) v(S_i) \\ = {} & \Delta t_n \frac{\eta(T - t_n, K) K^2}{2} v(K). \end{aligned} \tag{9.17}$$

Taking $v = e_\epsilon^n$ in (9.17), using (9.10) and the nonincreasing character of $\mathcal{V}_\epsilon$, we obtain that

$$\begin{aligned} &\frac{1}{2}\|e_\epsilon^n\|^2 + \frac{1}{2}\|e_\epsilon^n - e_\epsilon^{n-1}\|^2 + \Delta t_n \left(\frac{\eta_{\min}}{4} |e_\epsilon^n|_V^2 - \lambda \|e_\epsilon^n\|^2 \right) \\ &\le \frac{1}{2}\|e_\epsilon^{n-1}\|^2 + \Delta t_n \frac{K^2 \eta_{\max}}{2} |e_\epsilon^n(K)| \\ &\le \frac{1}{2}\|e_\epsilon^{n-1}\|^2 + \Delta t_n \frac{C K^2 \eta_{\max}}{2} |e_\epsilon^n|_V, \end{aligned} \tag{9.18}$$

since there exists a constant C such that, for any $v \in V$,

$$|v(K)| \leq C|v|_V. \tag{9.19}$$

Multiplying (9.18) by $\prod_{i=1}^{n-1}(1-2\lambda\Delta t_i)$ and summing the equations on n, we obtain

$$\begin{aligned}&\left(\prod_{m=1}^{n}(1-2\lambda\Delta t_m)\right)\|e_\epsilon^n\|^2+\frac{1}{2}\eta_{\min}\sum_{m=1}^{n}\Delta t_m\left(\prod_{i=1}^{m-1}(1-2\lambda\Delta t_i)\right)|e_\epsilon^m|_V^2\\&\leq CK^2\eta_{\max}\sum_{m=1}^{n}\Delta t_m\left(\prod_{i=1}^{m-1}(1-2\lambda\Delta t_i)\right)|e_\epsilon^m|_V.\end{aligned} \tag{9.20}$$

This proves that $\max_{0\leq n\leq N}\|e_\epsilon^n\|$ is bounded uniformly with respect to ϵ (and also with respect to η for η satisfying assumptions (9.8) and (9.9)). This implies that $(P_\epsilon^n)_{0\leq n\leq N}$ is bounded (in any norm since $(V_h)^{N+1}$ is a finite-dimensional space) uniformly with respect to ϵ (and also to η). It is also clear that the real numbers $\mathcal{V}_\epsilon(P_\epsilon^n(S_i)-P_\circ(S_i))$ are bounded uniformly with respect to ϵ.

We aim at proving that the sequence $(P_\epsilon^n)_{0\leq n\leq N}$ converges to $(P^n)_{0\leq n\leq N}$ as $\epsilon\to 0$. To do so, we extract a subsequence that converges, and we want to prove that the limit is precisely $(P^n)_{0\leq n\leq N}$. The main point consists in proving that the limit belongs to $(\mathcal{K})^{N+1}$. For this, we need to use a discrete maximum principle. As seen in Chapter 3, this amounts to looking for monotonicity properties of the matrix $\mathbf{M}+\Delta t_n\mathbf{A}^n$. The matrix $\mathbf{M}+\Delta t_n\mathbf{A}^n$ cannot be an M-matrix, since the diffusion coefficient $\frac{\eta S^2}{2}$ vanishes at $S=0$. However, it is very reasonable to assume that the block of $\mathbf{M}+\Delta t_n\mathbf{A}^n$ associated to the nodes S_i greater than a given value is an M-matrix. If $\underline{S}>0$ and if the block of $\mathbf{M}+\Delta t_n\mathbf{A}^n$ associated to the nodes S_i greater than $\underline{S}$ is an M-matrix, then we will prove that $P_\epsilon^n\in\mathcal{K}$ so the property will hold at the limit. For a matrix $\mathbf{A}$, we call $\mathbf{A}_\ell$ the block of $\mathbf{A}$ corresponding to row and column indices greater than or equal to ℓ.

Assumption 9.1. We assume that $\underline{S}>0$, i.e., $\rho>0$, and that the parameters h and $\frac{h^2}{\min_n\Delta t_n}$ are small enough so that the matrices $\mathbf{A}_l^n$ and $\mathbf{M}_l+\Delta t_n\mathbf{A}_l^n$ are tridiagonal irreducible M-matrices for all n, $1\leq n\leq N$, and l, $\rho\leq l<N_h$, and for all the functions η satisfying (9.8) and (9.9).

Proposition 9.1. *Let η satisfy* (9.8) *and* (9.9)*, and choose $\Delta t<\frac{1}{2\lambda}$, with λ given in* (9.10)*. Let Assumption* 9.1 *be satisfied. The sequence $(P_\epsilon^n)_{0\leq n\leq N}$ given by* (9.13), (9.14) *converges to $(P^n)_{0\leq n\leq N}$ given by* (9.11), (9.12) *in $(V_h)^{N+1}$ as ϵ tends to* 0.

Proof. For $\rho\leq i\leq N_h$ let us define $\mu_{\epsilon,i}^n$ to be the real number,

$$\begin{aligned}\mu_{\epsilon,i}^n&=\left(P_\epsilon^n-P_\epsilon^{n-1},w^i\right)+\Delta t_n a_{T-t_n}(P_\epsilon^n,w^i)\\&=rK\Delta t_n 1_{S_i\leq K}\frac{|\tilde{\Omega}_i|}{2}(\mathcal{V}_\epsilon(P_\epsilon^n(S_i)-P_\circ(S_i)).\end{aligned} \tag{9.21}$$

It is clear that the numbers $\mu_{\epsilon,i}^n$ are bounded uniformly with respect to ϵ. From the uniform bounds on P_ϵ^n and $\mu_{\epsilon,i}^n$, we know that we can find a sequence ϵ_k converging to 0 such that

$(P^n_{\epsilon_k})_{0\le n\le N}$ converges to $(\tilde{P}^n)_{0\le n\le N}$ in V_h^{N+1}, and $\mu^n_{\epsilon_k,i}$ converges to μ^n_i, $\rho \le i \le N_h$, $i \le 1 \le N$, as k tends to ∞. It is clear that $\tilde{P}^0 = P_\circ$, and that $\mu^n_i \ge 0$, $\rho \le i \le N_h$. Passing to the limit in (9.21), we obtain that

$$\left(\tilde{P}^n - \tilde{P}^{n-1}, w^i\right) + \Delta t_n a_{T-t_n}(\tilde{P}^n, w^i) = \mu^n_i \ge 0, \quad \rho \le i \le N_h,\ 1 \le n \le N. \quad (9.22)$$

By using (9.17) and the discrete maximum principle (because $\mathbf{M}_\rho + \Delta t_n \mathbf{A}^n_\rho$ is an M-matrix), it is possible to prove by induction on n that

$$P^n_{\epsilon_k} \ge P_\circ. \quad (9.23)$$

Passing to the limit, we obtain that

$$\tilde{P}^n(S_i) \ge P_\circ(S_i), \quad 0 \le i \le N_h. \quad (9.24)$$

Assume that for some i, $\rho \le i \le N_h$, and for some $\alpha > 0$, $\tilde{P}^n(S_i) > P_\circ(S_i) + \alpha$. Then for k large enough, $P^n_{\epsilon_k}(S_i) > P_\circ(S_i) + \frac{\alpha}{2}$. This and (9.21) yield that $\lim_{k\to\infty} \mu^n_{\epsilon_k,i} = 0$, i.e., $\mu^n_i = 0$. We have proved that

$$\mu^n_i(\tilde{P}^n(S_i) - P_\circ(S_i)) = 0, \qquad \rho \le i \le N_h,\ 1 \le n \le N. \quad (9.25)$$

Therefore $(\tilde{P}^n)_{0\le n\le N}$ satisfies (9.22), (9.24), (9.25) and $\tilde{P}^n(S_i) = P_\circ(S_i)$ for $i < \rho$. These properties are equivalent to (9.12). Since (9.12) has a unique solution, we have proved that $\tilde{P}^n = P^n$. The uniqueness of the limit shows that $\lim_{\epsilon\to 0} P^n_\epsilon = P^n$. □

Remark 9.1. *As proven in* [4], *the result stated in Proposition* 9.1 *holds with Assumption* 9.1 *replaced with the following weaker assumption.*

Assumption 9.2. *Let κ' be the integer such that $x_i > \frac{K}{2} \Leftrightarrow \kappa' \le i \le N_h$. We assume that the parameters h and $\frac{h^2}{\min_n \Delta t_n}$ are small enough so that the matrices $\tilde{A}^n_l$ and $\tilde{M}_l + \Delta t_n \tilde{A}^n_l$ are tridiagonal irreducible M-matrices for all n, $1 \le n \le N$, and l, $\kappa' \le l < N_h$, and for all the functions η satisfying* (9.8) *and* (9.9).

With this assumption, we have $P^n_\epsilon - P_\circ \ge -C\epsilon$ for a positive constant C.

9.3 Necessary Optimality Conditions

For ease of exposition only, we first simplify problem (9.3) by assuming that only one option (with strike K and maturity T) is observed; i.e., the family I has only one element. The simpler least squares problem reads as follows:

Find $\eta \in \mathcal{H}$ minimizing $J(\eta) + J_R(\eta)$, $J(\eta) = |P^N(S_\circ) - \bar{P}|^2$, where

$$\begin{aligned} &P^0 = P_\circ, \\ &\forall v \in \mathcal{K}_h, \quad (P^n - P^{n-1}, v - P^n) + \Delta t_n a_{T-t_n}(P^n, v - P^n) \ge 0, \quad 1 \le n \le N. \end{aligned} \quad (9.26)$$

It is possible to formulate the discrete variational inequality (9.5) as an identity involving P^n and the real numbers $\mu^n_i = (P^n - P^{n-1}, w^i) + \Delta t_n a_{T-t_n}(P^n, w^i)$, with additional constraints: $P^n \geq P_\circ$, $\mu_i^n \geq 0$, and $\mu_i^n(P^n(S_i) - P_\circ(S_i)) = 0$. It is then tempting to use the Lagrange machinery for the least squares problem. However, as observed by Bergougnioux and Mignot [15] for optimal control of obstacle problems, it is generally not possible to find a necessary optimality condition with as many Lagrange multipliers as there are constraints, because the additional constraints are not qualified and the Lagrange system that one would obtain has no solutions. So it is not easy to derive suitable optimality conditions from the variational inequality itself. Instead, following Ito and Kunish [75] and Hintermüller [73], it is possible to consider a least squares problem for the penalized version of the inequality in (9.26):

Find $\eta \in \mathcal{H}$ minimizing $J_\epsilon(\eta) + J_R(\eta)$, $J_\epsilon(\eta) = |P_\epsilon^N(S_\circ) - \bar{P}|^2$, where $P_\epsilon^0 = P_\circ$. For $1 \leq n \leq N$, $P_\epsilon^n - P_\circ \in \tilde{V}_h$, and for any $v \in \tilde{V}_h$,

$$\left(P_\epsilon^n - P_\epsilon^{n-1}, v\right) + \Delta t_n \left(a_{T-t_n}(P_\epsilon^n, v) - rK \sum_{\rho \leq i \leq \kappa} \frac{|\tilde{\Omega}_i|}{2} \mathcal{V}_\epsilon(P_\epsilon^n(S_i) - P_\circ(S_i))v(S_i) \right) = 0. \tag{9.27}$$

Proceeding as in Chapter 8, it is possible to find necessary optimality conditions for (9.27). To do so, we introduce the adjoint state $(y_\epsilon^n)_{1 \leq n \leq N}$, $y_\epsilon^n \in \tilde{V}_h$, such that for all $v \in \tilde{V}_h$,

$$\begin{aligned}
&\left(y_\epsilon^N, v\right) + \Delta t_N \left(a_0(v, y_\epsilon^N) - rK \sum_{\rho \leq i \leq \kappa} \frac{|\tilde{\Omega}_i|}{2} \mathcal{V}'_\epsilon(P_\epsilon^N(S_i) - P_\circ(S_i)) y_\epsilon^N(S_i) v(S_i) \right) \\
&\quad = 2(P_\epsilon^N(S_\circ) - \bar{P}) v(S_\circ), \\
&\left(y_\epsilon^n - y_\epsilon^{n+1}, v\right) + \Delta t_n \left(a_{T-t_n}(v, y_\epsilon^n) - rK \sum_{\rho \leq i \leq \kappa} \frac{|\tilde{\Omega}_i|}{2} \mathcal{V}'_\epsilon(P_\epsilon^n(S_i) - P_\circ(S_i)) y_\epsilon^n(S_i) v(S_i) \right) \\
&\quad = 0, \qquad 1 \leq n < N.
\end{aligned} \tag{9.28}$$

The existence and uniqueness of (9.28) are ensured if $\Delta t < \frac{1}{2\lambda}$, λ in (9.10). The existence of a solution to the least squares problem (9.27) is obtained by the same arguments as for Proposition 8.3.

We state the necessary optimality condition for (9.27) without any proof.

Proposition 9.2. *Let η_ϵ^* be a solution of (9.27), and define $P_\epsilon^{*,n}$ to be the corresponding state function and $y_\epsilon^{*,n}$ the adjoint state solution to (9.28) with $\eta = \eta_\epsilon^*$. For all $\eta \in \mathcal{H}$, denoting $\delta\eta = \eta - \eta_\epsilon^*$, we have*

$$\begin{aligned}
&\langle DJ_R(\eta_\epsilon^*), \delta\eta \rangle - \frac{1}{2} \sum_{n=1}^{N} \Delta t_n \left(\delta\eta(S, T - t_n) S \frac{\partial P_\epsilon^{*,n}}{\partial S}, S \frac{\partial y_\epsilon^{*,n}}{\partial S} \right) \\
&- \sum_{n=1}^{N} \Delta t_n \left(\left(\delta\eta(S, T - t_n) + \frac{1}{2} S \frac{\partial \delta\eta}{\partial S}(S, T - t_n) \right) S \frac{\partial P_\epsilon^{*,n}}{\partial S}, y_{\epsilon^*}^n \right) \geq 0,
\end{aligned} \tag{9.29}$$

which can also be written as

$$0 \le \langle DJ_R(\eta_\epsilon^*), \delta\eta\rangle$$

$$-\frac{1}{2}\sum_{n=1}^{N}\Delta t_n \sum_{i=\rho}^{N_h} S_i^2 \delta\eta(T-t_n, S_i) y_\epsilon^{*,n}(S_i) \left(\begin{array}{c} \dfrac{P_\epsilon^{*,n}(S_i) - P_\epsilon^{*,n}(S_{i-1})}{h_i} \\ +\dfrac{P_\epsilon^{*,n}(S_i) - P_\epsilon^{*,n}(S_{i+1})}{h_{i+1}} \end{array} \right). \tag{9.30}$$

Let $(\epsilon_i)_{i\in\mathbb{N}}$ be a sequence of penalty parameters tending to zero, and let $\eta_{\epsilon_i}^*$ be a minimizer of (9.27) with $\epsilon = \epsilon_i$. Since $\mathcal{H}$ is a compact subset of Y, it is possible to extract a subsequence, still denoted ϵ_i, such that $\eta_{\epsilon_i}^*$ converges to η^* in $\mathcal{H}$. Arguing exactly as in the proof of Proposition 9.1, it is possible to prove the following result.

Lemma 9.3. *Let Assumption* 9.1 *or Assumption* 9.2 *be satisfied. Let* (ϵ_i) *be a sequence of penalty parameters converging to* 0 *and let* $\eta_{\epsilon_i}^*$ *be a minimizer of* (9.27) *with* $\epsilon = \epsilon_i$. *Let* $\eta_{\epsilon_i}^*$ *converge to* η^* *in* $\mathcal{H}$ *as* $i \to \infty$; *then* $(P_{\epsilon_i}^{*,n})_{0\le n\le N}$ *converges to* $(P^{*,n})_{0\le n\le N}$, $P^{*,n} \in \mathcal{K}_h$, *the solution to*

$$\begin{aligned} &P^0 = P_\circ, \\ &\left(P^{*,n} - P^{*,n-1}, v - P^{*,n}\right) + \Delta t_n a_{T-t_n}^*(P^{*,n}, v - P^{*,n}) \ge 0, \qquad \forall v \in \mathcal{K}_h,\ 1 \le n \le N, \end{aligned} \tag{9.31}$$

where $a_t^*(v,w) = (\frac{\eta^*(S,t)}{2} S\frac{\partial v}{\partial S}, S\frac{\partial w}{\partial S}) + ((-r + \eta^*(S,t) + \frac{1}{2}S\frac{\partial \eta^*}{\partial S}(S,t))S\frac{\partial v}{\partial S}, w) + r(v,w)$. *Furthermore, for any* j, $\rho \le j \le N_h$, $\mu_{\epsilon_i,j}^{*,n} = rK\Delta t_n \frac{|\tilde\Omega_j|}{2} \mathcal{V}_{\epsilon_i}(P_\epsilon^{*,n}(S_j) - P_\circ(S_j))$ *converges to* $\mu_j^{*,n} = (P^{*,n} - P^{*,n-1}, w^j) + \Delta t_n a_{T-t_n}^*(P^{*,n}, w^j)$.

Proposition 9.4. *With the assumptions of Lemma* 9.3, *let* (ϵ_i) *be a sequence of penalty parameters converging to* 0, *and let* $\eta_{\epsilon_i}^*$ *be a solution of* (9.27) *with* $\epsilon = \epsilon_i$. *Let* η^* *be the limit of* $\eta_{\epsilon_i}^*$ *as* $i \to \infty$; *then* η^* *is a solution of* (9.26).

Proof. For $\eta \in \mathcal{H}$, $J_{\epsilon_i}(\eta_{\epsilon_i}^*) + J_R(\eta_{\epsilon_i}^*) \le J_{\epsilon_i}(\eta) + J_R(\eta)$ because $\eta_{\epsilon_i}^*$ is a minimizer for (9.27) with $\epsilon = \epsilon_i^*$. But $J_{\epsilon_i}(\eta_{\epsilon_i}^*) = |P_{\epsilon_i}^{*,N}(S_\circ) - \bar P|^2$ converges to $J(\eta^*) = |P^{*,N}(S_\circ) - \bar P|^2$ from Lemma 9.3, and $J_{\epsilon_i}(\eta)$ converges to $J(\eta)$ from Proposition 9.1. Therefore, from the continuity of J_R in $\mathcal{H}$, we see that, for all $\eta \in \mathcal{H}$,

$$J(\eta^*) + J_R(\eta^*) \le J(\eta) + J_R(\eta). \qquad \square$$

Let η^* be a solution of (9.26) that can be approached by $\eta_{\epsilon_i}^*$, a solution of (9.27), with $\epsilon = \epsilon_i$. In what follows, we drop the index i in ϵ_i in order to alleviate the notation. The aim is to find optimality conditions satisfied by η^* by passing to the limit in the optimality conditions obtained for (9.27). For this, the following assumption can be useful.

Assumption 9.3. Let $\underline{S}$ be positive. We assume that the parameters h and $\frac{h^2}{\min_n \Delta t_n}$ are small enough so that the matrices $(\mathbf{A}_l^n)^T$ and $(\mathbf{M}_l + \Delta t_n \mathbf{A}_l^n)^T$ are tridiagonal irreducible

M-matrices for all n, $1 \le n \le N$, and l, $\rho \le l < N_h$, and for all the functions η satisfying (9.8) and (9.9).

Theorem 9.5. *Let Assumption 9.1 or Assumption 9.2 be satisfied. Let η^* be a solution of (9.26) that can be obtained as the limit of a sequence $(\eta^*_\epsilon)_\epsilon$ of solutions of (9.27). Then there exist $(y^{*,n})_{1\le n\le N}$, $y^{*,n} \in \tilde{V}_h$, and real numbers α_i^n, $\rho \le i \le N_h$, $1 \le n \le N$, satisfying*

$$\begin{aligned}
&\forall v \in \tilde{V}_h, \quad \left(y^{*,N}, v\right) + \Delta t_N \left(a_0^*(v, y^{*,N}) + \langle \alpha^N, v\rangle\right) = 2(u^{*,N}(S_\circ) - \bar{P})v(S_\circ),\\
&\forall v \in \tilde{V}_h, \quad \left(y^{*,n} - y^{*,n+1}, v\right) + \Delta t_n \left(a_{T-t_n}^*(v, y^{*,n}) + \langle \alpha^n, v\rangle\right) = 0, \quad 1 \le n < N,
\end{aligned} \tag{9.32}$$

where α^n is the linear form on $\tilde{V}_h$ defined by

$$\langle \alpha^n, w^i\rangle = \alpha_i^n, \quad i = \rho, \dots, N_h, \tag{9.33}$$

with

$$\alpha_i^n (P^{*,n}(S_i) - P_\circ(S_i)) = 0, \qquad \rho \le i \le N_h, \quad 1 \le n \le N, \tag{9.34}$$

$$\alpha_i^n y^{*,n}(S_i) \ge 0, \qquad \rho \le i \le N_h, \quad 1 \le n \le N, \tag{9.35}$$

and

$$\mu_i^{*,n} y^{*,n}(S_i) = 0, \qquad \rho \le i \le N_h, \quad 1 \le n \le N, \tag{9.36}$$

such that for any $\eta \in \mathcal{H}$, noting by $\delta\eta = \eta - \eta^*$,

$$\begin{aligned}
&\langle DJ_R(\eta^*), \delta\eta\rangle - \frac{1}{2}\sum_{n=1}^{N} \Delta t_n \left(\delta\eta(S, T - t_n) S\frac{\partial P^{*,n}}{\partial S}, S\frac{\partial y^{*,n}}{\partial S}\right)\\
&-\sum_{n=1}^{N} \Delta t_n \left(\left(\delta\eta(S, T - t_n) + \frac{1}{2} S\frac{\partial \delta\eta}{\partial S}(S, T - t_n)\right) S\frac{\partial P^{*,n}}{\partial S}, y^{*,n}\right) \ge 0,
\end{aligned} \tag{9.37}$$

which can also be written as

$$\begin{aligned}
0 \le{}& \langle DJ_R(\eta^*), \delta\eta\rangle\\
&- \frac{1}{2}\sum_{n=1}^{N} \Delta t_n \sum_{i=\rho}^{N_h} S_i^2 \delta\eta(S_i, T - t_n) y^{*,n}(S_i) \left(\begin{array}{c} \dfrac{P^{*,n}(S_i) - P^{*,n}(S_{i-1})}{h_i} \\ + \dfrac{P^{*,n}(S_i) - P^{*,n}(S_{i+1})}{h_{i+1}} \end{array}\right).
\end{aligned} \tag{9.38}$$

If $P^{*,N}(S_\circ) - \bar{P} = 0$, then $y^{*,n} = 0$ and $\alpha^n = 0$ for $1 \le n \le N$.
If Assumption 9.3 is verified and $P^{*,N}(S_\circ) - \bar{P} \ne 0$, then

$$(P^{*,N}(S_\circ) - \bar{P}) y_i^{*,n} \ge 0, \qquad \rho \le i \le N_h, \quad 1 \le n \le N. \tag{9.39}$$

Proof. We take $v = y_\epsilon^{*,N}$ in the first line of (9.28), and $v = y_\epsilon^{*,n}$ in the second line. Using the nonincreasing character of $\mathcal{V}_\epsilon$, we obtain that

$$(1-\lambda\Delta t_N)\|y_\epsilon^{*,N}\|^2 + \Delta t_N \left(\begin{array}{l} \frac{\eta_{\min}}{4}|y_\epsilon^{*,N}|_V^2 \\ -rK\sum_{\rho\le i\le\kappa}\frac{|\tilde\Omega_i|}{2}\mathcal{V}'_\epsilon(P_\epsilon^{*,N}(S_i)-P_\circ(S_i))(y_\epsilon^{*,N}(S_i))^2 \end{array}\right)$$

$$\le 2|P_\epsilon^{*,N}(S_\circ)-\bar P||y_\epsilon^{*,N}(S_\circ)|,$$

$$\frac12(1-2\lambda\Delta t_n)\|y_\epsilon^{*,n}\|^2 + \Delta t_n \left(\begin{array}{l} \frac{\eta_{\min}}{4}|y_\epsilon^{*,n}|_V^2 \\ -rK\sum_{\rho\le i\le\kappa}\frac{|\tilde\Omega_i|}{2}\mathcal{V}'_\epsilon(P_\epsilon^{*,n}(S_i)-P_\circ(S_i))(y_\epsilon^{*,n}(S_i))^2 \end{array}\right)$$

$$\le \frac12\|y_\epsilon^{*,n+1}\|^2.$$

Using (9.19) and the uniform bounds on $(P_\epsilon^{*,n})_{0\le n\le N}$ with respect to ϵ and $\eta\in\mathcal{H}$, we obtain that $(y_\epsilon^{*,n})_{1\le n\le N}$ is uniformly bounded with respect to ϵ and η in $\mathcal{H}$, and furthermore that

$$\max_{1\le n\le N}\sum_{\rho\le i\le\kappa}\frac{|\tilde\Omega_i|}{2}\mathcal{V}'_\epsilon(P_\epsilon^{*,n}(S_i)-P_\circ(S_i))(y_\epsilon^{*,n}(S_i))^2\le C, \tag{9.40}$$

where C depends neither on ϵ nor on $\eta\in\mathcal{H}$.

From (9.28) and the uniform bound on $(y_\epsilon^{*,n})_{1\le n\le N}$, we infer that the real numbers

$$\alpha_{\epsilon,j}^n = -rK1_{j\le\kappa}\frac{|\tilde\Omega_j|}{2}\mathcal{V}'_\epsilon(P_\epsilon^{*,n}(S_j)-P_\circ(S_j))y_\epsilon^{*,n}(S_j),\quad \rho\le j, \tag{9.41}$$

are all bounded (in absolute value) by a constant independent of ϵ.

Therefore, we can extract a subsequence still called (ϵ) such that $P_\epsilon^{*,n}\to P^{*,n}$ in V_h, $y_\epsilon^{*,n}\to y^{*,n}$ in $\tilde V_h$, and $\alpha_{\epsilon,j}^n\to\alpha_j^n$ for all j,n, $\rho\le j\le N_h$, $1\le n\le N$, and passing to the limit in (9.28) (thanks to Lemma 9.3), we obtain (9.32).

Assume that for $i\le\kappa$, $P^{*,n}(S_i)\ge P_\circ(S_i)+\zeta$, $\zeta>0$. For ϵ small enough, $P_\epsilon^{*,n}(S_i)\ge P_\circ(S_i)+\frac{\zeta}{2}>P_\circ(S_i)+\epsilon$, which implies that $\alpha_{\epsilon,i}^n=0$.

We have proved that

$$P^{*,n}(S_i)>P_\circ(S_i)\Rightarrow\alpha_i^n=0, \tag{9.42}$$

or in equivalent manner (9.34). From the fact that $\alpha_{\epsilon,i}^n y_\epsilon^{*,n}(S_i)\ge 0$ we deduce (9.35).

The next thing to do is to study $\mu_{\epsilon,i}^{*,n}y_\epsilon^{*,n}(S_i)$ as $\epsilon\to 0$:

$$\begin{aligned}|\mu_{\epsilon,i}^{*,n}y_\epsilon^{*,n}(S_i)| &= rK\frac{|\tilde\Omega_i|}{2}\mathcal{V}_\epsilon(P_\epsilon^{*,n}(S_i)-P_\circ(S_i))|y_\epsilon^{*,n}(S_i)|\\ &= rK\frac{|\tilde\Omega_i|}{2}(\mathcal{V}_\epsilon(P_\epsilon^{*,n}(S_i)-P_\circ(S_i))-\mathcal{V}_\epsilon(\epsilon))|y_\epsilon^{*,n}(S_i)|\end{aligned}$$

because $\mathcal{V}_\epsilon(\epsilon) = 0$. Since $\mathcal{V}_\epsilon$ is convex, $0 \le \mathcal{V}_\epsilon(P_\epsilon^{*,n} - P_\circ) - \mathcal{V}_\epsilon(\epsilon) \le \mathcal{V}'_\epsilon(P_\epsilon^{*,n} - P_\circ)(P_\epsilon^{*,n} - P_\circ - \epsilon)$. However, $\mathcal{V}'_\epsilon(P_\epsilon^{*,n} - P_\circ) \ne 0 \Leftrightarrow P_\epsilon^{*,n} - P_\circ \le \epsilon$, and on the other hand $P_\epsilon^{*,n} - P_\circ \ge -C\epsilon$ (see the proof of Proposition 9.1 and Remark 9.1). Therefore, $|\mathcal{V}'_\epsilon(P_\epsilon^{*,n} - P_\circ)(P_\epsilon^{*,n} - P_\circ - \epsilon)| \le C\epsilon|\mathcal{V}'_\epsilon(P_\epsilon^{*,n} - P_\circ)|$. This implies that

$$\begin{aligned}
|\mu_{\epsilon,i}^{*,n} y_\epsilon^{*,n}(S_i)| &\le C\epsilon \frac{|\tilde{\Omega}_i|}{2} |\mathcal{V}'_\epsilon(P_\epsilon^{*,n}(S_i) - P_\circ(S_i))||y_\epsilon^{*,n}(S_i)| \\
&\le C\epsilon \left(\frac{|\tilde{\Omega}_i|}{2} |\mathcal{V}'_\epsilon(P_\epsilon^{*,n}(S_i) - P_\circ(S_i))| \right)^{\frac{1}{2}} \left(\frac{|\tilde{\Omega}_i|}{2} |\mathcal{V}'_\epsilon(P_\epsilon^{*,n}(S_i) - P_\circ(S_i))|(y_\epsilon^{*,n}(S_i))^2 \right)^{\frac{1}{2}} \\
&\le C\sqrt{\epsilon} \left(\frac{|\tilde{\Omega}_i|}{2} |\mathcal{V}'_\epsilon(P_\epsilon^{*,n}(S_i) - P_\circ(S_i))|(y_\epsilon^{*,n}(S_i))^2 \right)^{\frac{1}{2}},
\end{aligned}$$

where we have used the fact that $|\mathcal{V}'_\epsilon| \le \frac{2}{\epsilon}$. Finally, from (9.40) we obtain that

$$|\mu_{\epsilon,i}^{*,n} y_\epsilon^{*,n}(S_i)| \le C\sqrt{\epsilon},$$

which implies (9.36).

Finally, (9.37) is obtained by passing to the limit in (9.29).

Assume that $P^{*,N}(S_\circ) - \bar{P} \ne 0$ and, without limitations, that $P^{*,N}(S_\circ) - \bar{P} > 0$. Then for ϵ small enough, $P_\epsilon^{*,N}(S_\circ) - \bar{P} > 0$. If Assumption 9.3 is satisfied, then the matrix of the system of linear equations satisfied by $(y_{\epsilon,i}^{*,n})_{\rho \le i \le N_h}$ is an M-matrix, because $-rK\mathcal{V}'_\epsilon \ge 0$. Using the discrete maximum principle, we can prove by induction on $N - n$ that for all $1 \le n \le N$ and for all $\rho \le j \le N_h$, $y_{\epsilon,i}^{*,n} \ge 0$, and by passing to the limit that $y_i^{*,n} \ge 0$, $\rho \le i \le N_h$, $1 \le n \le N$. We have proved (9.39). □

Remark 9.2. *With the assumptions of Theorem 9.5, and if, at the optimum, we have the strict complementarity*

$$P^{*,n}(S_i) > P_\circ(S_i) \Leftrightarrow \mu_i^{*,n} = 0, \tag{9.43}$$

then the pairs $(y^{*,n}, \alpha^n)_{1 \le n \le N}$ are defined uniquely by (9.32), (9.34), (9.36). *In particular, $y^{*,n} \in \tilde{V}_h$ satisfies, for all i, $\rho \le i \le N_h$,*

$$\begin{aligned}
&\forall n, 0 \le n < N, \\
&P^{*,n}(S_i) = P_\circ(S_i) \Rightarrow y^{*,n}(S_i) = 0, \\
&P^{*,n}(S_i) > P_\circ(S_i) \Rightarrow \left(y^{*,n} - y^{*,n+1}, w^i\right) + \Delta t_n a_{T-t_n}^*(w^i, y^{*,n}) = 0, \\
&P^{*,N}(S_i) > P_\circ(S_i) \Rightarrow \left(y^{*,N}, w^i\right) + \Delta t_N a_0^*(w^i, y^{*,N}) = 2(P^{*,N}(S_\circ) - \bar{P})w^i(S_\circ).
\end{aligned}$$

In this case, the whole sequence $(y_\epsilon^{*,n})$ converges to $(y^{*,n})$.

Remark 9.3. *In the case where the solution $P^{*,n}$ exhibits a free boundary, i.e., there exist N real numbers γ_h^n, $1 \le n \le N$, such that for all $i, 0 \le i \le N_h$, $P^{*,n}(S_i) = P_\circ(S_i) \Leftrightarrow S_i \le \gamma_h^n$, then the lack of strict complementarity can occur only at $(t_n, \gamma_h^n)_{1 \le n \le N}$. From the a priori estimates on $y^{*,n}$, we can even say that if the strict complementarity condition is not true at $(t_n, \gamma_h^n)_{1 \le n \le N}$, then $|y^{*,n}(\gamma_h^n)| \le c\sqrt{h}$. Then imposing $y^{*,n}(\gamma_h^n) = 0$ does not yield a large error on the optimality condition.*

Summary on the Optimality Conditions. We go back to problem (9.3). We have proved the following result.

Theorem 9.6. *Let η^* be a minimizer of* (9.3) *which can be found as a limit of a sequence η^*_ϵ of minimizers for the penalized problem, and let $(P_i^{*,n})_{i\in I}$ be the solutions to* (9.4), (9.5) *with $\eta = \eta^*$. There exist $y_i^{*,n} \in \tilde{V}_h$ and $\alpha^n_{i,j} \in \mathbb{R}$, $1 \le n \le N_i$, $\rho \le j \le N_h$, $i \in I$, such that for all $v \in \tilde{V}_h$,*

$$
\begin{aligned}
&\left(y_i^{*,N_i}, v\right) + \Delta t_{N_i}\left(a_0^*(v, y_i^{*,N_i}) + \sum_{j=\rho}^{N_h} \alpha_{i,j}^{N_i} v(S_j)\right) = 2(P_i^{*,N_i}(S_\circ) - \bar{P}_i)v(S_\circ),\\
&\left(y_i^{*,n} - y_i^{*,n+1}, v\right) + \Delta t_n\left(a^*_{T_i - t_n}(v, y_i^{*,n}) + \sum_{j=\rho}^{N_h} \alpha_{i,j}^{n} v(S_j)\right) = 0, \quad 1 \le n < N,
\end{aligned}
\tag{9.44}
$$

with, for all j, n, $\rho \le j \le N_h$, $1 \le n \le N_i$,

$$
\alpha_{i,j}^n (P_i^{*,n}(S_j) - P_\circ(S_j)) = 0, \quad \mu_{i,j}^{*,n} y_i^{*,n}(S_j) = 0, \quad \alpha_i^n y_i^{*,n}(S_j) \ge 0,
$$

such that for any $\eta \in \mathcal{H}$, denoting $\delta\eta = \eta - \eta^*$,

$$
\begin{aligned}
0 \le\ & \langle DJ_R(\eta^*), \delta\eta \rangle \\
& - \frac{1}{2}\sum_{i\in I}\sum_{n=1}^{N_i} \Delta t_n \sum_{j=\rho}^{N_h} S_j^2 \delta\eta(S_j, T_i - t_n) y_i^{*,n}(S_j)
\left(
\begin{array}{c}
\dfrac{P_i^{*,n}(S_j) - P_i^{*,n}(S_{j-1})}{h_j} \\
+ \dfrac{P_i^{*,n}(S_j) - P_i^{*,n}(S_{j+1})}{h_{j+1}}
\end{array}
\right).
\end{aligned}
$$

9.4 Differentiability

In [4], the following result is proved.

Proposition 9.7. *Let Assumption* 9.1 *or Assumption* 9.2 *be satisfied for all η obeying* (9.8) *and* (9.9). *Let $\eta \in \mathcal{H}$ be such that the strict complementarity conditions*

$$
P_i^n(S_j) > P_{\circ,i}(S_j) \Leftrightarrow \mu_{i,j}^n = 0 \tag{9.45}
$$

are fulfilled for all $i \in I$ and for all j, $\rho \le j \le N_h$, where P_i^n is the solution to (9.4), (9.5), *and $\mu_{i,j}^n = (P_i^n - P_i^{n-1}, w^j) + \Delta t_n a_{T-t_n}(P_i^n, w^j)$. The functional J is differentiable at η, and for any admissible variation χ of η,*

$$
\begin{aligned}
\langle DJ(\eta), \chi \rangle = & -\frac{1}{2}\sum_{i\in I}\sum_{n=1}^{N} \Delta t_n \left(\chi(S, T - t_n) S \frac{\partial P_i^n}{\partial S}, S\frac{\partial y_i^n}{\partial S}\right) \\
& - \sum_{i\in I}\sum_{n=1}^{N} \Delta t_n \left(\left(\chi(S, T - t_n) + \frac{1}{2} S \frac{\partial \chi}{\partial S}(S, T - t_n)\right) S \frac{\partial P_i^n}{\partial S}, y_i^n\right),
\end{aligned}
\tag{9.46}
$$

where $y_i^n = y^n(\eta)_i \in \tilde V_h$, $\alpha_{i,j}^n \in \mathbb{R}$, $\rho \le j \le N_h$, *are the solution to, for all* $v \in \tilde V_h$,

$$\left(y_i^{N_i}, v\right) + \Delta t_{N_i}\left(a_0(v, y_i^{N_i}) + \sum_{j=\rho}^{N_h} \alpha_{i,j}^{N_i} v(S_j)\right) = 2(P_i^{N_i}(S_\circ) - \bar P_i)v(S_\circ),$$

$$\left(y_i^n - y_i^{n+1}, v\right) + \Delta t_n\left(a_{T_i - t_n}(v, y_i^n) + \sum_{j=\rho}^{N_h} \alpha_{i,j}^n v(S_j)\right) = 0, \quad 1 \le n < N,$$

with

$$\alpha_{i,j}^n(P_i^n(S_j) - P_\circ(S_j)) = 0, \quad \mu_{i,j}^n y_i^n(S_j) = 0, \quad \alpha_i^n y_i^n(S_j) \ge 0.$$

Another way of writing (9.46) *is*

$$\begin{aligned}&\langle DJ(\eta), \chi\rangle \\ &= -\frac{1}{2}\sum_{i\in I}\sum_{n=1}^{N} \Delta t_n \sum_{j=\rho}^{N_h} S_j^2 \chi(S_j, T - t_n) y_i^n(S_j)\left(\begin{array}{c}\dfrac{P_i^n(S_j) - P_i^n(S_{j-1})}{h_j} \\ + \dfrac{P_i^n(S_j) - P_i^n(S_{j+1})}{h_{j+1}}\end{array}\right).\end{aligned} \tag{9.47}$$

9.5 Algorithm

We describe the simplest possible projected descent method in the space Y, where the descent direction is computed thanks to the considerations above. The degrees of freedom of a function $\chi \in Y$ are the values of χ at some nodes of a grid and we call them $(\Lambda_\ell^*(\chi))_{1\le\ell\le L}$ (Λ_ℓ^* is the linear form on Y which maps χ to its value at a given node). We endow Y with the basis $(\Lambda_\ell(\chi))_{1\le\ell\le L}$ defined by $\Lambda_\ell^*(\Lambda_k) = \delta_{\ell k}$, and we define the inner product $(\sum_{\ell=1}^L a_\ell\Lambda_\ell, \sum_{\ell=1}^L b_\ell\Lambda_\ell)_Y = \sum_{l=1}^L a_\ell b_\ell$.

Algorithm.

- Choose $\eta \in \mathcal{H}$, $\epsilon > 0$ and $\rho > 0$, set $e = +\infty$.
- **While** $e > \epsilon$ **do**
  1. Compute $(P_i)_{i\in I}$ by (9.4), (9.5), by using, for example, one of the algorithms proposed in §6.5 and $J(\eta) + J_R(\eta)$, $J(\eta) = \sum_{i\in I} |P_i^{N_i}(S_\circ) - \bar P_i|^2$.
  2. For all $i \in I$, compute $(y_i^n)_{1\le n\le N_i}$, $y_i^n \in \tilde V_h$ satisfying (9.44).
  3. Compute $\zeta \in Y$ such that for all $\chi \in Y$,

$$\begin{aligned}&(\zeta, \chi)_Y \\ &= -\frac{1}{2}\sum_{i\in I}\sum_{n=1}^{N_i} \Delta t_n \sum_{j=\rho}^{N_h} S_j^2 \chi(S_j, T_i - t_n) y_i^n(S_j)\left(\begin{array}{c}\dfrac{u_i^n(S_j) - u_i^n(S_{j-1})}{h_j} \\ + \dfrac{u_i^n(S_j) - u_i^n(S_{j+1})}{h_{j+1}}\end{array}\right).\end{aligned} \tag{9.48}$$

4. Set $\tilde{\eta} = \pi_{\mathcal{H}}(\eta - \rho(\text{grad} J_R(\eta) + \zeta))$, $e = \|\tilde{\eta} - \eta\|$, $\eta = \tilde{\eta}$, where $\pi_{\mathcal{H}}$ is the projection on $\mathcal{H}$.

- **end_do**

The complete justification of the algorithm above is still an open question because it is not proved that $-\text{grad} J_R(\eta) - \zeta$ is always a descent direction. However, from Proposition 9.7 and Remark 9.3, we know that most often ζ is exactly $\text{grad} J(\eta)$: in this case, the algorithm coincides with a projected gradient method.

In the numerical tests below, we have used variants of this algorithm (an interior point algorithm due to Herskovits [68]—a quasi-Newton algorithm which can handle general constraints), which have proved very robust. In particular, we never experienced breakdown caused by the fact that the direction ζ is not a descent direction.

Parallelism. The algorithm above can be parallelized in a very natural way on a distributed memory machine with N_p processors, because the computations of the pairs (P_i, y_i), $i \in I$, are independent of each other. We split I in $I = \cup_{k=1}^{N_p} I_k$ in order to balance the amount of work among the processors, the processor labeled k being responsible for the sums over $i \in I_k$ in $J(\eta)$ and (9.48). Note that the complexity of the computation of P_i, y_i depends on i, so load balancing is not straightforward. The data for η and ζ are replicated on the N_p processors. The processor labeled k computes its own contribution to $J(\eta)$ and to (9.48), i.e., the sums over $i \in I_k$, in an independent manner; then communications are needed for assembling the sums over $i \in I$ in $J(\eta)$ and in (9.48).

For programming, we have used C++ with the message passing library `mpi`.

The following is a simple version of the code for computing the adjoint state y_i.

ALGORITHM 9.1. Backward loop for the adjoint problem.

```
template <class Vecto, class Tab, class Mat>
void  euler_scheme<Vecto,Tab, Mat>::backward_loop_iv(const int init_time,
const Tab& eta, Tab& y,  const Tab &p, const Vecto & ob,const int strike)
                                              //    p :  price of the put
                                                //    y :  adjoint state
                                              //    ob:  obstacle  (K-x)_+
                         //    strike:  the index of the strike in the mesh
{
  int ti=t.size()-1;
  dt=t[ti]-t[ti-1];
                                   //   builds the matrix A^T at final time
  disc->Build_Imp_Euler_LHS_Matrix_T(dt,eta[ti],LhsmatT);
                         //   modifies the adjoint system in the exercise zone
  for (int i=0;i<p[ti].size();i++)
    if(p[ti][i]==ob[i]&&  disc->meshpoints()[i]<2
       * disc->meshpoints()[strike])
      {
        LhsmatT(i,i)=1.;
        if (i< p[ti].size()-1)
          LhsmatT(i,i+1)=0.;
        if (i>0)
          LhsmatT(i,i-1)=0;
        y[ti][i]=0;
```

```
      }
                                        //   solves for adjoint state at final time
  gauss(LhsmatT,y[ti]);
                                                             //   loop on t
  for(int tit=t.size()-1;tit>init_time+1;tit--)
    {
      dt=t[tit-1]-t[tit-2];
      disc->rate()=rate[tit-1];
                                                        //   builds the matrix A^T
      disc->Build_Imp_Euler_LHS_Matrix_T(dt,eta[tit-1],LhsmatT);
                           //    modifies the adjoint system in the exercise zone
      for (int i=0;i<p[tit-1].size();i++)
        if(p[tit-1][i]==ob[i]&& disc->meshpoints()[i]<2
           *  disc->meshpoints()[strike])
          {
            LhsmatT(i,i)=1.;
            if (i< p[tit-1].size()-1)
              LhsmatT(i,i+1)=0.;
            if (i>0)
              LhsmatT(i,i-1)=0;
          }
                           //   builds the right-hand side of the adjoint problem
      for (int i=0; i<y[0].size();i++)
        y[tit-1][i]+=Massmat(i,i)*y[tit][i];
      for (int i=0; i<y[0].size()-1;i++)
        y[tit-1][i]+=Massmat(i,i+1)*y[tit][i+1];
      for (int i=1; i<y[0].size();i++)
        y[tit-1][i]+=Massmat(i,i-1)*y[tit][i-1];
                          //    modifies the right-hand side in the exercise zone
      for (int i=0;i<p[tit-1].size();i++)
        if(p[tit-1][i]==ob[i]&& disc->meshpoints()[i]<2
           *disc->meshpoints()[strike] )
          y[tit-1][i]=0;
                                                //   solves for the adjoint state
      gauss(LhsmatT,y[tit-1]);
    }
  y[init_time]=0;
}
```

9.6 Results

9.6.1 Testing the Method with a Target Volatility

In order to test the method, we choose a given target squared volatility function η_g, and we compute the prices of a series of American puts (with ten different maturities and ten different strikes) by solving (9.4), (9.5) with $\eta = \eta_g$. The chosen (squared) volatility surface

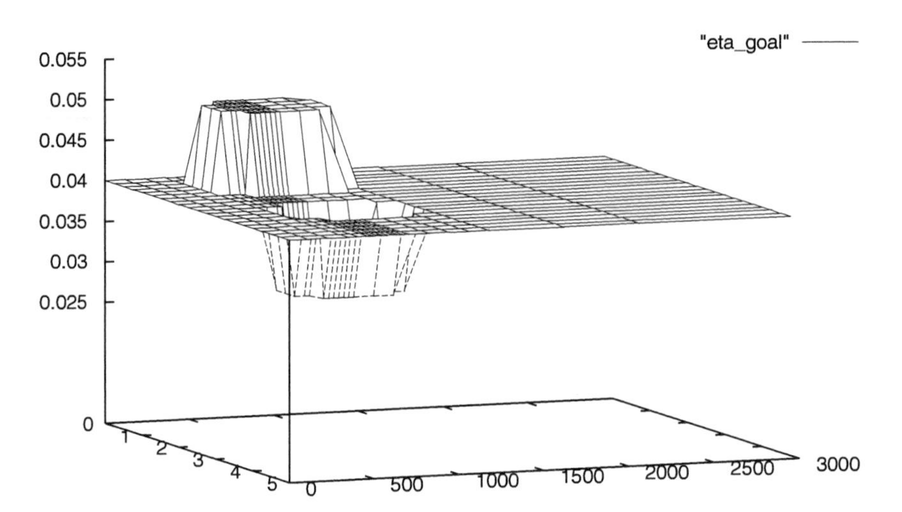

Figure 9.1. *The squared volatility surface to be found by the calibration algorithm.*

is plotted in Figure 9.1. It is the interpolation on a Cartesian grid of the piecewise constant function

$$(S,t) \mapsto 0.04 + 0.01(1_{\omega_0} - 1_{\omega_1}),$$

$$[9pt]\omega_0 = \left\{ \left(8\frac{S}{\bar{S}} - 2\right)^2 + \left(8\frac{t}{T} - 1.6\right)^2 \le 1 \right\},$$

$$\omega_1 = \left\{ \left(8\frac{S}{\bar{S}} - 2\right)^2 + \left(8\frac{t}{T} - 4.8\right)^2 \le 1 \right\},$$

and $\bar{S}$ and T are given below. We are aware that such a volatility is not realistic. The reason for choosing it is that the exercise prices of the options are not monotone functions of time. Note also that the function η_g is piecewise affine, and that $0.03 \le \eta_g \le 0.05$. The other parameters of the computation are

- interest rate: 0.06;
- observed price: $S_\circ = 590$;
- $\bar{S} = 5 \times S_\circ$;
- the maturities are $T_i = 0.5i$, $1 \le i \le 10$; we set $T = 5$;
- the strikes are at $0.85S_\circ$, $0.9S_\circ$, $0.95S_\circ$, $S_\circ$, $1.05S_\circ$, $1.1S_\circ$, $1.2S_\circ$, $1.3S_\circ$, and $1.4S_\circ$;
- the constraints on η are $0.005 \le \eta \le 0.1$.

By running the calibration program, we aim at recovering the surface plotted in Figure 9.1.

Finite Elements with a Uniform Mesh. In the first test, the squared volatility is described by piecewise bilinear functions (Q_1) of S and t, with a uniform 10×10 mesh. The discrete space for η has dimension 100.

The Tychonoff functional is of the form

$$J_R(\eta) = \int_t \int_S a(S)(\eta(S,t) - 0.04)^2 + \int_t \int_S b(S) \left(\frac{\partial \eta}{\partial t}(S,t) \right)^2 + \int_t \int_S c(S) \left(S \frac{\partial \eta}{\partial S}(S,t) \right)^2 + \int_t \int_S d(S) \left(S \frac{\partial^2 \eta}{\partial t \partial S}(S,t) \right)^2,$$

where a, b, c, and d are positive functions to be chosen properly. In particular, these functions play an important role in the out-of-money region, where we do not have any information.

The mesh for u is obtained by refining the mesh for η. It is uniform in the t variable and 25 times finer than the mesh for η. It is nonuniform in the S variable, with a refinement in the money region.

The optimization loop is initialized with a uniform volatility: $\eta = 0.04$. At convergence, we obtain the surface plotted in Figure 9.2. We see that the volatility obtained by the algorithm is very close to the goal. It is interesting to plot the relative error between the observed prices and the prices computed with the volatility η at convergence: In Figure 9.3, we see that the relative error does not exceed 10^{-3} and is more often of the order 10^{-5}. In Figure 9.4, we plot the graphs showing the convergence of the optimization loop: In the top panel, we see the value of $J(\eta)$ as a function of the number of iterations, in logarithmic scale. We see that J is reduced by a factor 10^7. In the bottom panel, the norm of $\zeta + \operatorname{grad} J_R(\eta)$, with ζ given by (9.48), is displayed. We see the typical behavior of the Newton algorithm, with a quadratic rate near convergence. At convergence, $\zeta + \operatorname{grad} J_R(\eta)$ is 0, because the constraints on η are not active.

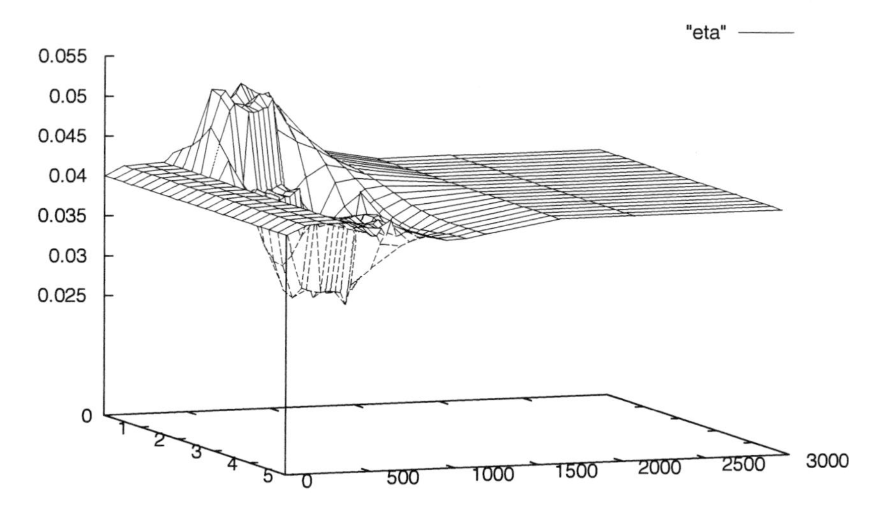

Figure 9.2. *The squared volatility surface obtained by running the calibration program.*

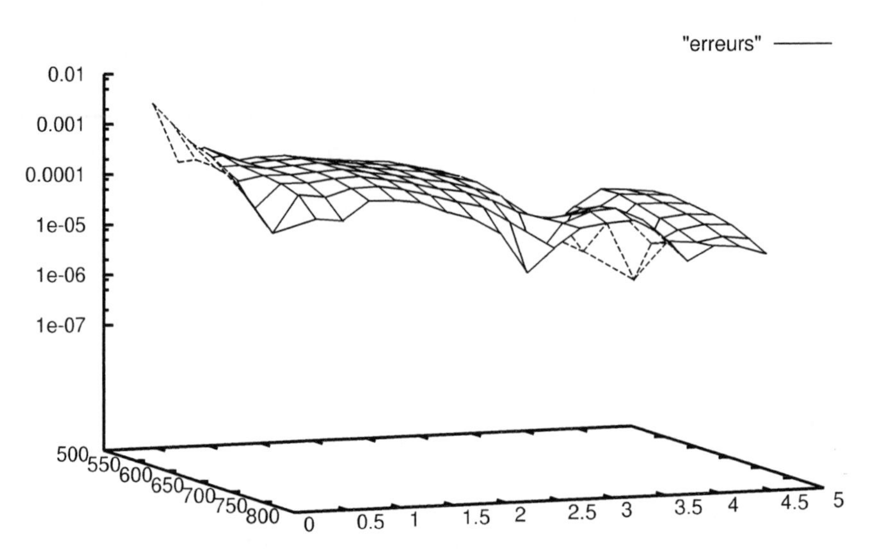

Figure 9.3. *Relative errors between the observed prices and those obtained with η found after running the calibration program.*

It is also possible to use nonuniform meshes for the volatilities (see [4]). However, with finite elements, the number of unknowns for the volatility grows very rapidly, and the quasi-Newton method becomes rapidly expensive in terms of memory and CPU time.

Bicubic Splines. We run the same test case, but now, the squared volatility is described by bicubic splines. More precisely, η is chosen as the sum of

- a piecewise affine function of S which takes two constant values in the regions $S \leq 0.5S_\circ$ and $S \geq 1.5S_\circ$, and which is linear in the region $|S - S_\circ| \leq 0.5S_\circ$;
- a bicubic spline in the region $|S - S_\circ| \leq 0.5S_\circ$, $|t - T/2| < T/2$, whose value and derivatives vanish on $|S - S_\circ| = 0.5S_\circ$ and on $|t - T/2| = T/2$. In this experiment, there are 10×10 control points.

Therefore, the number of degrees of freedom for the volatility is $10 \times 10 + 2$. The optimization loop is initialized with a uniform volatility: $\eta = 0.04$. At convergence, we obtain the surface plotted in Figure 9.5. We see that the volatility obtained by the algorithm is not as close to the goal as when discretized by finite elements. The reason for that is clearly that η_g is piecewise affine and cannot be represented exactly by bicubic splines. Nevertheless, the main qualitative features of the volatility surface are recovered. The accuracy of the method is best seen when we plot the relative errors between the observed prices and the prices computed with the volatility η at convergence: In Figure 9.6, we see that the relative error does not exceed 5×10^{-3} and is most often smaller than 10^{-3}. In Figure 9.7, we plot the graphs monitoring the optimization loop: we see that the convergence is much faster than with finite elements.

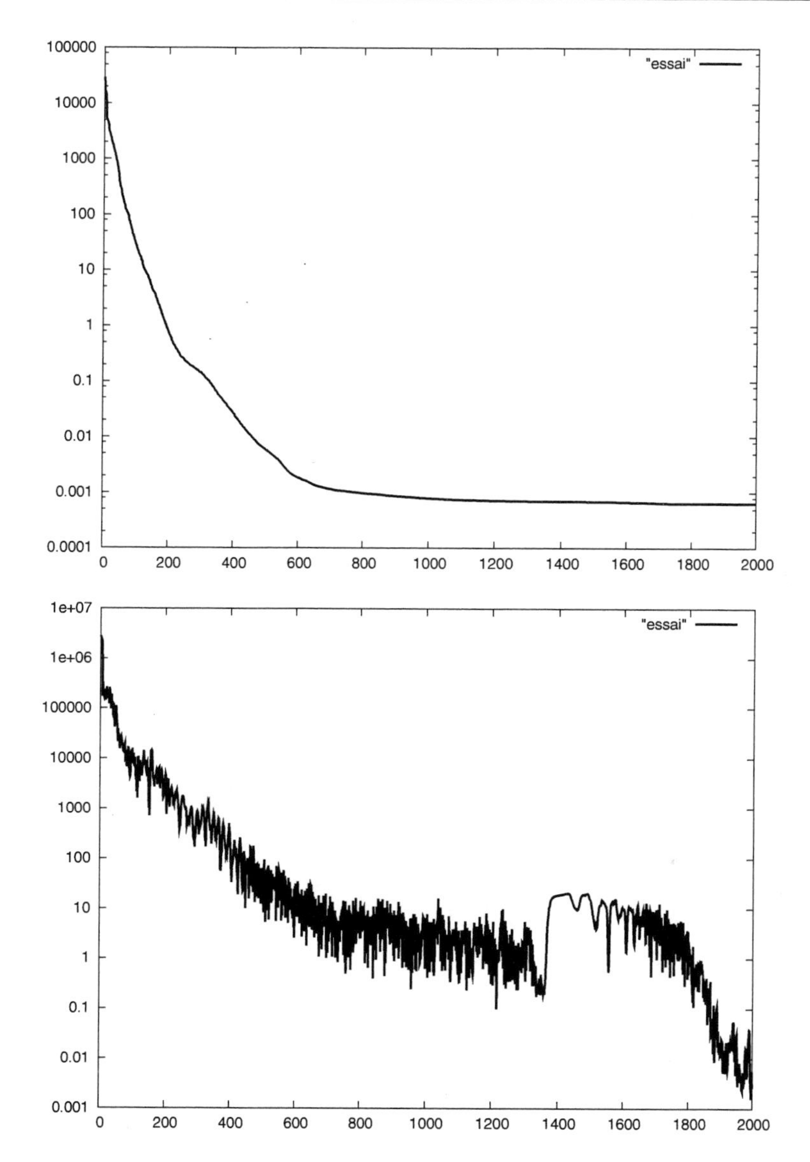

Figure 9.4. *$J(\eta)$ as a function of the number of iterations and the norm of $\zeta +$* grad $J_R(\eta)$, *with ζ given by* (9.48) *as a function of the number of iterations.*

9.6.2 Results with American Puts on the FTSE 100 Index

In this paragraph, we consider American puts on the FTSE 100 index. The data correspond to June 6, 2001. We thank José Da Fonseca for providing us with the data.

The price of the underlying asset is $x_\circ = 5890$. The American puts correspond to four different maturities: 0.122, 0.199, 0.295, and 0.55 years. We set $T = 0.55$. The interest rate r varies with time, so r is replaced by $r(t)$ in (9.6), and this function is known. For

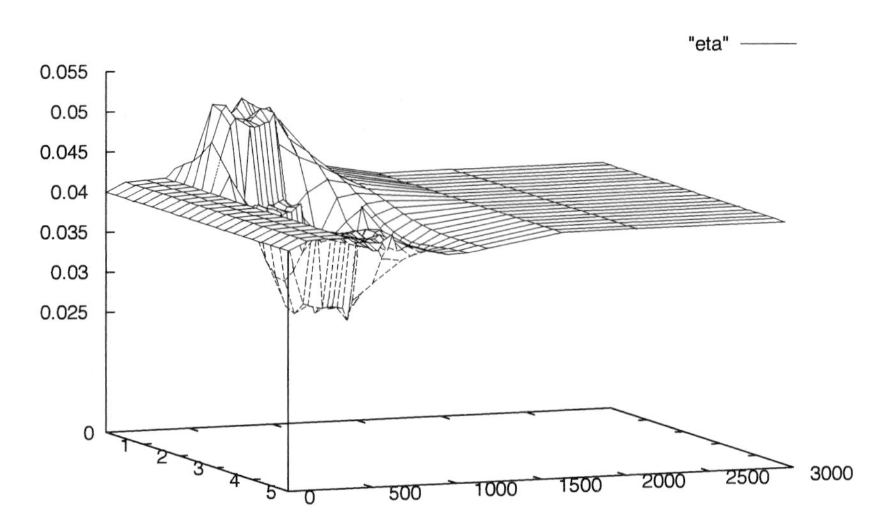

Figure 9.5. *The squared volatility surface obtained by running the calibration program.*

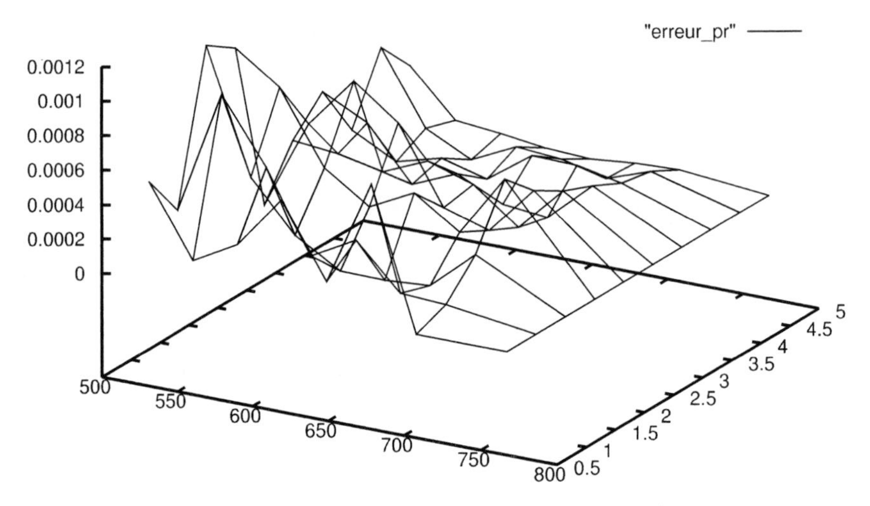

Figure 9.6. *Relative errors between the observed prices and those obtained with η found after running the calibration program.*

these maturities, the prices of the observed options versus strike are plotted in Figure 9.8. The aim is to find the volatility surface from these prices. The volatility is discretized by functions that are the sum of

- a piecewise affine function in the S-variable which is constant in the regions $S < 1000$ and $S > 9000$ and affine in the region $1000 < S < 9000$;
- a bicubic spline in the region $1000 < S < 9000$, $|t - T/2| < T/2 + 0.1$, whose value and derivatives vanish on the boundary of this rectangle. The control points of the spline are plotted in Figure 9.9, where the time variable is $T - t$. We see that the

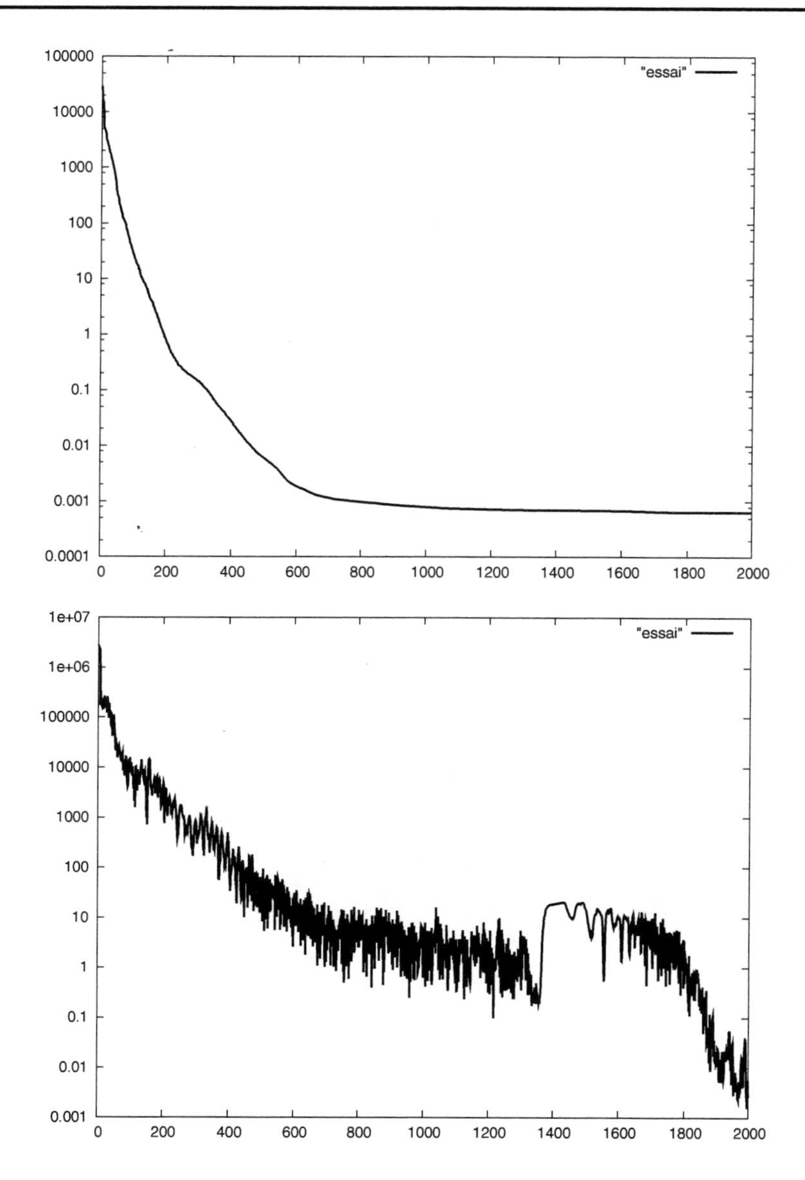

Figure 9.7. *$J(\eta)$ as a function of the number of iterations and the norm of $\zeta + \text{grad} J_R(\eta)$, with ζ given by* (9.48) *as a function of the number of iterations.*

control points are not uniformly distributed: the mesh is refined for small times t and at the money region.

The grid for u is nonuniform with 745 nodes in the S-direction and 210 nodes in the t-direction. For simplicity, the grid is chosen in such a way that the points $(T_i, K_i)_{i \in I}$ coincide with some grid nodes.

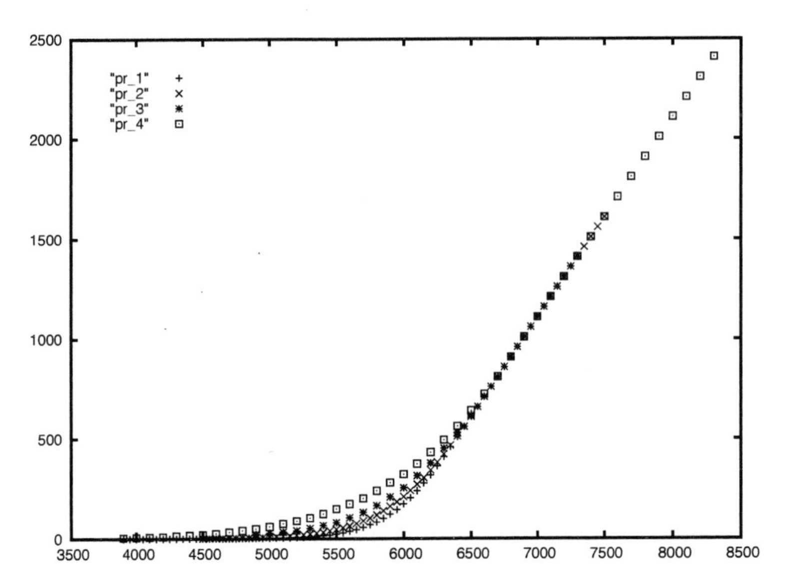

Figure 9.8. *The data for the inverse problem: the prices of a family of American puts on the FTSE* 100 *index.*

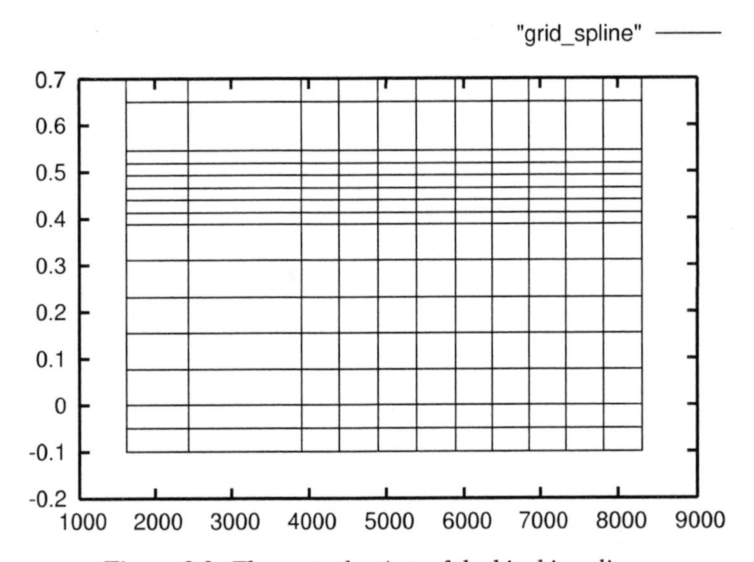

Figure 9.9. *The control points of the bicubic splines.*

The (squared) volatility obtained at convergence is displayed in Figure 9.10: the surface has a smile shape. The relative errors between the observed prices and those computed at convergence are plotted in Figure 9.11, top. They are larger than those observed for the test problem in §9.6.1. However, we have to realize that the available observed prices are themselves given with a roundoff error, which is exactly 0.5. In Figure 9.11, bottom, we

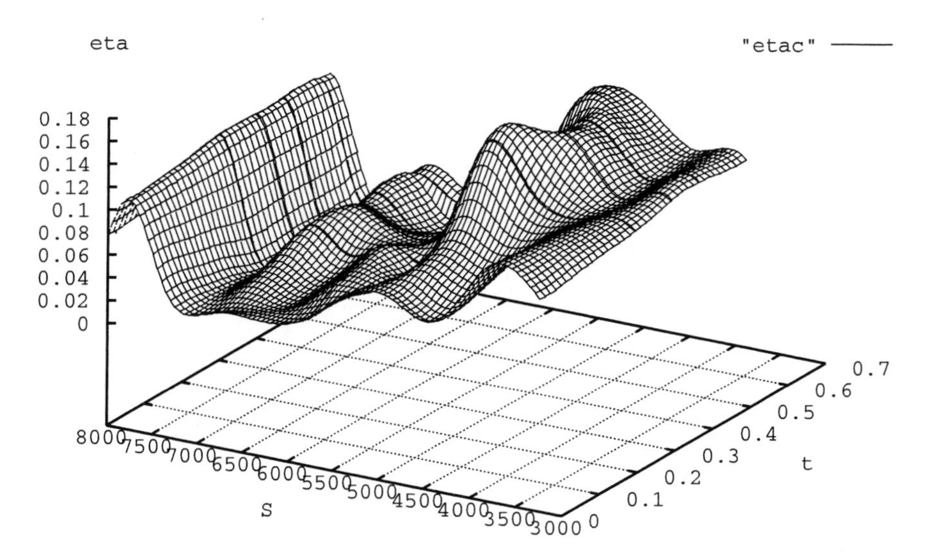

Figure 9.10. *The squared volatility surface obtained by running the calibration program.*

have plotted the relative roundoff error on the observed prices. Doing so, we see that the relative errors on the prices at convergence are of the same order as the roundoff error on the observed prices. Therefore, it is very natural that the optimization program cannot improve on this level of error.

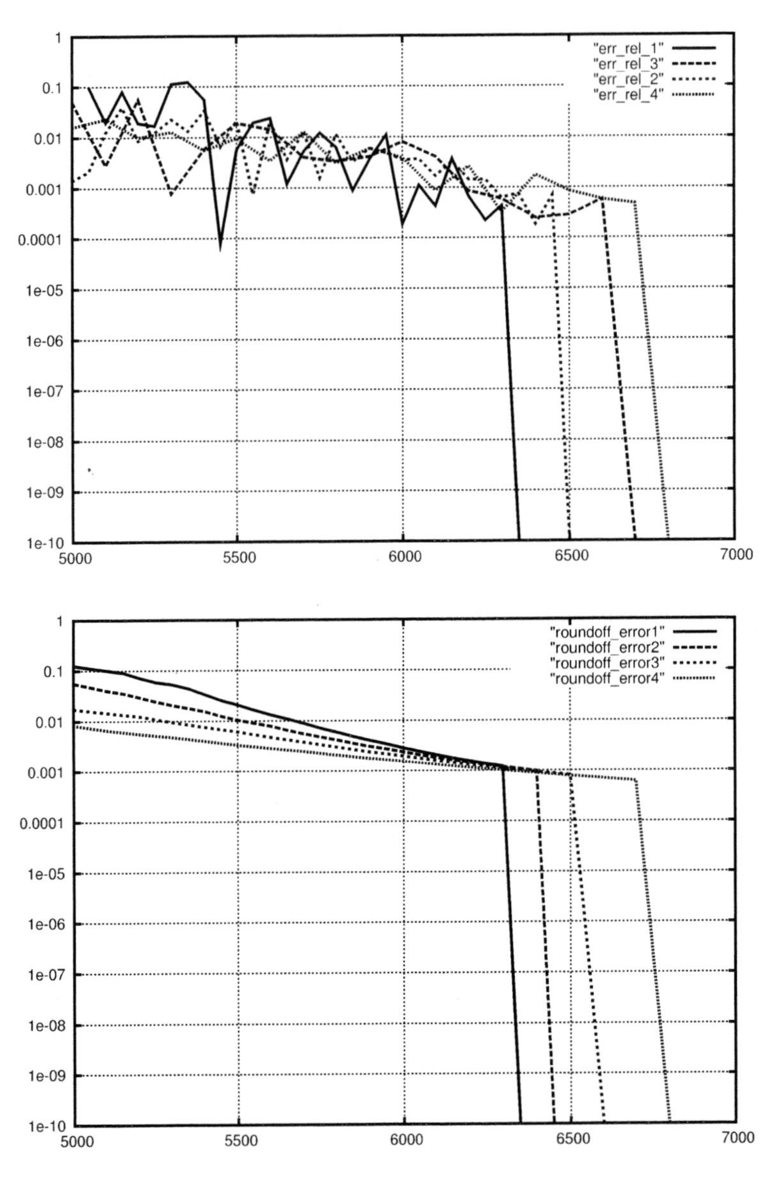

Figure 9.11. *Top: relative errors between the observed prices and those obtained with η found after running the calibration program. A curve corresponds to a given maturity. Bottom: relative roundoff error on observed prices. The two errors are of the same order.*

Bibliography

[1] Milton Abramowitz and Irene A. Stegun, editors. *Handbook of Mathematical Functions with Formulas, Graphs, and Mathematical Tables*. Dover, New York, 1992.

[2] Yves Achdou. An inverse problem for a parabolic variational inequality arising in volatility calibration with American options. *SIAM J. Control Optim.*, 43:1583–1615, 2005.

[3] Yves Achdou, Bruno Franchi, and Nicoletta Tchou. A partial differential equation connected to option pricing with stochastic volatility: Regularity results and discretization. *Math. Comp.*, 74:1291–1322, 2005.

[4] Yves Achdou and Olivier Pironneau. A numerical procedure for the calibration of American options. To appear in *Appl. Math. Finance*.

[5] Yves Achdou and Nicoletta Tchou. Variational analysis for the Black and Scholes equation with stochastic volatility. *M2AN Math. Model. Numer. Anal.*, 36:373–395, 2002.

[6] L.B.G. Andersen and R. Brotherton-Ratcliffe. The equity option volatility smile: An implicit finite difference approach. *J. Comput. Finance*, 1:5–32, 1998.

[7] M. Avellaneda, M. Friedman, C. Holmes, and D. Samperi. Calibrating volatility surfaces via relative entropy minimization. *Appl. Math. Finance*, 4:37–64, 1997.

[8] Owe Axelsson. *Iterative Solution Methods*. Cambridge University Press, New York, 1994.

[9] Vlad Bally. An elementary introduction to Malliavin calculus. *INRIA Research Report*, 4718(February), 2003.

[10] Vlad Bally, L. Caramello, and A. Zanette. Pricing and hedging American option by Monte-Carlo methods using a Malliavin calculus approach. *INRIA Research Report*, 4804(April), 2003.

[11] Guy Barles, Julien Burdeau, Marc Romano, and Nicolas Samsœn. Estimation de la frontière libre des options américaines au voisinage de l'échéance. *C. R. Acad. Sci. Paris Sér.* I *Math.*, 316:171–174, 1993.

[12] A. Bensoussan. On the theory of option pricing. *Acta Appl. Math.*, 2:139–158, 1984.

[13] Alain Bensoussan and Jacques-Louis Lions. *Impulse Control and Quasivariational Inequalities.* μ. Gauthier-Villars, Montrouge, 1984. Translated from the French by J. M. Cole.

[14] Amal Bergam, Christine Bernardi, and Zoubida Mghazli. A posteriori analysis of the finite element discretization of a nonlinear parabolic equation. *Math. Comput.*, 74: 1097–1116, 2005.

[15] Maïtine Bergounioux and Fulbert Mignot. Optimal control of obstacle problems: Existence of Lagrange multipliers. *ESAIM Control Optim. Calc. Var.*, 5:45–70, 2000.

[16] F. Black and M. Scholes. The pricing of options and corporate liabilities. *J. Pol. Econ.*, 81:637–659, 1973.

[17] Phelim P. Boyle and Draviam Thangaraj. Volatility estimation from observed option prices. *Decis. Econ. Finance*, 23:31–52, 2000.

[18] Dietrich Braess. *Finite Elements: Theory, Fast Solvers, and Applications in Solid Mechanics.* Cambridge University Press, Cambridge, 2001. Second Edition.

[19] Michael J. Brennan and Eduardo S. Schwartz. The valuation of the American put option. *J. Finance*, 32:449–462, 1977.

[20] S.C. Brenner and R. Scott. *The Mathematical Theory of Finite Element Methods.* Volume 15 of *Texts in Applied Mathematics.* Springer-Verlag, New York, 1994.

[21] H. Brezis. *Analyse Fonctionnelle, Théorie et Applications.* Masson, Paris, 1983.

[22] Mark Broadie and Paul Glasserman. Pricing American-style securities using simulation. *J. Econom. Dynam. Control*, 21:1323–1352, 1997.

[23] Hans-Joachim Bungartz and Michael Griebel. Sparse grids. *Acta Numerica*, 13:1–123, 2004.

[24] P. Carr and D. Madan. Option valuation using the fast Fourier transform. *J. Comput. Finance*, 2:61–73, 1998.

[25] Thierry Cazenave and Alain Haraux. *An Introduction to Semilinear Evolution Equations.* The Clarendon Press, Oxford University Press, New York, 1998. Translated from the 1990 French original by Yvan Martel and revised by the authors.

[26] Terence Chan. Pricing contingent claims on stocks driven by Lévy processes. *Ann. Appl. Probab.*, 9:504–528, 1999.

[27] P.G. Ciarlet. *The Finite Element Method for Elliptic Problems.* North–Holland, Amsterdam, 1978. Revised Edition: SIAM, Philadelphia, PA, 2002.

[28] P.G. Ciarlet. Basic error estimates for elliptic problems. In *Handbook of Numerical Analysis*, Vol. II, 17–351. North–Holland, Amsterdam, 1991.

[29] T.F. Coleman, Y. Li, and A. Verma. Reconstructing the unknown local volatility function. *J. Comput. Finance*, 2:77–102, 1999.

[30] R. Cont and P. Tankov. *Financial Modelling with Jump Processes.* Chapman and Hall, Boca Raton, FL, 2003.

[31] Rama Cont and Peter Tankov. Nonparametric calibration of jump-diffusion option pricing models. *J. Comput. Finance*, 7:1–49, 2004.

[32] Rama Cont and Ekaterina Voltchkova. Finite Difference Methods for Option Pricing in Jump-Diffusion and Exponential Lévy Models. Rapport Interne 513, CMAP, Ecole Polytechnique, 2003.

[33] Rama Cont and Ekaterina Voltchkova. Integro-differential Equations for Option Prices in Exponential Lévy Models. Rapport Interne 547, CMAP, Ecole Polytechnique, 2004.

[34] R. Courant. Variational methods for the solution of problems of equilibrium and vibrations. *Bull. Amer. Math. Soc.*, 49:1–23, 1943.

[35] R. Courant, K. Friedrichs, and H. Lewy. On the partial difference equations of mathematical physics. *IBM J. Res. Develop.*, 11:215–234, 1967.

[36] J.C. Cox and M. Rubinstein. *Option Market.* Prentice–Hall, Englewood Cliffs, NJ, 1985.

[37] Michael G. Crandall, Hitoshi Ishii, and Pierre-Louis Lions. User's guide to viscosity solutions of second order partial differential equations. *Bull. Amer. Math. Soc. (N.S.)*, 27:1–67, 1992.

[38] Ionut Danaila, Frédéric Hecht, and Olivier Pironneau. *Simulation Numérique en C++.* Dunod, Paris, 2003.

[39] Carl de Boor. *A Practical Guide to Splines.* Volume 27 of *Applied Mathematical Sciences.* Springer-Verlag, New York, 2001. Revised Edition.

[40] S. DelPino and O. Pironneau. *FreeFem3D.* www.ann.jussieu.fr.

[41] B. Dupire. Pricing with a smile. *Risk*, 7:18–20, 1994.

[42] Ernst Eberlein. Application of generalized hyperbolic Lévy motions to finance. In *Lévy Processes*, 319–336. Birkhäuser Boston, Boston, MA, 2001.

[43] Ernst Eberlein and Karsten Prause. The generalized hyperbolic model: Financial derivatives and risk measures. In *Mathematical Finance—Bachelier Congress, 2000 (Paris)*, Springer Finance, 245–267. Springer-Verlag, Berlin, 2002.

[44] Ernst Eberlein and Sebastian Raible. Term structure models driven by general Lévy processes. *Math. Finance*, 9:31–53, 1999.

[45] Kenneth Eriksson, Don Estep, Peter Hansbo, and Claes Johnson. Introduction to computational methods for differential equations. In *Theory and Numerics of Ordinary and Partial Differential Equations (Leicester, 1994)*, Adv. Numer. Anal., IV, 77–122. Oxford University Press, New York, 1995.

[46] Kenneth Eriksson and Claes Johnson. Adaptive finite element methods for parabolic problems. I. A linear model problem. *SIAM J. Numer. Anal.*, 28:43–77, 1991.

[47] Kenneth Eriksson and Claes Johnson. Adaptive finite element methods for parabolic problems. II. Optimal error estimates in $L_\infty L_2$ and $L_\infty L_\infty$. *SIAM J. Numer. Anal.*, 32:706–740, 1995.

[48] Wendell H. Fleming and Raymond W. Rishel. *Deterministic and Stochastic Optimal Control*. Volume 1 of *Applications of Mathematics*. Springer-Verlag, New York, 1975.

[49] Wendell H. Fleming and H. Mete Soner. *Controlled Markov Processes and Viscosity Solutions*. Volume 25 of *Applications of Mathematics*. Springer-Verlag, New York, 1993.

[50] P.A. Forsyth and K.R. Vetzal. Quadratic convergence for valuing American options using a penalty method. *SIAM J. Sci. Comput.*, 23:2095–2122, 2002.

[51] Jean-Pierre Fouque, George Papanicolaou, and K. Ronnie Sircar. *Derivatives in Financial Markets with Stochastic Volatility*. Cambridge University Press, Cambridge, 2000.

[52] Eric Fournié, Jean-Michel Lasry, Jérôme Lebuchoux, and Pierre-Louis Lions. Applications of Malliavin calculus to Monte-Carlo methods in finance. II. *Finance Stoch.*, 5:201–236, 2001.

[53] Eric Fournié, Jean-Michel Lasry, Jérôme Lebuchoux, Pierre-Louis Lions, and Nizar Touzi. Applications of Malliavin calculus to Monte Carlo methods in finance. *Finance Stoch.*, 3:391–412, 1999.

[54] Leopoldo P. Franca, Sérgio L. Frey, and Thomas J. R. Hughes. Stabilized finite element methods. I. Application to the advective-diffusive model. *Comput. Methods Appl. Mech. Engrg.*, 95:253–276, 1992.

[55] A. Friedman. *Partial Differential Equations of Parabolic Type*. Prentice–Hall, Englewood Cliffs, NJ, 1964.

[56] Avner Friedman. Regularity theorems for variational inequalities in unbounded domains and applications to stopping time problems. *Arch. Rational Mech. Anal.*, 52:134–160, 1973.

[57] Avner Friedman. Stochastic games and variational inequalities. *Arch. Rational Mech. Anal.*, 51:321–346, 1973.

[58] K.O. Friedrichs. The identity of weak and strong extensions of differential operators. *Trans. Amer. Math. Soc.*, 55:132–151, 1944.

[59] M. Galassi, J. Davies, J. Theiler, B. Gough, G. Jungman, M. Booth, and F. Rossi. *The GNU Scientific Library*. www.gnu.org/software/gsl.

[60] Paul Glasserman. *Monte Carlo Methods in Financial Engineering*. Volume 53 of *Applications of Mathematics*. Springer-Verlag, New York, 2004.

[61] Roland Glowinski, Jacques-Louis Lions, and Raymond Trémolières. *Numerical Analysis of Variational Inequalities*. Volume 8 of *Studies in Mathematics and Its Applications*. North–Holland, Amsterdam, 1981. Translated from the French.

[62] Gene H. Golub and Charles F. Van Loan. *Matrix Computations*. The Johns Hopkins University Press, Baltimore, MD, 1989. Second Edition.

[63] M. Griebel. Adaptive sparse grid multilevel methods for elliptic PDEs based on finite differences. *Computing*, 61:151–179, 1998.

[64] M. Griebel and P. Oswald. Tensor-product-type subspace splittings and multilevel iterative methods for anisotropic problems. *Adv. Comput. Math.*, 4:171–206, 1995.

[65] M. Griebel, M. Schneider, and C. Zenger. A combination technique for the solution of sparse grid problems. In *Proceedings of the IMACS International Symposium on Iterative Methods in Linear Algebra*, 263–281, Elsevier, Amsterdam, 1992.

[66] Andreas Griewank. *Evaluating Derivatives: Principles and Techniques of Algorithmic Differentiation*. Volume 19 of *Frontiers in Applied Mathematics*. SIAM, Philadelphia, PA, 2000.

[67] J. Michael Harrison and Stanley R. Pliska. Martingales and stochastic integrals in the theory of continuous trading. *Stochastic Process. Appl.*, 11:215–260, 1981.

[68] J. Herskovits. Feasible direction interior-point technique for nonlinear optimization. *J. Optim. Theory Appl.*, 99:121–146, 1998.

[69] Magnus R. Hestenes and Eduard Stiefel. Methods of conjugate gradients for solving linear systems. *J. Research Nat. Bur. Standards*, 49:409–436, 1952.

[70] S. Heston. A closed-form solution for options with stochastic volatility with application to bond and currency options. *Rev. Financial Studies*, 6:327–343, 1993.

[71] N. Hilber, A.M. Matache, and C. Schwab. Sparse Wavelets Methods for Option Pricing under Stochastic Volatility. Technical report, Seminar for Applied Mathematics, ETH Zurich, 2004.

[72] M. Hintermüller, K. Ito, and K. Kunisch. The primal-dual active set strategy as a semismooth Newton method. *SIAM J. Optim.*, 13:865–888, 2002.

[73] Michael Hintermüller. Inverse coefficient problems for variational inequalities: Optimality conditions and numerical realization. *M2AN Math. Model. Numer. Anal.*, 35:129–152, 2001.

[74] J.C. Hull and A. White. The pricing of options on assets with stochastic volatilities. *J. Finance*, 42:281–300, 1987.

[75] K. Ito and K. Kunisch. Optimal control of elliptic variational inequalities. *Appl. Math. Optim.*, 41:343–364, 2000.

[76] Kazufumi Ito and Karl Kunisch. Semi-smooth Newton methods for variational inequalities of the first kind. *M2AN Math. Model. Numer. Anal.*, 37:41–62, 2003.

[77] N. Jackson, E. Süli, and S. Howison. Computation of deterministic volatility surfaces. *App. Math. Finances*, 2:5–32, 1998.

[78] Patrick Jaillet, Damien Lamberton, and Bernard Lapeyre. Variational inequalities and the pricing of American options. *Acta Appl. Math.*, 21:263–289, 1990.

[79] I. Karatzas and S. E. Shreve. *Brownian Motion and Stochastic Calculus*. Springer-Verlag, New York, 1991. Second Edition.

[80] Joseph B. Keller and Dan Givoli. Exact nonreflecting boundary conditions. *J. Comput. Phys.*, 82:172–192, 1989.

[81] C.T. Kelley. *Iterative Methods for Linear and Nonlinear Equations*. Volume 16 of *Frontiers in Applied Mathematics*. SIAM, Philadelphia, PA, 1995.

[82] D. Kinderlehrer and G. Stampacchia. *An Introduction to Variational Inequalities and Their Application*. Academic Press, New York, 1980.

[83] R. Lagnado and S. Osher. Reconciling differences. *Risk*, 10:79–83, 1997.

[84] R. Lagnado and S. Osher. A technique for calibrating derivative security pricing models: Numerical solutions of an inverse problem. *J. Comput. Finance*, 1:13–25, 1997.

[85] D. Lamberton and B. Lapeyre. *Introduction au Calcul Stochastique Appliqué à la Finance*. Ellipses, Paris, 1997.

[86] Damien Lamberton. Critical price for an American option near maturity. In *Seminar on Stochastic Analysis, Random Fields and Applications (Ascona, 1993)*. Volume 36 of *Progress in Probability*, 353–358. Birkhäuser, Basel, 1995.

[87] B. Lapeyre and A. Sulem. *Simulation of Financial Models: Mathematical Foundations and Applications*. Cambridge University Press, Cambridge, to appear.

[88] P.D. Lax and R.D. Richtmyer. Survey of the stability of linear finite difference equations. *Comm. Pure Appl. Math.*, 9:267–293, 1956.

[89] Jacques-Louis Lions. *Quelques Méthodes de Résolution des Problèmes aux Limites non Linéaires*. Dunod, Paris, 1969.

[90] Jacques-Louis Lions and Enrico Magenes. *Problèmes aux Limites non Homogènes et Applications*, Vols. I and II. Dunod, Paris, 1968.

[91] Yuh-Dauh Lyuu. *Financial Engineering and Computation: Principles, Mathematics, Algorithms*. Cambridge University Press, Cambridge, 2002.

[92] D. Madan. Financial modeling with discontinuous price processes. In O.E. Barndorff-Nielsen, T. Mikosh, and S. Resnick, editors, *Lévy Processes—Theory and Applications*. Birkhäuser Boston, Boston, MA, 2001.

[93] A.-M. Matache, P.-A. Nitsche, and C. Schwab. Wavelet Galerkin Pricing of American Options on Lévy Driven Assets. 2003. Research Report SAM 2003-06.

[94] A.-M. Matache, C. Schwab, and T.P. Wihler. Fast Numerical Solution of Parabolic Integro-differential Equations with Applications in Finance. Technical report no. 1954, IMA University of Minnesota, Minneapolis, MN, 2004.

[95] A.-M. Matache, T. von Petersdoff, and C. Schwab. Fast deterministic pricing of Lévy driven assets. *M2AN Math. Model. Numer. Anal.*, 38:37–72, 2004.

[96] Stephen F. McCormick, editor. *Multigrid Methods*. Volume 3 of *Frontiers in Applied Mathematics*. SIAM, Philadelphia, PA, 1987.

[97] Robert C. Merton. Theory of rational option pricing. *Bell J. Econom. Management Sci.*, 4:141–183, 1973.

[98] A. Pazy. *Semigroups of Linear Operators and Applications to Partial Differential Equations*. Volume 44 of *Applied Mathematical Sciences*. Springer-Verlag, New York, 1983.

[99] Huyên Pham. Optimal stopping of controlled jump-diffusion processes: A viscosity solution approach. *J. Math. Systems*, 8:1–27, 1998.

[100] O. Pironneau and F. Hecht. `freefem++`, www.freefem.org.

[101] Elijah Polak. *Optimization: Algorithms and Consistent Approximations*. Volume 124 of *Applied Mathematical Sciences*. Springer-Verlag, New York, 1997.

[102] M.H. Protter and H.F. Weinberger. *Maximum Principles in Differential Equations*. Springer-Verlag, New York, 1984. Corrected reprint of the 1967 original.

[103] P.-A. Raviart and J.-M. Thomas. *Introduction à l'analyse numérique des équations aux dérivées partielles*. Masson, Paris, 1983.

[104] A. Revuz and M. Yor. *Continuous Martingale Calculus*. Springer-Verlag, New York, 1990.

[105] Robert D. Richtmyer and K. W. Morton. *Difference methods for initial-value problems*. Interscience Tracts in Pure and Applied Mathematics, No. 4. John Wiley & Sons, New York, London, Sydney, 1967. Second Edition.

[106] Y. Saad. *Iterative Methods for Sparse Linear Systems*. PWS Publishing Company, Boston, MA, 1996. Second Edition: SIAM, Philadelphia, PA, 2003.

[107] Y. Saad and M.H. Schultz. GMRES: A generalized minimal residual algorithm for solving nonsymmetric linear systems. *SIAM J. Sci. Statist. Comput.*, 7:856–869, 1986.

[108] Dominick Samperi. Calibrating a diffusion pricing model with uncertain volatility: Regularization and stability. *Math. Finance*, 12:71–87, 2002.

[109] Dominik Schötzau and Christoph Schwab. *hp*-discontinuous Galerkin time-stepping for parabolic problems. *C. R. Acad. Sci. Paris Sér.* I *Math.*, 333:1121–1126, 2001.

[110] E. Stein and J. Stein. Stock price distributions with stochastic volatility: An analytic approach. *Rev. Financial Studies*, 4:727–752, 1991.

[111] G. Strang and G.J. Fix. *An Analysis of the Finite Element Method.* Prentice–Hall, Englewood Cliffs, NJ, 1973.

[112] Vidar Thomée. *Galerkin Finite Element Methods for Parabolic Problems.* Volume 25 of *Springer Series in Computational Mathematics.* Springer-Verlag, Berlin, 1997.

[113] H.A. van der Vorst. Bi-CGSTAB: A fast and smoothly converging variant of Bi-CG for the solution of nonsymmetric linear systems. *SIAM J. Sci. Statist. Comput.*, 13:631–644, 1992.

[114] T. von Petersdoff and C. Schwab. Numerical solutions of parabolic equations in high dimensions. *M2AN Math. Model. Numer. Anal.*, 38:93–128, 2004.

[115] T. Werder, K. Gerdes, D. Schötzau, and C. Schwab. *hp*-discontinuous Galerkin time stepping for parabolic problems. *Comput. Methods Appl. Mech. Engrg.*, 190:6685–6708, 2001.

[116] Paul Wilmott, Jeff Dewynne, and Sam Howison. *Option Pricing*. Oxford Financial Press, Oxford, 1993.

[117] Paul Wilmott, Sam Howison, and Jeff Dewynne. *The Mathematics of Financial Derivatives: A Student Introduction*. Cambridge University Press, Cambridge, 1995.

[118] C. Zenger. Sparse grids. In W. Hackbusch, editor, *Parallel Algorithms for Partial Differential Equations*. Volume 31 of *Notes on Numerical Fluid Mechanics*. Vieweg, Braunschweig/Wiesbaden, 1991.

[119] R. Zvan, P.A. Forsyth, and K.R. Vetzal. Penalty methods for American options with stochastic volatility. *J. Comput. Appl. Math.*, 91:199–218, 1998.

Index

BOWLING GREEN STATE UNIVERSITY
DISCARDED
LIBRARY